Spatial Data Analysis
Theory and Practice

Spatial Data Analysis: Theory and Practice provides a broad-ranging treatment of the field of spatial data analysis. It begins with an overview of spatial data analysis and the importance of location (place, context and space) in scientific and policy-related research. Covering fundamental problems concerning how attributes in geographical space are represented to the latest methods of exploratory spatial data analysis and spatial modelling, it is designed to take the reader through the key areas that underpin the analysis of spatial data, providing a platform from which to view and critically appreciate many of the key areas of the field. Parts of the text are accessible to undergraduate and master's level students, but it also contains sufficient challenging material that it will be of interest to geographers, social scientists and economists, environmental scientists and statisticians, whose research takes them into the area of spatial analysis.

ROBERT HAINING is Professor of Human Geography at the University of Cambridge. He has published extensively in the field of spatial data analysis, with particular reference to applications in the areas of economic geography, medical geography and the geography of crime. His previous book, *Spatial Data Analysis in the Social and Environmental Sciences* (Cambridge University Press, 1993) was well received and cited internationally.

Spatial Data Analysis

Theory and Practice

ROBERT HAINING

University of Cambridge

CAMBRIDGE
UNIVERSITY PRESS

PUBLISHED BY THE PRESS SYNDICATE OF THE UNIVERSITY OF CAMBRIDGE
The Pitt Building, Trumpington Street, Cambridge, United Kingdom

CAMBRIDGE UNIVERSITY PRESS
The Edinburgh Building, Cambridge CB2 2RU, UK
40 West 20th Street, New York, NY 10011-4211, USA
477 Williamstown Road, Port Melbourne, VIC 3207, Australia
Ruiz de Alarcón 13, 28014 Madrid, Spain
Dock House, The Waterfront, Cape Town 8001, South Africa

http://www.cambridge.org

First published 2003
Reprinted 2004, 2005 (twice)

Printed in the United Kingdom at the University Press, Cambridge

Typeface Lexicon No. 2 10/14 pt and Lexicon No. 1   *System* LATEX 2_ε [TB]

A catalogue record for this book is available from the British Library

Library of Congress Cataloguing in Publication data

Haining, Robert P.
Spatial data analysis: theory and practice / Robert Haining.
   p.   cm.
Includes bibliographical references and index.
ISBN 0-521-77319-9 – ISBN 0-521-77437-3 (pbk.)
1. Spatial analysis (Statistics)   2. Geology – Statistical methods – Data processing.   I. Title.
QA278.2 .H345   2003
001.4'22–dc   2002031242

ISBN 0 521 77319 9 hardback
ISBN 0 521 77437 3 paperback

To my wife, Rachel, and our children,
Celia, Sarah and Mark

Contents

Preface

Interest in analysing spatial data has grown considerably in the scientific research community. This reflects the existence of well-formulated questions or hypothesis in which location plays a role, of spatial data of sufficient quality, of appropriate statistical methodology.

In writing this book I have drawn on a number of scientific and also policy-related fields to illustrate the scale of interest – actual and potential – in analysing spatial data. In seeking to provide this overview of the field I have given a prominent place to two fields of research: Geographic Information Science (GISc) and applied spatial statistics.

It is important as part of the process of understanding the results of spatial data analysis to define the relationship between geographic reality and how that reality is captured in a digital database in the form of a data matrix containing both attribute data and data on locations. The usefulness of operations on that data matrix – revising or improving an initial representation (e.g. spatial smoothing), testing hypotheses (e.g. does this map pattern contain spatial clusters of events?) or fitting models (e.g. to explain offence patterns or health outcomes in terms of socio-economic covariates) – will depend on how well the reality that is being represented has been captured in the data matrix. Awareness of this link is important and insights can be drawn from the GISc literature.

I have drawn on developments in spatial statistics which can be applied to data collected from continuous surfaces and from regions partitioned into sub-areas (e.g. a city divided into wards or enumeration districts). In covering this material I have attempted to draw out the important ideas whilst directing the reader to specialist sources and original papers. This book is not an exhaustive treatment of all areas of spatial statistics (it does not cover point processes), nor of all areas of spatial analysis (it does not include cartographic modelling).

Implementing a programme of spatial data analysis is greatly assisted if supporting software is available. Geographic information systems (GIS) software are now widely used to handle spatial data and there is a growing quantity of software some of it linked to GIS for implementing spatial statistical methods. The appendix directs the reader to some relevant software.

Readership

This book brings together techniques and models for analysing spatial data in a way that I hope is accessible to a wide readership, whilst still being of interest to the research community.

Parts of this book have been tried out on year 2 geography undergraduates at the University of Cambridge in an eight-hour lecture course that introduced them to certain areas of geographic information science and methods of spatial analysis. The parts used are chapters 1, 2, sections 3.1, 3.2.1, 3.2.3, 3.2.4(a) from chapter 3, selected sections from chapter 4 (e.g. detecting errors and outliers, areal interpolation problems), selected sections from chapter 7 (section 7.1.3, map smoothing) and some selected examples on modelling and mapping output using the normal linear regression model. In associated practicals simple methods for hot spot detection are applied (the first part of section 7.3.1(a)) together with logistic regression for modelling (along the lines of section 11.2.2(a)).

Parts of the book have been tried out on postgraduate students on a one year M.Phil. in Geographic Information Systems and Remote Sensing at Cambridge. One 16-hour course was on general methods of spatial analysis but particularly for data from continuous surfaces. In addition to some of the foundation material covered in chapters 1 to 4 there was an extended treatment of the material in section 4.4.2 with particular reference to kriging with Gaussian data (including estimation and modelling of the semi-variogram taken from chapter 10 and the references therein). A second 16-hour course dealt with exploratory spatial data analysis and spatial modelling with reference to the analysis of crime and health data. This focused on area data. The material in chapter 7 was included with an introduction provided by the conceptual frameworks described in chapter 5. The part of the course on modelling took selected material from chapter 9 and drew on examples referred to in that chapter and chapter 11.

Acknowledgements

This book has taken shape over the last two years at the University of Cambridge but has its roots in teaching and research that go back over many years most significantly to my time at the University of Sheffield. In one sense at least the book dates back to the early 1970s and a one-off lecture given by Michael Dacey at Northwestern University on spatial autocorrelation. That lecture was my introduction to the problems of analysing spatial data. Michael Goodchild invited me to spend some time at the NCGIA in Santa Barbara in the later 1980s and this too proved very formative.

I am grateful to friends and colleagues over the years with whom I have worked. The University of Sheffield had the foresight in the mid 1990s to invest in a research centre – the Sheffield Centre for Geographic Information and Spatial Analysis. This opened up opportunities for me to work on a range of different problems both theoretical and applied and fostered numerous collaborations both within the University and with local agencies. I would like to thank in particular Max Craglia, Ian Masser and Steve Wise in working with me to establish SCGISA and with whom I have undertaken many projects and had many interesting discussions.

I have had the benefit of working with many excellent researchers and in particular I would like to acknowledge Judith Bush, Paul Brindley, Vania Ceccato, Sue Collins, Andrew Costello, Young-Hoon Kim, Jingsheng Ma, Xiaoming Ning, Paola Signoretta and Dawn Thompson. At Cambridge I am working with Jane Law and together we are learning to apply Bayesian methodology to crime and health data, using the WinBUGS program. The examples in chapters 10 and 11 owe a great deal to her hard work. Jane and I are also working to encourage interest in these methods in agencies in the Cambridge region.

Sections on error propagation, missing-data estimation and spatial sampling have benefited from research collaborations with Giuseppe Arbia, Bob

Bennett, Dan Griffith, Luis Flores and Jinfeng Wang. Some of the visualization material and exploratory data analysis has benefited from two ESRC projects undertaken with Steve Wise. I have worked with Max Craglia on several projects with strong policy dimensions, notably two recent Home Office projects. Some of the material on spatial modelling has benefited from collaboration with Eric Sheppard and Paul Plummer on modelling price variation in interdependent markets. My interest in applications of spatial analysis methods to problems in the areas of health studies have been stimulated by projects with Marcus Blake, Judith Bush and Dawn Thompson; also with David Hall and Ravi Maheswaran at Sheffield and recently with Andy Cliff with whom I have done some work on American measles data. Martin Kulldorff has kindly given me advice on the use of his scan test. My more recent interest in the application of spatial analysis methods to data in criminology, as well as drawing my attention to relevant literature, owe much to advice from Tony Bottoms, Andrew Costello and Paul Wiles. Thanks also to the many people who have drawn my attention to a wide range of relevant literature – apologies for not including them all by name.

Parts of this book have been tested on undergraduate and postgraduate students at the University of Cambridge. My thanks to them for sitting through the 'first draft'. My thanks also to three anonymous readers who saw the first part of this book and made many excellent suggestions.

My thanks to Phil Stickler in the Cartography Laboratory at the University of Cambridge for drawing up the figures. Thanks also to the editorial and production guidance of Tracey Sanderson, Carol Miller and Anne Rix at Cambridge University Press.

Thanks to the following for allowing me to use their data in the examples: Dawn Thompson and Sheffield Health for the breast cancer screening data; South Yorkshire Police for several crime data sets, including the burglary and victimized offender data sets; Sheffield children's services unit for the data on children excluded from school; James Reid for the updated ward boundary data for Cambridgeshire; Sara Godward of the Cancer Intelligence Unit for the Cambridgeshire lung cancer data, Andy Cliff for the US measles data.

Finally my thanks to my mother and father who have given me such encouragement over the years. This book is dedicated in particular to Rachel, my wife and 'best friend', for all her support and not least her willingness and enthusiasm to upsticks and try something new and different on occasions too numerous to count.

Copyright acknowledgements

Some figures in this book display boundary material which is copyright of the Crown, the ED-LINE consortium. Ordnance Survey data supplied by EDINA Digimap (a JISC supplied service) were also used. Some of the figures in this book are reproduced with the kind permission of the original publishers. These are as follows and with full references in the bibliography.

Figure 3.5: Kluwer Academic Publishers. From *Geo-ENV II Geostatistics for Environmental Applications*, edited by J. Gomez-Hernandez, A. Soares and R. Frodevaux (1998) Savelieva et al. Conditional stochastic co-simulations of the Chernobyl fallout, fig. 14, p. 463.

Figure 4.3: Pion Limited, London. From M. Tranmer and D.G. Steel (1998) Investigating the ecological fallacy. *Environment and Planning*, A, **30**, fig. 1, p. 827 and fig. 2, p. 830.

Figure 6.2: Taylor and Francis Limited, London. From M. Craglia, R. Haining and P. Wiles (2000) A comparative evaluation of approaches to urban crime pattern analysis. *Urban Studies* **37**, fig. 5, p. 725.

Figure 6.3: Oxford University Press. From *Journal of Public Health Medicine* (1994) R. Haining, S. Wise and M. Blake. Constructing regions for small area health analysis, fig. 3, p. 433.

Figure 6.4: Kluwer Academic Publishers. From *Mathematical Geology*, 20 (1989) M. Oliver and R. Webster. A geostatistical basis for spatial weighting in multivariate classification, fig. 3, pp. 15–35.

Figure 6.7: Kluwer Academic Publishers. From G. Verly et al. (1984) *Geostatistics for Natural Resources Characterization*. N. Cressie towards resistant geostatistics, fig. 8, p. 33.

Figures 6.8 to 6.12: Springer-Verlag. From *Journal of Geographical Systems*, **2** (2000) R. Haining et al. Providing scientific visualization for spatial data analysis, pp. 121–40.

Figure 7.1: John Wiley and Sons Limited. From *Statistics in Medicine* (1999) K. Kafadar. Simultaneous smoothing and adjusting mortality rates in US counties. Figs. 2(a)–2(d). Thanks also to Dr Kafadar for providing the original digital version of these figures.

Figure 7.3 and 7.7: Routledge. From *GIS and Health*, edited by A. Gatrell and M. Loytonen. M. Kulldorff. Statistical methods for spatial epidemiology, figs. 4.1 and 4.2.

Figure 7.8: John Wiley and Sons Limited. From *Statistics in Medicine* (1988) R. Stone. Investigations of excess environmental risks around putative sources. Fig. 3.

Figures 8.1 and 8.2: Ohio State University Press, Columbus, Ohio. From *Geographical Analysis* (1991) R. Haining. Bivarate correlation with spatial data Figs. 2, 3 and 5

Figure 8.3: International Biometric Society. From *Biometrics* (1997). A. Cerioli Modified test of independence in 2×2 tables with spatial data, Fig. 2, pp. 619–28.

Figure 10.1: Ohio State University Press, Columbus, Ohio. From *Geographical Analysis* (1994) D. Griffith et al. Heterogeneity of attribute sampling error in spatial data sets, Fig. 1, p. 31a.

Figure 11.2: Taylor and Francis Limited, London. From *Urban Studies* (2001) M. Craglia et al. Modelling high intensity crime areas in English cities p. 1931.

Introduction

0.1   About the book

This book is about methods for analysing quantitative spatial data. 'Spatial' means each item of data has a geographical reference so we know where each case occurs on a map. This spatial indexing is important because it carries information that is relevant to the analysis of the data. The book is aimed at those studying or researching in the social, economic and environmental sciences. It details important elements of the methodology of spatial data analysis, emphasizes the ideas underlying this methodology and discusses applications. The purpose is to provide the reader with a coherent overview of the field as well as a critical appreciation of it.

There are many different types of spatial data and different forms of spatial data analysis so it is necessary to identify what is, and what is not, covered here. We do so by example:

1 *Data from a surface*. The data that are recorded have been taken from a set of *fixed* (or given) locations on a continuous surface. The continuous surface might refer to soil characteristics, air pollution, snow depth or precipitation levels. The attribute being measured is typically continuous valued. Note that for some of these variables (e.g. snow depth) a point observation is sufficient whilst for others (e.g. air pollution) an areal support or block is necessary in order to provide a measure for the attribute value.

The attribute need not be continuous valued and could be categorical. Land use constitutes a continuous surface. The surface might be divided into small parcels or blocks and land-use type recorded for each land parcel.

The data may originate from a sample of points (or small blocks) on the surface. The data may originate from exhaustively dividing up the surface into tracts and recording a representative value for the attribute for each tract. Once the data have been collected the location of each observation is treated as fixed.

2 *Data from objects*. In this case, data refer to point or area objects that are located in geographic space. An individual may be given a location according to their place of residence. At some scale of analysis the set of retail outlets in a town can be represented by points; even towns scattered across a region may be thought of as a set of points. Attributes may be continuous valued or discrete valued, quantitative or qualitative. Objects may be aggregated into larger groupings – for example populations aggregated by census tracts. Now attribute values are representative of the aggregated population. Again, once the data have been collected, the locations of the points or areas or aggregate zonings are treated as fixed.

The purpose of analysis may be to describe the spatial variation in attribute values across the study area. The next step might be to explain the spatial pattern of variation in terms of other attributes. Description might involve identifying interesting features in the data, including detecting clusters or concentrations of high (or low) values and the next step might be to try to understand why certain areas of the map have a concentration of high (or low) values. In some areas of spatial analysis such as geostatistics the aim may be to provide estimates or predictions of attribute values at unsampled locations or to make a map of the attribute on the basis of the sample data.

That the locations of attribute values are treated as fixed is in contrast to classes of problems, not treated here, where it is the location of the points or areas that are the outcome of some process and where the analyst is concerned to describe and explain the location patterns. These are referred to as *point pattern* data and *object* data (Cressie, 1991, p. 8). To give an example: suppose data were available on the location of all retail outlets of a particular type across an urban area. In addition, for each site, attribute data have been recorded including the price for a particular commodity. The methods of this book would be appropriate for describing and explaining the variation in prices across the set of retail outlets treating their locations as fixed. If we are interested in describing and explaining the location pattern of the individual retail outlets within the urban area, as the outcome of a point location process, then this falls outside the domain of the methods here. Note however that if we are willing to view the location problem through a partitioning of the urban area into a set of *fixed* areas that have been defined independently of

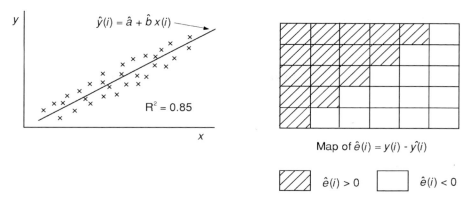

Figure 0.1 Evidence in assessing the adequacy of fit of a regression model

the distribution of retail sites, then the number of retail sites (0, 1, 2, . . .) in each area becomes an attribute and the methods of data analysis in this book are relevant.

The methods of this book are appropriate for analysing the variation in, for example, disease, crime and socio-economic data across a set of areal units such as census tracts or fixed-point sites. They are also appropriate where a sensor has recorded data across a study area in terms of a rectangular grid of small areas (pixels) from which land use or other environmental data have been obtained.

There are two aspects to variation in a spatial data set. The first is *variation* in the data values disregarding the information provided by the locational index. The second is *spatial* variation – the variation in the data values across the map. *Describing* these two aspects of variation calls for two different terminologies and involves different strategies. *Explaining* variation – that is finding a model that will account for the variation in an attribute – could, as an outcome, also provide a good explanation of its *spatial* variation. It is also possible that a model that apparently does well in describing attribute variation leaves important aspects of its spatial variation unexplained. For example all the cases that are very poorly fitted by the model might be in one part of the map. This would arise in regression analysis if all the cases that have the largest positive residuals are in one part of the map and all the cases that have the largest negative residuals are in another. In figure 0.1 the goodness-of-fit statistic ($R^2 \times 100$) equals 85%. X has accounted for 85% of the variation in the response variable (Y). This suggests an adequate model. But there is a strong spatial structure to the pattern of positive and negative residuals ($\hat{e}(i)$). The analyst will conclude that the model is in need of further development if parameters of interest are to be properly estimated or hypotheses properly tested and will need to consider strategies for achieving this.

0.2   What is spatial data analysis?

The term 'spatial analysis' has a pedigree in geography that can be traced back to at least the 1950s and for an overview of historical developments at that time see Berry and Marble (1968, pp. 1–9). Spatial analysis is a term widely used in the Geographical Information *Systems* (GIS) and Geographical Information *Science* (GISc) literatures. A definition of spatial analysis is that it represents a collection of techniques and models that explicitly use the spatial referencing associated with each data value or object that is specified within the system under study. Spatial analysis methods need to make assumptions about or draw on data describing the spatial relationships or spatial interactions between cases. The results of any spatial analysis are not the same under re-arrangements of the spatial distribution of values or reconfiguration of the spatial structure of the system under investigation (Chorley, 1972; Haining, 1994, p. 45).

Spatial analysis has three main elements. First it includes *cartographic modelling*. Each data set is represented as a map and map-based operations (or implementing map algebras) generate new maps. For example buffering is the operation of identifying all areas on a map within a given distance of some spatial object such as a hospital clinic, a well, or a linear feature such as a road. Overlaying includes logical operations (.AND.; .OR.; .XOR.) and arithmetic (+; −; ×; /) operations. The logical overlay denoted by .AND. identifies the areas on a map that simultaneously satisfy a set of conditions on two or more variables (Arbia et al., 1998). The arithmetic overlay operation of addition sums the values of two or more variables area by area (Arbia et al., 1999).

Second, spatial analysis includes forms of *mathematical modelling* where model outcomes are dependent on the form of spatial interaction between objects in the model, or spatial relationships or the geographical positioning of objects within the model. For example, the configuration of streams and the geography of their intersections in a hydrological model will have an effect on the movement of water through different areas of a catchment. The geographical distribution of different population groups and the distribution of their density in a region may have an influence on the spread of an infectious disease whilst the location of topographical barriers may have an influence on the colonization of a region by a new species. Finally, spatial analysis includes the development and application of statistical techniques for the proper analysis of spatial data and which, as a consequence, make use of the spatial referencing in the data. This is the area of spatial analysis that we refer to here as *spatial data analysis*.

There are many features to spatial data that call for careful consideration when undertaking statistical analysis. Although the analysis of spatial dependence is a critical element in spatial data analysis and central for example in specifying sampling designs or undertaking spatial prediction, an excessive attention to just that aspect of spatial data can lead the analyst to ignore other issues. For example: the effect of an areal partition on the precision of an estimator or the wider set of assumptions and data effects that determine whether a model can be considered adequate for the purpose intended. In this sense spatial data analysis is a *subfield* of the more general field of data analysis. In defining the skills and concepts necessary for undertaking a proper analysis of spatial data there is, then, an important role for areas of statistical theory developed to handle other types of, non-spatial, data. In adopting this rather broader definition of spatial data analysis a link is maintained to the wider body of statistical theory and method.

0.3   **Motivation for the book**

This book is a descendant of *Spatial Data Analysis in the Social and Environmental Sciences* (1990) which dealt with the same types of spatial data. The earlier book reviewed models for describing and explaining spatial variation and discussed the role of robust methods of fitting partly in response to the perceived nature and quality of spatial data. That book described both exploratory spatial data analysis and spatial modelling. Exploratory spatial data analysis (ESDA) includes amongst other activities the identification of data properties and formulating hypotheses from data (Good, 1983). It provides a methodology for drawing out useful information from data. The findings from exploratory analysis provide input into spatial modelling. Modelling involves specification, parameter estimation and inference (testing hypotheses, computing confidence intervals, assessing goodness of fit) through which the analyst hopes to estimate parameters of interest and test hypotheses. In assessing a model, the tools and methods of ESDA may again play a useful role, leading to further iterations of model specification, estimation and inference.

Over the last decade or so there have been a number of developments, theoretical and practical, which have had important implications for the conduct of spatial data analysis (Haining, 1996). We briefly sketch some of these developments by way of illustration which also provides some motivation for the timing of this book.

One of the first research agendas set by the United States National Center for Geographic Information and Analysis (NCGIA) after its founding in the second half of the 1980s was in the area of spatial data accuracy (Goodchild and Gopal, 1989). Research in this area particularly into the nature of error in spatial data and how such error may propagate as a result of performing different types of map operations like overlaying or buffering has important implications for the conduct of data analysis. It helps to define the limits to what may be concluded from the analysis of spatial data (Arbia et al., 1999). This remains an area of special importance at a time when there are ever-growing volumes of spatially referenced data produced both by government agencies and the private sector. This research focus on data accuracy and data quality in turn led to a focus on issues of spatial representation (when geographic reality is translated into digital form) and what terms such as 'quality', 'accuracy' and 'error' mean in the case of spatial data.

Exploratory spatial data analysis (ESDA) methods were not widely used in the late 1980s, although Cressie (1984) had written about methods for exploring geostatistical data and Openshaw and colleagues had developed a 'geographical analysis machine' for looking for clusters of events in inhomogeneous spatial populations (Openshaw et al., 1987). There has been considerable interest in this area since that time together with new research into visualization tools and associated software to support ESDA. Notable in this respect has been the pioneering work of statisticians including Haslett and colleagues (e.g. Haslett et al., 1990, 1991; Unwin et al., 1996) and geographers including Monmonier and MacEachren (e.g. Monmonier and MacEachren, 1992). One aspect of ESDA that has attracted interest is the development of local statistics (based on using spatial subsets of the data) in order to detect local properties and describe spatial heterogeneity. The complex nature of spatial variation has long been a subject of comment. The development of local statistics to compliment the array of familiar 'whole map' or global statistics that provide descriptions of the average properties of a map is in part a response to the recognition of the heterogeneous nature of spatial variation (e.g. Getis and Ord, 1992, 1996; Anselin, 1995; Fotheringham et al., 2000). New techniques for the detection of spatial clusters of events (such as clusters of a particular disease or crime) have been developed which represent additions to the spatial analysis toolkit (Besag and Newell, 1991; Kulldorff and Nagarwalla, 1995).

In the area of statistical modelling of spatial data, Bayesian approaches attracted attention in the 1990s, in part because of the availability of numerical methods within new software for fitting a wide range of models (Gilks et al., 1996). Prior to the early 1990s much spatial modelling was based on spatial modifications to the linear regression model in which, for example, spatial dependence was modelled through the response variable. There were few applications of Bayesian methods (for an exception see Hepple, 1979). Bayesian methods have introduced other ways for modelling the effects of spatial dependence.

Over the last decade there have been important advances in software. The requirement to write ones own software with the attendant anxiety of making subtle (and not so subtle) programming errors, always acted as a brake on the utilization of spatial analysis methods particularly outside statistics. Bailey and Gatrell (1995) provided software as part of their book although this software was largely for teaching purposes. There have been considerable advances in making software available including much that can be downloaded free off the web. This book is not linked to any one piece of spatial analysis software but the appendix gives a list of software that can be used to implement many of the methods discussed in this book.

Geographic Information Systems (GIS) are software systems for capturing, storing, managing and displaying spatial data. The ability to directly capture events (such as the location of the offence by an officer attending the crime scene) on to a GIS has important implications for the rapid and timely accumulation of spatial data and the conduct of analysis. One of their most important capabilities for the purpose of spatial data analysis is that they provide a platform for integrating different data sets that may not necessarily be referenced to the same spatial framework. The problem of *how* to link data sets that derive from incompatible spatial frameworks (for example linking pixel-based environmental data and enumeration district-based population data) has attracted considerable interest. Less attention however seems to have been paid to the *consequences* of such linkage on the conduct of analysis and the interpretation of results, given the errors and uncertainties that such linkage necessarily induces in the database. Commercial GIS (e.g. ArcGIS Geostatistical Analyst) now provides some spatial analysis capability including statistical analysis. This opens up at least in some areas of research the possibility for a seamless environment for the storage, management, display and also *analysis* of spatially referenced statistical data.

The breadth of disciplinary interest in spatial data analysis is evident from earlier books in the field (Ripley, 1981; Upton and Fingleton, 1985; Anselin, 1988; Haining, 1990; Cressie, 1991; Bailey and Gatrell, 1995) and edited volumes such as Fotheringham and Rogerson (1994), Fischer et al. (1996) and Longley and Batty (1996). The continued vitality of the field over the last decade is illustrated by the growing number of applications and the increasing number of journals that have carried theme issues (see, e.g., special issues of *Papers in Regional Science*, 1991 (3); *Computational Statistics*, 1996 (4); *International*

Regional Science Review, 1997 (1&2); *The Statistician*, 1998 (3); *Journal of Real Estate Finance and Economics*, 1998 (1); *Statistics in Medicine*, 2000 (19, parts 17 and 18)).

The emergence of an area of quantitative research is due to the availability of good-quality data, the emergence of well-formulated hypotheses that can be expressed in mathematical terms, the availability of appropriate mathematical and statistical tools and techniques and the availability of technology for facilitating analysis. This diagnosis seems to apply to geographical and environmental epidemiology and to health services research (Cuzick and Elliott, 1992) where the availability of geo-coded health data and the growth of geographical databases and the development of new statistical techniques has generated considerable research activity. The expanding use of spatial analysis methods reflects the significance of place and space in theorizing disciplinary subfields such as the interest in area contextual effects in explaining health behaviours.

Developments in criminology seem to reflect a similar pattern: the collection of geo-coded offence, offender and victim data and the development of local statistics and their availability through spatial analysis software such as provided by the United States' National Institute of Justice. Such work is given further impetus through theorizing the role of spatial relationships and the context of place in shaping offence, offender and victim geographies (Bottoms and Wiles, 1997).

0.4   Organization

Chapter 1, discusses the relevance of spatial data analysis in selected areas of scientific and policy-related research and provides motivation for the rest of the book. Chapter 2, discusses the nature of spatial data and the relationship between the spatial data matrix, the foundation of all the analysis in this book, and the geographical reality it seeks to capture. This chapter draws heavily on the geographical information science literature. This leads to a discussion of data quality issues and methods of quantifying spatial dependence. This book is not just about the issues for the analysis of spatial data raised by spatial dependence but, as already noted, it is an important property that influences many stages and aspects of data analysis.

Chapter 3 discusses sources of spatial data and concentrates in particular on spatial sampling. There is a section on obtaining simulated data from spatial models. Chapter 4 looks at the implications of different aspects of data quality for the conduct of spatial data analysis. The emphasis is on techniques that address some of the problems often encountered with spatial data such as missing values, data on incompatible areal systems and inference problems associated with ecological (spatially aggregated) data.

Chapters 5 to 7 deal with exploratory spatial data analysis. Chapter 5 is a short chapter describing different conceptual models of spatial variation that might be used to underpin a programme of exploratory data analysis in the sense of specifying in a quite informal way what spatial structures might be looked for in a data set. Chapter 6 deals with visual methods for exploring spatial data whilst chapter 7 describes numerical methods for identifying data properties concentrating on spatial smoothing, clustering methods and map comparison methods. Splitting in this way makes the discussion of ESDA more manageable, in my view, but it is important to remember that visual and numerical methods are complementary.

Chapter 8 describes some of the implications for carrying out statistical tests when data are not independent. Tests of differences of means, bivariate correlation tests and chi-square tests on spatial data are discussed. These topics are approached through the concept of the 'effective sample size' which means calculating the amount of information about the process contained in the dependent set of data. The 'nuisance' aspect of spatial dependence (in the statistical sense) is that positive spatial dependence reduces the amount of information the analyst has for making inferences about the population. This reduction is relative to the amount of information the analyst would have if the observations were independent.

Chapters 9 to 11 discuss the modelling of spatial data. Chapter 9 describes statistical models for spatial variation when the data are from a continuous surface and when data refer to areal aggregates or point or area objects. Just as chapter 5 describes models that underpin ESDA, chapter 9 describes the range of descriptive and explanatory models that underpin formal data analysis where the aim is to estimate parameters of interest and test hypotheses. Models for representing spatial variation are mentioned and occasionally used in earlier chapters. The reader is encouraged, after completing chapter 2, to dip into section 9.1 especially for material on spatial covariance and spatial autoregressive models.

Chapter 10 discusses and provides examples of descriptive spatial modelling where the aim is to find a model to represent the variation in a response variable. The coverage includes trend surface models with independent errors and trend surface models with spatially correlated errors. Models for describing spatial variation in discrete-valued regional variables are also treated. Bayesian methods for disease mapping are applied. Chapter 11 discusses and provides examples of explanatory modelling using regression where the spatial variation in a response is modelled in terms of covariates. This

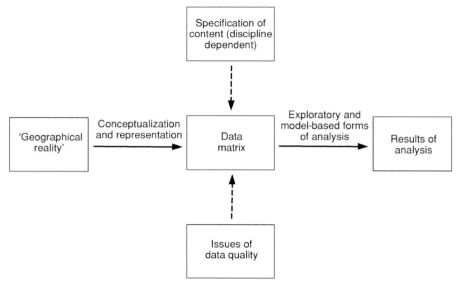

Figure 0.2 Overall structure of the book

chapter includes a review of different methodologies for modelling in non-experimental sciences. There is an appendix on available software.

Figure 0.2 represents the overall structure of the book.

0.5   The spatial data matrix

Underlying all the analyses here is a data matrix. We now indicate the content of that matrix as well as introducing some notation.

Let $Z_1, Z_2, \ldots, Z_k$ refer to k variables or attributes and $\mathbf{S}$ to location. The type of spatial data set to be considered in this book can be represented as:

$$
\begin{array}{c}
\overbrace{}^{\text{Data on the } k \text{ variables}} \quad \overbrace{\phantom{\mathbf{s}(1)}}^{\text{Location}} \\[-2pt]
\left[\begin{array}{ccccc}
z_1(1) & z_2(1) & \cdots & z_k(1) & \mathbf{s}(1) \\
z_1(2) & z_2(2) & \cdots & z_k(2) & \mathbf{s}(2) \\
\vdots & \vdots & \vdots & \vdots & \vdots \\
z_1(n) & z_2(n) & \cdots & z_k(n) & \mathbf{s}(n)
\end{array}\right]
\begin{array}{l}
\text{Case 1} \\
\text{Case 2} \\
\vdots \\
\text{Case } n
\end{array}
\end{array}
\tag{0.1}
$$

The use of the lower case symbol on Z and $\mathbf{S}$ denotes an actual data value whilst the number inside the brackets, 1, 2, . . . etc, references the particular case. Attached to every case (i) is a location $\mathbf{s}(i)$. In a later chapter we shall be more specific about how the location of a case is referenced and what other information on $\mathbf{s}(1), \ldots, \mathbf{s}(n)$ may need to be recorded in order to undertake analysis – such as which other sites (cases) represent spatial neighbours of any given site (case). At this stage, however, and since we are only interested in two-dimensional space, it is sufficient to note that there will be occasions when the referencing will involve two co-ordinates. Together these fix the location of the case with respect to two axes that are at right angles to one another (orthogonal). So, the bold font for $\mathbf{s}$ signals that this is a vector and may contain more than one number for the purpose of identifying the spatial location of the case: for example $\mathbf{s}(i) = (s_1(i), s_2(i))$. In this book we only look at methods that treat the locations as fixed – we will not be looking at problems where there is randomness associated with the locations of the cases.

The structure (0.1) can be shortened to the form:

$$\{z_1(i), z_2(i), \ldots, z_k(i) \mid \mathbf{s}(i)\}_{i=1,\ldots,n} \tag{0.2}$$

and when no confusion arises the notation outside the curly brackets will be dropped.

In addition to possessing a spatial reference, data also have, at least implicitly, a temporal reference. The type of data set specified by 0.2 might be re-expressed in the form:

$$\{z_1(i, t), z_2(i, t), \ldots, z_k(i, t) \mid \mathbf{s}(i), t\}_{i=1,\ldots,n} \tag{0.3}$$

where t denotes time. However, all data values are meant to refer to the same point in time which is why t will be suppressed in the notation. The implications of this assumption will need careful consideration in any particular analysis because it is not always possible to have data on different attributes referring to the same time period. Population censuses for example are only taken every 10 years in the UK. At this stage we simply note that this is not a book about analysing space–time variation, except inasmuch as we might compare results arising from separate analyses of two or more time periods.

On various occasions throughout the book, depending on the context, the variables or attributes $Z_1, Z_2, \ldots, Z_k$ will be divided into groups and labelled differently. In the case of data modelling, the variable whose variation is to be modelled will be denoted Y. In regression Y is called the *response* or *dependent* or *endogenous* variable. The variables used to explain the variation in the response are called *explanatory* or *independent*, or *exogenous* or *predictor* variables and are usually labelled differently such as $X_1, X_2, \ldots, X_k$.

The context for spatial data analysis

1

Spatial data analysis: scientific and policy context

Seen from the perspective of the scientist or the policy maker, analytical techniques are a means to an end: for the scientist the development of rigorous, scientifically based understanding of events and processes; for the policy maker the strategic and tactical deployment of resources informed by the application of scientific method and understanding. This chapter describes various areas that raise questions calling for the analysis of spatial data.

The chapter is organized as follows. Section 1.1 identifies how location and spatial relationships enter generically into scientific explanation and section 1.2 briefly discusses how they enter into questions in selected thematic areas of science and general scientific problem solving. Section 1.3 considers the ways in which geography and spatial relationships are important in the area of policy making. Section 1.4 gives some examples of how problems and misinterpretations can arise in analysing spatial data.

1.1   Spatial data analysis in science

All events have space and time co-ordinates attached to them – they happen somewhere at sometime. In many areas of experimental science, the exact spatial co-ordinates of where experiments are performed do not usually need to enter the database. Such information is not of any material importance in analysing the outcomes because all information relevant to the outcome is carried by the explanatory variables. The individual experiments are independent and any case indexing could, without loss of information relevant to explaining the outcomes, be exchanged across the set of cases.

The social and environmental sciences are observational not experimental sciences. Outcomes have to be taken as found and the researcher is not usually able to experiment with the levels of the explanatory variables nor to replicate. In subsequent attempts to model observed variation in the response variable, the design matrix of explanatory variables is often fixed both in terms of what variables have been measured and their levels. It follows that at later modelling stages model errors include not only the effects of measurement error and sampling error but also various forms of possible misspecification error.

In many areas of observational science, recording the place and time of individual events in the database will be important. First, the social sciences study processes in different types of places and spaces – the structure of places and spaces may influence the unfolding of social and economic processes; social and economic processes may in turn shape the structure of places and spaces. Schaeffer (1953) provides an early discussion of the importance of this type of theory in geography and Losch (1939) in economics. Second, recording where events have occurred means it becomes possible to link with data in other databases – for example linking postcoded or address-based health data and socio-economic data from the Census. A high degree of precision might be called for in recording location to ensure accurate linkage across databases.

Spatial data analysis has a role to play in supporting the search for scientific explanation. It also has a role to play in more general problem solving because observations in geographic space are dependent – observations that are geographically close together tend to be alike, and are more alike than those which are further apart. This is a generic property of geographic space that can be exploited in problem-solving situations such as spatial interpolation. However this same property of spatial dependence raises problems for the application of 'classical' statistical reference theory because data dependence induces data redundancy which affects the information content of a sample ('effective sample size').

1.1.1   Generic issues of place, context and space in scientific explanation

(a)   Location as place and context

Location enters into scientific explanation when geographically defined areas are conceptualized as collections of a particular mix of attribute values. Ecological analysis is the analysis of spatially aggregated data where the object of study is the spatial unit. In other circumstances the object of study might comprise individuals or households. Analysis may then need to include not only individual-level characteristics but also area-level or ecological attributes that might impact on individual-level outcomes.

'Place' can be used to further scientific understanding by providing variability in explanatory variables. The diversity of places in terms of variable values consitutes a form of 'natural' laboratory. Consider the case of air pollution levels across a large region which contains many urban areas with contrasting economic bases and as a consequence measurable differences in levels and forms of air pollution. Data of this type combined with population data can be used for an ecological analysis of the relationship between levels of air pollution at the place of residence and the incidence of respiratory conditions in a population, controlling for the effects of possible 'confounders' (e.g. age, deprivation and lifestyle). The Harvard 'six cities' study used the variability in air pollution levels across six cities in the USA to examine the relationship between levels of fine particle matter in the atmosphere and the relative risk of disease (Dockery et al., 1993).

Explaining spatial variation needs to disentangle 'compositional' and 'contextual' influences. Geographical variations in disease rates may be due to differences between areas in the resident population in terms of say age and material well being (the compositional effect). Variation may also be due to differences between areas in terms of exposure to factors that might cause the particular disease or attributes of the areas that may have a direct or indirect effect on people's health (the contextual effect).

Contextual properties of geographical areas may be important in a number of areas of analysis. Variation in economic growth rates across a collection of regional economies may be explained in terms of the variation in types of firms and firm properties (the compositional effect). It may be due to the characteristics of the regions that comprise the environments within which the firms must operate (the contextual effect). Regional characteristics might include the tightness of regional labour markets, the nature of regional business networks, wider institutional support and the level of social capital as measured by levels of trust, solidarity and group formation within the region (Knack and Keefer, 1997). The contextual effect may operate at several scales or levels. Hedonic house price models include the price effects of neighbourhood quality and also the quality of *adjacent* neighbourhoods (Anas and Eum, 1984). Brooks-Gunn et al. (1993) in their study of adolescent development comment: 'individuals cannot be studied without consideration of the multiple ecological systems in which they [the adolescents] operate' (p. 354). The contextual effect of 'place' can operate at a hierarchy of scales from the immediate neighbourhood up to regional scales and above. Neighbourhoods influence behaviour, attitudes, values and opportunities and the authors review four theories about how neighbourhoods may affect child development. Contagion theory stresses the power of peer group influences to spread problem behaviour. Collective socialization theory emphasizes how neighbourhoods provide role models and monitor behaviour. Competition theory emphasizes the effects on child development of competing for scarce neighbourhood resources whilst relative deprivation theory stresses the effects on child development of individuals evaluating themselves against others. Pickett and Pearl (2001) provide a critical review of multilevel analyses that have examined how the socio-economic context provided by different types of neighbourhood, after controlling for individual level circumstances, can affect health outcomes. Jones and Duncan (1996) describe generic contextual effects in geography.

The introduction of 'place' raises the generic problem of how to handle scale effects. 'Place' can refer to areal objects of varying sizes – even within the same analysis. In most areas of the social sciences properties of areas are scaled up from data on individuals or smaller subareas (including point locations) by the arithmetic operation of averaging – that is by implicitly assuming additivity. This seems to be a consequence of the nature of area-level concepts in the social sciences (e.g. social cohesion, social capital and social control; material deprivation) which allows analysts to adopt any reasonable operational convention. In environmental science a similar form of change of scale problem arises in change of support problems where data measured on one support (e.g. point samples) are converted to another (e.g. a small area or block) through weighted averaging. But not all change of scale problems in environmental science are linear and can be handled in this way, as discussed for example in Chilès and Delfiner (1999, pp. 593–602) in the case of upscaling permeability measurements. There is detailed discussion of upscaling and downscaling problems and methods in environmental science in Bierkens et al. (2000).

(b)   Location and spatial relationships

The second way location enters into scientific explanation is through the 'space' view. This emphasizes how objects are positioned with respect to one another and how this relative positioning may enter explicitly into explaining variability. This derives from the interactions between the different places that are a function of those spatial relationships. This generic conception of location as denoting the disposition of objects with respect to one another introduces relational considerations such as distance (and direction), gradient or neighbourhood and configuration or system-wide properties which may play a role in the explanation of attribute variability. The roles that these influences may play in any explanation are ultimately dependent on place attributes and in particular on the interactions that are generated as a consequence of these place attributes and their spatial distribution. We consider different ways spatial relationships construct or configure space: through *distance* separation, by generating *gradients* and by inducing an area-wide *spatial organization*.

Distance can be defined through different metrics – for example straight line physical distance, time distance (how long it takes to travel from *A* to *B*), cost distance, perceived distance. Distance can be defined in terms of networks of relationships and in qualitative terms: near to, far from, next to, etc. Distance becomes part of a scientific explanation when attribute variability across a set of areas is shown to be a consequence of how far areas are from a particular region that possesses what may be a critical level of some causal factor. The geography of economic underdevelopment reflects variation in levels of absolute disadvantage in terms of endowments, including lack of natural resources, poor land quality and disease. However it also appears to reflect distance from the core economic centres because distance affects prices and flows of new technology (Gallup et al. 1999; Venables, 1999). The incidence of cancer of the larynx might be linked to certain types of emissions and disease counts by area might be linked to distance from a particular noxious facility (Gatrell and Dunn, 1995). The measurement of distance might need to allow for such characteristics as prevailing wind direction and topographic attributes that could affect the direction of spread and amount of dilution of the emissions. In situations where outcomes are a product of interaction between individuals or groups then the level of an attribute in one area may influence (and be influenced by) levels of the same attribute in other nearby areas. High levels of an infectious disease in one area may through social contact and the greater risk of an infected individual contacting a non-infected individual lead to high levels in other nearby areas. Proximity also acts as a surrogate for the frequency with which individuals visit an area and become exposed to a highly localized causal agent. In various ways the relative proximity of areas, providing a surrogate for the intensity of different types of social contact, becomes integral to how geographic space becomes a consideration in accounting for the spatial variability of the incidence of the disease.

A gradient is a local property of a space, for example how similar or how different two neighbouring areas are in terms of variable characteristics. Measured surface water at a location after a rainstorm reflects not only the water retention characteristics of the location but also neighbourhood conditions that affect runoff levels and hence surface water accumulation rates. The economic gradient between two adjacent areas as measured by unemployment rates or average household income levels may influence crime rates, inducing an effect in both neighbourhoods that is not purely a consequence of the characteristics of the two respective neighbourhoods. Rather it reflects the fact that two areas of such contrasting economic circumstances are close together (Bowers and Hirschfield, 1999). Block (1979) remarked in the context of property crime: 'it is clear that neighbourhoods in which poor and middle class families live in close proximity are likely to have higher crime rates than other neighbourhoods' (p. 52). This was ascribed to a sharpened sense of frustration on the part of the have-nots combined with routine activity and opportunity theories that describe motivated offender behaviour. Johnstone (1978) encountered a similar neighbourhood effect in a study of adolescent delinquency.

The overall spatial organization of attributes of the study region may be important. In some instances the overall spatial distribution, how a totality of events in an area are distributed in relation to each other, may influence outcomes and overall, system-wide, properties. In the surface water example, levels of accumulation at a location will reflect not only local conditions and neighbourhood conditions but will also be affected by the overall configuration of wider system attributes such as the size, shape and topography of the catchment. Explanations of trading levels between two areas may be based not only on the economic characteristics of the two regions (which affects what they can supply and levels of demand) and their distance apart (which affects transport costs) but also on the nature of 'intervening opportunities' for trade. This can produce different levels of trade between pairs of regions that in terms of economic characteristics and distance apart are otherwise identical (Stouffer in Isard, 1960, p. 538). Faminow and Benson (1990) discuss how the spatial structure of markets changes the nature of tests for market integration.

Health may be related to social relativities rather than absolute standards of living (Wilkinson, 1996). The spatial distribution of material deprivation within a city, the extent to which deprived populations are spatially concentrated or scattered and thus experience different forms of relative rather than absolute deprivation may have an influence on the overall health statistics for a city (Gatrell, 1998). The geography of deprivation may influence the sorts of social comparisons people make. This in turn may influence their health via psychological factors and health-related behaviours (MacLeod et al., 1999). To what extent is persistent inter-generational poverty amongst certain ethnic groups in the USA a consequence of their spatial concentration in certain types of ghettos, spatially enlarged by processes of selective migration and characterized by high levels of poverty and long-term unemployment (Wilson, 1997)? Are areas with high levels of violent drug-related crime embedded in deprived areas of a city which are extensive enough to create special problems for policing (Craglia et al., 2000)?

The importance of spatial relativities in explaining attribute variation is scale dependent – that is the role of such relativities is dependent on the scale of the spatial unit through which events are observed and measured, in relation to the underlying processes. What may be a relational property in understanding why particular houses are burgled in an individual level analysis (for example, whether there are street lights outside the house or not) becomes a property of the place in an ecological analysis (quality of street lighting). If there is some crime displacement from areas where street lighting is good to neighbouring areas where street lighting is poor this will not be evident in the data if the spatial scale of the analysis is such as to average areas of contrasting street lighting or is larger than the scale at which any displacement effect occurs. Moving up the spatial scale of analysis, what may call for the inclusion of relational properties when analysis is in terms of urban census tracts may be analysed as a pure place effect at county or state levels of analysis. What will be an economic spillover of consumer expenditure from one area to another if the areas are small will be a local multiplier if the scale of the geographic areas exceed the scale of consumer travel behaviour. There may be neighbourhood effects in voting behaviour at the tract level as a result of interaction linked to 'the communication process, bandwagon effects, reference group behaviour, or other forms of "symbolic interactionism"' (Dow et al., 1982, p. 170). When comparing voting behaviours across larger regions, such effects are likely to become absorbed within the aggregate measure or become a contextual effect linked to variation in intra-area social interaction. At this scale other variables, such as socio-economic attributes, may assume greater significance.

1.1.2   Spatial processes

Certain processes, referred to as 'spatial processes' for short, operate in geographic space, and four generic types are now discussed: diffusion processes, processes involving exchange and transfer, interaction processes and dispersal processes.

A *diffusion* process is where some attribute is taken up by a population and, at any point in time, it is possible to specify which individuals (or areas) have the attribute and which do not. The mechanism by which the attribute spreads through the population depends on the attribute itself. Conscious or unconscious acquisition or adoption may depend on inter-personal contact, communication or the exerting of influence and pressure, as in the case of voting behaviour or the spread of political power (Doreian and Hummon, 1976; Johnston, 1986). In the case of an infectious disease, like influenza, the diffusion of the disease may be the result of contact between infected and non-infected but susceptible individuals or the dispersal of a virus as in the case of a disease like foot and mouth in livestock (Cliff et al., 1985). The density and spatial distribution of the population in relation to the scale at which the mechanism responsible for the spread operates will have an important influence on how the attribute diffuses and its rate of diffusion.

Urban and regional economies are bound together by processes of mutual commodity exchange and income transfer. Income earned in the production and sale of a commodity at one place may be spent on goods and services elsewhere. Through such processes of *exchange and transfer* the economic fortunes of different cities and regions become inter-linked. The binding together of local spatial economies through wage expenditure, sometimes called wage diffusion, and other 'spillover' effects may be reflected in the spatial structure of the level of per capita income (Haining, 1987).

A third type of process involves *interaction* in which outcomes at one location influence and are influenced by outcomes at other locations. The determination of prices at a set of retail outlets in an area may reflect a process of price action and reaction by retailers in that market. Whether retailer A responds to a price change by another (B) depends on the anticipated effect of that price shift on levels of demand at A. This may influence whether any price reaction at A needs to fully match the price shift at B or not. The closer the retail competitor at B is the more likely it is that A will need to respond in full (Haining, 1983; Plummer et al., 1998). Such interaction seems to be affected by the spatial distribution of sellers, including their density and clustering (Fik, 1988, 1991).

In a diffusion process the attribute spreads through a population and the individuals in the population have a fixed location. The final type of process, a *dispersal* process, represents the dispersal of the population itself. Such processes of dispersal may involve, for example, the dispersal of seeds from a parent plant or the spread of physical properties like atmospheric or maritime pollution or the spread of nutrients in a soil.

1.2   Place and space in specific areas of scientific explanation

The need for rigorous methods for spatial data analysis will be felt most strongly in those areas of thematic science where geographic space has entered directly into theorizing or theory construction. It will also be felt in areas of study where the identification of any regularities in spatial data is taken to signal something of substantive interest that justifies closer investigation. The next subsection discusses definitions and this is followed by a few brief examples.

1.2.1   Defining spatial subdisciplines

The recognition of the importance of location in a thematic discipline is signalled when subfields are defined prefixed with words such as 'geographical', 'spatial', 'environmental' or 'regional': geographical and environmental epidemiology (Elliott et al., 1992, 2000), spatial archaeology (Clarke, 1977) and spatial archaeometry, environmental criminology (Brantingham and Brantingham, 1991), regional economics (Richardson, 1970; Armstrong and

Taylor, 2000). Geography has systematic subfields which may overlap with the above with labels like: medical geography, historical geography, the geography of crime, economic geography. To the extent that there are real differences between these two approaches, geography as a synthetic discipline is often most interested in understanding particular places, drawing on the ideas and theories of the thematic disciplines (to which geographers themselves may contribute) in order to construct explanations or develop case studies. On the other side the thematic fields draw on place and space for the reasons discussed above – to develop understanding of the processes underlying disease incidence, pre-historic societies, the occurrence of crime and victimization, wealth creation.

Epidemiology distinguishes between geographical and environmental epidemiology. Geographical epidemiology focuses on the description of the geography of disease at different scales, ecological studies and the effects of migration on disease incidence (English, 1992). It is concerned with examining the factors associated with spatially varying levels of incidence, prevalence, mortality and recovery rates of a disease after controlling for age and sex. Environmental epidemiology seeks to model area-specific relative risk, after controlling for population characteristics and socio-economic confounders, arising from exposure to environmental risk factors such as naturally occurring radiation, air pollution or contaminated water. The study of geographical patterns and relationships help our understanding of the causes of disease, if not directly then at least by suggesting hypotheses that may then be pursued by other forms of investigation.

Swartz (2000) in his review defines environmental criminology as concerned with micro-level research which focuses on 'individual locations, and attempts to explain the relationship between site-specific physical features, social characteristics and crime' (p. 40). This is distinguished from the 'ecological tradition' in criminology which is 'confined to relatively large aggregations of people and space' (p. 40). Bottoms and Wiles (1997) use the term environmental criminology which they define as: 'the study of crime, criminality and victimisation as they relate, first to particular places, and secondly to the way that individuals and organisations shape their activities spatially and in so doing are in turn influenced by place-based or spatial factors' (p. 305). The term environment is used more broadly than in epidemiology and the definition allows for both the micro level and ecological levels of spatial analysis.

Clarke (1977) defines spatial archaeology as 'the retrieval of information from archaeological spatial relationships and the study of the spatial consequences of former hominid activity patterns within and between features and structures and their articulation within sites, site systems and their

environments' (p. 9). Clarke identifies the key features of the subfield as the retrieval of useful archaeological information from the examination of the geography of archaeological data; the examination of archaeological data at a range
of different geographical scales; the use of the map as a key tool in the process of
extracting information. Hodder (1977) identifies the key stages of spatial analysis in archaeology as going from mapping, to the construction of summary
descriptions of mapped distributions to the identification of map properties
and local anomalies. Geo-coding data which have been collected from different field surveys and other disparate data sources provides a particularly useful
way to link and cross check data sets. When combined with appropriate spatial
analysis techniques this may assist with classification and the identification of
heritage areas (for an example in the case of dialect studies see Wilhelm and
Sander, 1998).

The field of regional economics as defined by Richardson (1970, p. 1) is concerned with the role of 'space, distance and regional differentiation' in economics. It has been broadly concerned with two classes of problem. Location
theory focuses on explaining the location of economic activity and why particular activities are located where they are. The field of study originated with the
work of Von Thunen who in the 19th century considered the problem of the
location of agricultural production. This area of regional economics developed
through the work of a succession of 19th-century and later theorists concerned
principally with industrial location theory. The other main area of study is the
regional economy and is concerned with explanations of economic growth at
the regional scale, the causes of poor economic performance at the regional
level and associated policy prescriptions.

Five areas of thematic science have been selected to illustrate the role of place
and space within them.

1.2.2   Examples: selected research areas

(a)   Environmental criminology

Early work in environmental criminology examined the links between
urbanization, industrialization and crime and how and why different urban–
industrial places generated different crime patterns. There is interest in the
criminological implications of the shift towards the post-industrial city. The
decline of traditional shopping areas and the changing nature of the inner
city, the creation of new out-of-town shopping centres and new forms of residential housing with new forms of occupancy are generating new offence
geographies (Bottoms and Wiles, 1997). Changes in the use of space within
an urban area together with new patterns of mobility and new life styles are

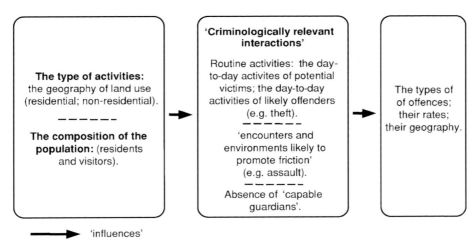

Figure 1.1 Wikström's (1990) model for the geography of offences (after Bottoms and Wiles, 1997)

inducing changes in offence patterns and the emergence of new geographical concentrations of offences (Ceccato et al., 2002). Wikström's (1990) model of where offences occur is based on variations in land use within the urban area and the forms of social interaction taking place within the urban area (figure 1.1). Offences take place where criminal opportunities intersect with areas that are known to the offender because of their routine use of that space.

The lack of 'neighbourhood organization' or social cohesion or the co-existence of certain types of social organization and disorganization within a neighbourhood are possible explanatory variables in understanding where high offence rates occur and where offenders come from (Shaw and McKay, 1942; Bursik and Grasmick, 1993; Wilson, 1997; Hirschfield and Bowers, 1997). Explanations of where offenders come from often lay emphasis on area-level attributes and emphasize housing type and neighbourhood socialization processes that may provide too few sanctions on juvenile delinquent behaviour in certain areas (Bronfenbrenner, 1979; Martens, 1993). Wikström and Loeber (2000) identify neighbourhood socio-economic context as having a direct impact on the late onset of offending for certain groups of young offenders. Sampson et al. (1997) identify the role of collective efficacy, defined as a combination of social cohesion and a willingness for individuals to act on behalf of the common good to explain area-level variation in victimization rates.

Area-level contextual influences linked to social organization and processes of informal social control within neighbourhoods play a role in explanations of the causes of offending and victimization. These influences when analysed at an aggregate level are measured at the area level. Variables include:

socio-economic variables, demographic variables linked to family structure and residential mobility, ethnicity variables measuring the degree of ethnic heterogeneity found in an area and environmental variables linked to the nature and, in the case of housing, the density of the built form. By contrast there appear to be few analyses that introduce spatial processes or spatial relationships explicitly into explanations in criminology. Messner et al. (1999) suggest that the distribution of violent crime in the USA may be linked to the dynamics of youth gangs so that the geography of youth violence may be the expression of a spatially contagious process linked to social networks and other forms of communication. Cohen and Tita (1999) suggest that on the question of whether there is a diffusion process going on: 'the jury is still out' (p. 376).

Swartz (2000) distinguishes micro from ecological (or macro) traditions of analysis within criminology. A shift to the micro scale requires that socio-economic, demographic and environmental variables, which are still relevant at this scale of analysis, are defined appropriately. Now, for example, measures of the quality of the environment at the micro scale (e.g. in terms of lighting, street width, presence of cul-de-sacs) become important. And they are important in both the opportunities they may offer for crime and for the effect they may have on the formation of social networks. However, perhaps the more significant shift, in the context of spatial analysis, is that the city can now be treated in a more fragmented, local way. New modes of analysis assume particular importance such as the detection and investigation of crime 'hot spots' and other forms of localized patternings of offences, victims and offenders. As spatially fine-grained offence and victim data have become available through police recording systems so there has been an increase in micro-scale analyses in criminology.

(b)   Geographical and environmental (spatial) epidemiology

Geographical studies examining disease variations do not generally shed an unambiguous light on the causes of disease because exposure to a risk factor and disease outcome are not measured on the same individuals. Environmental risk factors and neighbourhood contextual effects may have quite small impacts which are overwhelmed by individual circumstances or lifestyle factors. In ecological analyses, regression and correlation techniques are used to explore and test for relationships between attributes, dose levels and disease outcomes but it can be difficult to separate out compositional from contextual effects. International scale studies can often provide the most insight because differences on a global scale can be large, such as in the case of the link between exposure to sunlight and the incidence of rickets (English, 1992). The important consideration is not the geographic scale, per se, but whether there is adequate variation in the risk factor and the populations are sufficiently distinct in terms of their exposure levels (Lloyd, 1995).

In small-area studies exposure levels may be more homogeneous but the interpretation of geographical variation is made difficult by the effects of population movements and migration, the size of the population at risk and errors in population estimates (English, 1992). Area estimates of the level of a risk factor, such as an areal estimate of the level of air pollution, are used to impute levels of exposure experienced by individuals. This may be the most cost-effective way of examining the impact of environmental risk factors. Measuring exposure at an individual level is often both costly and potentially unreliable. Analyses can be strengthened by selecting subgroups of the population (such as by age or race) and by controlling for potentially confounding variables such as socio-economic factors (Jolley et al., 1992).

However geographical studies can suggest causal hypotheses so that within epidemiology as a whole, geographical and environmental epidemiology represents a form of exploratory analysis (see chapters 6 and 7). Cuzick and Elliott (1992) classify the several types of small-area studies: investigations of clusters where there is no putative source of a risk factor; investigations of incidence rates around possible point sources of a risk factor of a given type; investigations of clustering as a general phenomenon; ecological studies; mapping disease rates. Epidemiologists appear to be divided on the value of these different small-area studies. The search for a sound methodology to undertake cluster detection has led to numerous techniques appearing in the medical and statistical literature, whilst at the same time drawing criticism that their contribution to establishing links between risk factors and health outcomes has been fairly limited. Swerdlow (1992) cites studies where the levels of raised incidence of malignant nasal conditions were traced to occupational hazards identified in areas of England with local boot and shoe and furniture making. Small-area studies may also be helpful in pointing to the specific source of an outbreak when the risk factors are understood, as in the case of an outbreak of toxoplasmosis in Greater Victoria, Canada which was linked through mapping to one reservoir in the water distribution system (Bowie et al., 1997).

The study of infectious diseases raises questions about the origins of an outbreak, how it develops through time, the geographical form and extent of its spread and the conditions under which a small outbreak may turn into a major epidemic in which a large proportion of the population becomes infected. Predicting the course and geographic spread of an infectious disease is critical to trying to control it, but each of the individual questions raises wider questions about the role of place and space, and these have influenced mathematical modelling and empirical investigations of infectious diseases (Bailey, 1975).

For example, certain characteristics of places have been identified as important in understanding the origins of an outbreak. In the case of common infectious diseases like measles, the origins of outbreaks have been linked to urban centres of sufficient size and in which the disease is endemic with an epidemic occurring when the conditions for spread are right (Bartlett, 1957, 1960).

The Hamer–Soper model is basic to deterministic and stochastic modelling of the course of an epidemic. Although there are important variants the focus is on transition rates (in the case of deterministic models for large populations) or transition probabilities (in the case of stochastic models for small populations) which are specified for each of the three states of an individual. *Susceptibles* are individuals not yet infected but who are members of the population at risk; *infecteds* are individuals with the disease and at a stage when they might pass it on to a susceptible; *removals* are individuals who have been vaccinated or had the disease and are no longer infectious nor susceptible. Early work assumed a population with homogeneous mixing so that all individuals were assumed to have the same-sized acquaintance and kinship circles. For example, the transition rate or probability for a susceptible to become infected in a given interval of time was modelled as proportional to the numbers of infecteds and susceptibles and the length of the time interval. From the set of transitions it was then possible to derive threshold conditions under which a small outbreak would become an epidemic (see Bailey, 1975 for a review).

The multi-region version of the Hamer–Soper model in Cliff et al. (1993, p. 363) was used to model measles outbreaks in Iceland and allowed homogeneous mixing within regions. However inter-regional transmission of the disease was the result of inhomogeneous mixing. Infection was passed from region j to region i through an inter-regional transition process. This process was a function of the number of susceptibles in region i and the number of infecteds in j, with a parameter that was modelled as an inverse function of the distance between the centroids of the regions.

The multi-region Hamer–Soper model limits the number of parameters to be estimated, by assuming that the inhomogeneous mixing between the N regions, rather than generate $N(N - 1)$ parameters, is a function of distance so that only a single parameter needs to be estimated. Large numbers of parameters create problems for model estimation and inference, and the models could become still more complex if it becomes necessary to add more information to capture the internal characteristics of the regions. The model adopts a top–down approach to the analysis of complex systems, partitioning the study area into a pre-determined number of regions or zones.

Other modelling approaches have adopted a bottom–up approach representing the process in terms of a large population of individuals. In these models it is interaction at the micro level that defines the dynamics and the geography of the spread of the epidemic. An early example of this is Bailey (1967) who studied the spread of a disease on a lattice of individuals, each classified as either a susceptible, an infected or a removal, and where the spread starts from a single infected individual at the centre of the lattice. The model is a stochastic model of disease spread. At any given time, a susceptible only has a non-zero probability of becoming an infected if spatially adjacent to an infected. In a model of this type susceptibles change their states according to local (neighbourhood) transition rules. The model contains no mechanism for the infection to 'jump' and in particular there can be no transmission between spatially separated populations since there is no migration. This is an early example of the application of cellular automata theory (Coucelis, 1985; Phipps, 1989). System-wide properties emerge from micro-scale interactions. Bailey analysed the threshold conditions under which an outbreak would become a pandemic. He used a regular lattice for his simulations, but more complex spatial inhomogeneities can be incorporated through the spatial configuration of the population, as Hagerstrand (1967) employed in his models of innovation diffusion – an even earlier example of this type of modelling.

(c)   Regional economics and the new economic geography
    The subdiscipline of regional economics is positioned at the intersection of geography and economics and overlaps with the field of regional science. The nature of regional science and its original links with economics and geography can be gauged from Isard (1960). The current emphasis within both these areas of research can be judged from journals including the *Journal of Regional Science, Papers of the Regional Science Association, International Regional Science Review* and *Regional Studies*. The field of regional economics is principally concerned with regional problems and the analysis of economic activity at the subnational scale. Research in this area focuses on case studies and the mathematical modelling of economic growth at regional scales – models which have to reflect the different economic circumstances applying at the regional as opposed to the national scale (Armstrong and Taylor, 2000).
    Early approaches to understanding regional growth differences focused on the role of the export sector and led to an approach to modelling based on pre-defined regions between which factors of production would move as well as flows of goods in response to levels of regional demand. Regional econometric and input–output modelling were characterized by a top–down approach in which inter- and intra-regional relationships were specified usually in terms of large numbers of parameters. One purpose was to develop regional forecasting models to track how economic change in particular sectors in particular regions

would transmit effects to other sectors in other regions through the export sector.

A long-standing interest in regional economics is the extent to which there is convergence or divergence in per capita income growth rates between regions in the same market (Barro and Sala-i-Martin, 1995; Armstrong and Taylor, 2000). Inter-regional and inter-sectoral flows of labour and capital, responding to wage and profit differences, were seen as important in inducing convergence. However, migration of inputs, drawing on neo-classical arguments, is only one type of spatial mechanism that could induce convergence. Baumol (1994) identifies the role of technology transfers and the spatial feedback effects arising from productivity growth. In addition to these spatial mechanisms there are other geographical aspects to the modelling. These include the effects of spatial heterogeneity (regional differences in resource endowments, labour quality, local government and institutional policies) and the effects of local spillovers for example (Rey and Montouri, 1999; Rey, 2001; Moreno and Trehan, 1997; Conley, 1999).

Economists 'new economic geography' is concerned with regional growth and with understanding how the operation of the economy at regional scales affects national economic performance (Krugman, 1995; Porter, 1998) and trade (Krugman, 1991). This field, according to Krugman (1998), has served 'the important purpose of placing geographical analysis squarely in the economic mainstream' (p. 7), although its content and overall direction has drawn criticism from some economic geographers (Martin and Sunley, 1996; Martin, 1999).

Porter's theory, whilst not cast in formal terms, is concerned with the positive externalities (the contextual benefits) that a firm enjoys by being located where the environment confers competitive advantage on its operations. The theoretical underpinings to this advantage are captured in 'Porter's diamond', a conceptual model consisting of four components: factor conditions, demand conditions, firm strategy and the role of related and supporting industries. Geographical proximity strengthens and intensifies the interactions within the diamond and Porter (1998, p. 154) argues that competitive advantage accrues most effectively to a firm from a combination of the right system-wide (or national) conditions combined with intensely local conditions that foster industry clusters and geographical agglomerations.

A central feature of Krugman's modelling is the 'tug of war between forces that tend to promote geographical concentration and those that tend to oppose it – between "centripetal" and "centrifugal" forces' (Krugman, 1998 p. 8). The former includes external economies such as access to markets, and natural advantages. The latter includes external diseconomies such as congestion and pollution costs, land rents and immobile factors. Models are general equilibrium and spatial structure, for example an uneven distribution of economic activity across locations emerges from assumptions about market structure and the maximizing behaviour of individuals. At the centre of new economic geography models is a view of the space economy as a complex, self-organizing, adaptive structure: complex in the sense of large numbers of individual producers and consumers; self organizing through 'invisible-hand-of-the-market' processes; adaptive in the sense of consumers and producers responding to changes in, for example, tastes, lifestyles and technology. Where neo-classical theory is based on diminishing returns in which any spatial structure (such as the creation of rich and poor regions) is self cancelling (through convergence), the new economic geography is based on increasing returns from which spatial structure is an emergent property (Waldrop, 1992). Model outputs are characterized by bifurcations so that shifts from one spatial structure to another can result from smooth shifts in underlying parameters.

(d)   Urban studies

Krugman's deterministic models appear to share common ground with multi-agent models used in urban modelling. In multi-agent models active autonomous agents interact *and* change location as well as their own attributes. Individuals are responding not only to local but also global or system-wide information. Again, spatial structure in the distribution of individuals is an emergent property, and multi-agent models, unlike those of the regional approach to urban modelling developed in the 1970s and 1980s, are not based on pre-defined zones and typically use far fewer parameters (Benenson, 1998).

These stochastic models have been used to simulate the residential behaviour of individuals in a city. They have evolved from cellular automata modelling approaches to urban structure (see section 1.2.2(b)). They describe a dynamic view of human interaction patterns and spatial behaviours that contrasts with the more static relational structures found in cellular automata theory (Benenson, 1998; Xie, 1996). In Benenson's model the probability of a household migrating is a function of the local economic tension or cognitive dissonance they experience at their current location. These tensions are measured by the difference between their economic status or their cultural identity and the average status of their neighbours. The probability of moving to any vacant house is a function of the new levels of economic tension or cognitive dissonance they would experience at the new location. If the household is forced to continue to occupy its current location, cultural identity can change.

A point of interest with both multi-agent and cellular automata models is how complex structures, and changes to those structures can arise from quite simple spatial processes and sparse parameterizations (White and Engelen, 1994; Portugali et al., 1994; Batty, 1998; Benenson, 1998). The inclusion of spatial interaction can lead to fundamentally different results on the existance and stability of equilibria that echo phase transition behaviour in some physical processes (Follmer, 1974; Haining, 1985). It is the possibility of producing spatial structure in new parsimonious ways (rather than assuming regional structures), together with the fact that the introduction of spatial relationships into familiar models can yield new and in some cases surprising insights, that underlies at least some of the current interest in space in certain areas of thematic social science. This interest, as Krugman (1998) for example points out, is underpinned by new areas of mathematics that make it possible to model these systems. In addition modern computers make it possible to simulate models that are not amenable to other forms of analysis.

Local-scale interactions between fixed elementary units, whether these are defined in terms of individuals or small areas, can affect both local properties and system-wide properties as illustrated by cellular automata theory. This effect is also demonstrated through certain models of intra-urban retailing where pricing at any site responds to pricing strategies at competitive neighbours. This can yield fundamentally different price geographies depending on the form of the profit objective and the spatial structure of the sites in relation to the choice sets of consumers (Sheppard et al., 1992; Haining et al., 1996). Multi-agent modelling adds another, system-wide level to the set of interactions, allowing individuals to migrate around the space and change type as a function of local circumstances, global conditions and local conditions in other parts of the region. However, all these forms of modelling raise questions about how model expectations should be compared with observed data for purposes of model validation. One aspect involves comparing the spatial structure generated by model simulations with observed spatial structures and this calls directly for methods of spatial data analysis (Cliff and Ord, 1981).

(e)   Environmental sciences

Wegener (2000) provides a classification by application area of the large range of spatial models in environmental sciences drawing on Goodchild et al. (1993) and Fedra (1993). Atmospheric spatial modelling includes general circulation models and diffusion models for the dispersion of air-borne pollutants. Hydrological models includes surface water and ground water modelling. Land process spatial modelling includes models for surface phenomena such as plant growth or soil erosion and models for subsurface phenomena such as geological models and models of subsurface contamination (through waste disposal or infiltration). Biological and ecological spatial modelling includes vegetation and wildlife modelling – models of forest growth, fish-yield models, models for the spread of diseases through natural or farmed populations and models for the effect of resource extraction (like fishing) on stock levels. Finally the classification includes integrated models which involve combinations of the above groups such as atmospheric modelling and the transport of air-borne infectious diseases to livestock populations. To the earlier classification, Wegener adds environmental planning models such as those for noise in an urban area.

Space in environmental modelling is often continuous and spatial relationships are defined in terms of distance – either straight line or in terms of network structure as in the case of rivers in a catchment. Biological and ecological models of the spread of disease may introduce problems of short- and long-distance migration of the modelled populations. This needs to be accommodated in order to represent population mixing within the spatial model. Examples of different types of spatial modelling in the environmental sciences can be found for example in Goodchild et al. (1993) and Fotheringham and Wegener (2000).

Table 1.1 provides a summary of different ways place and space enter generically into the construction of scientific explanation in the examples cited in this section. The table identifies the different generic classes and selects an illustrative example for each. The two 'views' of geography are not mutually exclusive, as illustrated in the bottom row of the table.

1.2.3   Spatial data analysis in problem solving

There is a similarity to nearby attribute values in geographic space and Tobler (quoted for example in Longley et al., 2001) has referred to this property as the 'First Law of Geography'. Fisher (1935) noted in the context of designing agricultural field trials: 'patches in close proximity are commonly more alike, as judged by yield of crops, than those which are further apart' (p. 66). Processes that determine soil properties operate at many different spatial scales from large-scale earth movements that are responsible for the distribution of rock formations to the small-scale activities of earth worms. The consequence is a surface of values that displays spatial dependence in variable values and may contain different scales of spatial variation (Webster, 1985).

The same is true even when values represent aggregates with respect to an areal partition. That socio-economic characteristics tend to be similar between adjacent areas has often been noted (see Neprash, 1934, for an early observation of this). As Stephan (1934) remarked: 'data of geographic units are tied together . . . we know by virtue of their very social character, persons, groups

Table 1.1 *A summary of the generic treatment of geography in scientific explanation (see text for details)*

	Location as place and context		Location as relationships between places				
	Individual level (individual units, e.g. people or households, as the objects of analysis).	Ecological level (spatial aggregates of individual units as the objects of analysis).	'Top–Down' inter-regional models.	Distance influences.	Neighbourhood (e.g. local gradient) influences.	'Bottom–Up' interaction models.	Neighbourhood + system-wide (e.g. configuration) influences.
Classification	Relationship between individual-level response and individual-level characteristics, exposures and the contextual effect of area(s).	Relationship between a response and compositional effects and exposure effects.	Relationship between regional economic growth rates and aggregate characteristics of firms and area measures of social capital.	Relationship between a response and compositional effects and spatial contextual effects.			
Examples	Relationship between individual experiences of victimization and personal characteristics, neighbourhood characteristics and higher-level spatial contextual influences.	Relationship between rates of a disease and environmental exposures after allowing for confounders, and compositional effects.	(1) Hamer–Soper model of epidemics. (2) Regional econometric models. (3) Regional models of urban structure.	Disease incidence as a function of distance from a possible source of pollution.	(1) Cellular automata. (2) Differences in deprivation levels between adjacent areas as a factor in understanding crime rates.		(1) Multi-agent models. (2) Krugman models.

Spatial variation in offender rates as a function of aggregate household attributes, local neighbourhood attributes (place and context) adjacent neighbourhood attitudes, neighbourhood attitudes opportunities to offend (space). Spatial variation in uptake rates of a health service as a function of aggregate household attributes, neighbourhood attitudes (place and context) and physical access to service as a function of location (space).

and their characteristics are inter-related and not independent' (p. 165). There is strong correlation between the adjacent pixels on a remotely sensed image of land cover. The fact that events close together in geographic space tend to be more alike than those further apart can be exploited to handle a number of scientific and technical problems. We briefly list some examples. In all cases a knowledge of the underlying spatial dependence can be exploited to tackle the problem.

[1] A researcher wants to sample a surface to estimate a parameter (such as the average level of a variable) to a pre-specified level of precision. The aim is to do so as efficiently as possible – that is by devising a sampling plan that will ensure the desired level of precision without taking an unnecessarily large sample size (Dunn and Harrison, 1993).

[2] Samples have been taken across a surface and the analyst wishes to interpolate the value at a location that has not been sampled. Or the analyst wants to draw a map of surface variation which also reflects the uncertainty associated with the different parts of the map as a consequence of the sampling plan adopted (Isaaks and Srivastava, 1989).

[3] A spatial database has been assembled for a set of variables. The database contains values that are missing 'at random', in the sense that there is no underlying reason (such as suppression for confidentiality) why the particular values are missing. The analyst wants to obtain estimates of the missing values (Griffith et al., 1989). A variant of this problem is as follows. The analyst wishes to use the same database to model the relationship between a response variable and a set of explanatory variables. Rather than discard every case for which there is one or more missing values which could seriously reduce the data set, it is possible to fit the model using all the collected data whilst making allowance for the missing values (Little and Rubin, 1987).

[4] A group of artefacts have been found as a result of archaeological field research. It is not clear which can be classified as belonging to the same type. Information on the attributes of the artefacts combined with information on the location where they were found may be used to provide a classification (Barcelo and Pallares, 1998).

The spatial nature of data, in particular the spatial dependence in the data, can be exploited to help solve technical problems of the sort described. One of the main problems is to represent that spatial dependence in order to use it in the estimation problem.

There is another group of problems where the 'replication' provided by having observations on many geographical areas or the fact that each area is embedded within a larger set of other areas can be exploited to yield solutions to certain problems.

[1] One role for ecological inference is to use ecological data to learn about the behaviour of individuals within aggregates. Suppose there are n spatial units each with a cross-tabulation for which only marginal totals (row and column sums) are known. The objective is to make inferences about the cells of each of the tables. This problem may arise for example analysing data on race (black or white) and voting behaviour (voted or did not vote), where the totals of each race voting are known as are the totals of who voted and who did not. The real interest however lies in the cells within the cross tabulations, for example the proportion of voting-age black people who vote in a given electoral area or precinct – values which are unknown (King, 1997).

[2] A survey has allowed estimates of the unemployment rate to be obtained for each of a number of small areas in a region. It is known that, whilst the small-area estimators that have been used are unbiased estimators of the small-area unemployment levels, they have low precision (high estimator variance) because of the small number of samples falling in each area. Low precision makes it difficult to detect real differences between the small areas. However, the estimator for the regional level of unemployment has a much higher precision, but as an estimator for any of the small-area levels of unemployment it will be biased.

In both of the problems cited, an approach lies through applying methods in which the small-area estimator 'borrows information' or 'borrows strength'. In the first problem a statistical model may be specified that draws on information from all the other n individual tables in the data set in order to estimate cell values. In the second, the small-area estimator can borrow information from the region-wide estimator. In some applications the procedure of borrowing information focuses on a geographically defined neighbourhood around the small area and spatial dependence is built into the estimation procedure (Mollie, 1996).

1.3   Spatial data analysis in the policy area

The policy maker is concerned with the strategic and tactical deployment of resources. A framework for such deployment is as follows: identify the areas of need according to specified criteria and set objectives, target the intervention, manage and monitor the intervention and evaluate the outcomes in relation to the objectives set. There are many forms of resource targeting and *geographical* targeting is one of them. Geographical targeting means directing resources at specific areas. These might be large areas such as regions entitled to bid for funds under different objectives within the European Union's Structural Funds programme or small areas such as local community-based initiatives taken by the police in partnership with local community groups to reduce crime in a neighbourhood. Projects have different time horizons from the short-term tactical to the long-term strategic.

Implementating a programme of intervention at the local or regional level initiates a process involving many different participants, requiring the sharing and integration of relevant data sets, including geographically referenced data sets, underpinned by relevant analytical tools. Where modelling is an element of any analysis, the inclusion of variables that can be manipulated by policy instruments is usually an important element in model specification.

'Intelligence-led' intervention is informed by different types of knowledge from understanding general processes operating at different spatial scales to highly localized knowledge of places and circumstances and the likely effectiveness of different courses of action. McGuire (2000) illustrates this from the perspective of the New York City Police Department's computerized crime statistics (COMPSTAT) process which has been credited with playing a significant role in crime reduction in New York in the mid to late 1990s. In the UK, the 1998 Crime and Disorder Act has placed a statutory requirement on police to undertake crime and disorder audits with local partners and to produce strategies based on this work. The focus is particularly on high-volume crime such as burglary and car theft. This Act together with the availability of geo-coded offence, offender and victim data have played a significant role in the growth of crime mapping and analysis (Hirschfield and Bowers, 2001). Gordon and Womersley (1997) discuss the ways GIS mapping capability and spatial analysis can support local health service planning.

The distinction between tactical and strategic deployment of resources is important for methodological reasons. Tactical deployment is often focused on a very narrow and specific set of objectives. This might include dealing with a sudden upsurge in street robberies in an area of a city or, in the case of a health authority, a sudden outbreak of a disease in a particular area or the identification of an area of abnormally raised incidence. Data sets underlying the formulation of a tactical response usually refer to short periods of time. There may be a need for rapid data collection followed by relevant data processing, perhaps 'hot spot' analysis, to support the case that something unusual is happening, where it is happening and to prioritize amongst competing demands (Craglia et al., 2000).

In the context of disease, 'cluster investigations, initiated in response to reports of apparent disease excess in a locality, are often demanded by public concern, but are difficult to interpret . . . a balance (has) to be struck between generating unwarranted public concerns and identifying genuine health risks as early as possible' (Wilkinson, 1998, p. 185). In the case of health, controversy may surround whether a cluster is statistically significant and whether it really does signal something of substantive significance that calls for special investigation or intervention – particularly when the number of cases is small and no cause can be identified. In such cases it may not be clear what action ought to be taken, if any. Wilkinson takes the view that the use of geographical 'surveillance' techniques – GIS-based systems for computing disease rates and applying statistical tests to look for clusters – will grow as the technology advances but that 'the interpretation of their output . . . requires expert judgement and considerable circumspection' (p. 186).

Strategic deployment of resources is based on long-term data series and on analyses that have identified if not causes then at least strong associations between, for example, socio-economic and environmental attributes, crime, disease or ill health. Strategic deployment in the case of a health or police authority may be associated with a (re)focusing of mission arising from what are seen as shortcomings in current levels of performance in relation to priorities. From this may follow decisions on implementation that result in new geographical patterns of strategic resource targeting that may distinguish between those elements of spatial variation due to compositional effects and those due to area-level contextual effects. The Acheson (1998) Enquiry into Inequalities in Health reviewed the 'evidence on inequalities in health in England . . . as a contribution to the development of the Government's strategy for health, to identify areas for policy development likely to reduce these inequalities' (p. xi). One of the recommendations was 'a review of data needs to monitor inequalities in health and their determinants at a national and local level' (p. 120). The recently completed ESRC Health Variations Programme focused on understanding different aspects of health variation in the UK, including geographical variation and the importance of place (ESRC, 2001).

In the UK the strategic deployment of resources for tackling social exclusion focuses on neighbourhood renewal and requires the geographical coordination of many small-area data sets (crime, health, education, housing, employment). Craglia et al. (2002) report an example of a children in need audit by enumeration district in Sheffield that involves bringing together many different local data sets. The next step is the co-ordination of the corresponding services with the aim of narrowing the gap between the most deprived areas and others (Minister for the Cabinet Office, 1999).

In the context of policing Swartz (2000) remarks that ecological or macro-level data analysis (see section 1.2.1) is likely to be of most interest to politicians who are concerned with the levels of strategic resource allocation across the different areas of the country. The link between deprivation and ill health has provided a justification for weighting resource allocation to Regional Health Authorities in England to reflect geographical variation in deprivation (see, e.g., Martin et al., 1994). There are dangers with ecological analyses (see chapter 4) and Fieldhouse and Tye (1996) warn that directing expenditures at areas with high levels of deprivation may not be targetting a high proportion of deprived people.

National politicians are unlikely to be interested in insights from highly localized, micro-level, spatial data analysis but those with responsibility for local problems and local-scale resource allocation will be (Swartz, 2000). Local government, grass roots organizations and neighbourhood associations as well as local police officials may find insights from micro-scale analyses helpful in identifying persistent high crime areas as well as helpful in pointing the way in terms of how resources might best be targeted. Once the areas are clearly delimited, 'the solution to the problem might be as simple as improving street lighting . . . or as complex as improving the living conditions of local residents' (Swartz, 2000, p. 44).

The technique of geographic profiling is 'a strategic information management system designed to support serial violent crime investigations' (Rossmo, 2000, p. 211). In contrast to 'hot spot' techniques that are used to analyse volume crimes like burglary and car theft at the local scale, geographic profiling is used to narrow down the range of possible anchor points (e.g. home address, work address) of a criminal engaged in serial rape or sexual assault. Geographic profiling uses a spatial analysis technique, informed by assumptions about offender activity spaces. The latter is based on the work of Brantingham and Brantingham (1991) and geographic profiling is applied to local data sets to produce 'a probability surface showing the likelihood of offender residence within the hunting area' (Rossmo, 2000, p. 197).

The neighbourhood is an important scale for strategic resource targetting. The neighbourhood environment may be important in influencing or even shaping individual, preventative health behaviours and attitudes towards health (Sooman and Macintyre, 1995). Neighbourhood characteristics (social, material, physical) are likely to be important in determining the extent to which elderly members of a population are able to cope independently.

Examples of the need to be sensitive to individual-level characteristics and neighbourhood characteristics are not limited to health interventions. Wikström and Loeber (2000) note in the context of preventative strategies aimed at reducing youth offending that it is important to base strategy on 'knowledge about the interaction of "kinds of individuals" in "kinds of contexts"' which 'have higher potentials to be effective than strategies that pinpoint either the individual or the context' (p. 1111). Strategy aimed at addressing problems of disadvantage needs to separate out those aspects of an underlying problem that relate to individual circumstances and which might be addressed using individually targetted support from those aspects that relate to area-level group effects and which should be addressed by area-level programmes.

The case for localized geographical targeting is strengthened when micro-level data analysis is underpinned by an understanding of likely cause. The link between coronary heart disease and certain types of diet is well established so that when areas of high incidence have been identified within a city it is often clear what needs to be included in any form of intervention. In other circumstances where specific causes may not be understood, such as in the case of low uptake of a screening programme, if 'cold spot' analysis identifies such areas in a city, it may be necessary to undertake individual-level surveys within the areas to try to discover why.

1.4   Some examples of problems that arise in analysing spatial data

This final section provides a few examples where attention paid to the spatial nature of the problem may guard against drawing unwarranted conclusions or following inefficient procedures. The properties of spatial data have an impact on many aspects of data analysis from description and exploration of data sets through to modelling.

1.4.1   Description and map interpretation

Before drawing conclusions from mapped data about geographical variation it is important to try to ensure that values are 'equally robust'. Some elements of spatial variability may be an artefact of the data arising from errors that have propagated through a data set as a result of carrying out arithmetic or logical operations on data contaminated by error (Arbia et al., 1999). It may have arisen as a consequence of the small number problem (Gelman and Price, 1999). In mapping disease rates by area, sample sizes may be small, sampling variability large so that structures in the data, particularly extreme rates but also spatial trends, may be a statistical artefact. As fine-grained spatial data becomes more readily available the risk of making this type of error increases. Mollie (1996) provides an illustration using cancer data for French *departements* of how the most extreme rates tend to be associated with areas with the smallest populations.

1.4.2   Information redundancy

The presence of spatial dependence means the information content of a sample used to estimate a population parameter is less than would be the case if the n observations were independent. In the terminology of Clifford and Richardson (1985) the 'effective' sample size is less than the number of cases sampled because data points near to one another carry 'duplicate' information about the parameter. Statistical testing needs to recognize the degrees of freedom that are actually available for carrying out tests using spatial data where observations are not independent.

Because data points close to one another carry duplicate information this needs to be recognized in designing sampling plans (Dunn and Harrison, 1993). This feature of spatial data as well as the specific configuration of the observed data points needs to be recognized in spatial interpolation (Isaaks and Srivastava, 1989).

1.4.3   Modelling

A regression model may provide a good fit but a map of the residuals (the differences between the observed values of the response and the model predictions or fits) may show clear evidence of spatial structure that is confirmed by a test statistic. This violates one of the assumptions of regression which undermines the validity of inferences drawn from the model. In addition, notwithstanding the goodness of fit of the model, such a result suggests the model can be improved perhaps by inclusion of new covariates in the model specification.

1.5   Concluding remarks

This chapter has considered the importance of spatial data analysis in a number of different areas, distinguishing between the role of spatial data analysis as a tool of science and as a tool of the policy maker. Spatial data analysis is concerned with studying patterns in variables and associations between variables but in the absence of time series data it is usually not possible to identify causal relationships; this means neither the components of any causal system nor the directions of causation in those cases where there is ambiguity about the direction of causation. At best spatial analysis can point to possible causal relationships which can then be followed up by other methods.

Large quantities of fine-grained geographically referenced data are available, and can be linked on a common spatial reference with the help of a geographic information system. Geographic variability can today be mapped and analysed at a level of detail and spatial extent that in the past was difficult to undertake. The availability of such geo-coded data provides an opportunity to construct new views of old problems and to explore views of new spatial data prompting new understanding and new insights. Such a claim rests at least in part on the quality of the data and later chapters (particularly chapter 2) will discuss issues of data quality. However at this point it is worth remarking on two further points. First, fine-grained geo-coded data means precise information on location but clearly the locational reference must be relevant to the area of enquiry. Place of residence is important in analysing household burglary patterns but place of residence may be less useful in analysing geographies of chronic diseases with long latency times. Second, the greater the level of spatial detail the higher the level of noise there is likely to be in the data and the greater the need to draw on methods that distinguish between 'noise' and 'signal' in pattern analysis and in the analysis of relationships.

2

The nature of spatial data

This chapter discusses the nature of spatial data. All the analytical techniques in this book use a space-(time-) attribute data matrix (see the introduction). This matrix is the end product of a process of construction that starts from a conceptualization of geographical reality. What is it necessary to know about the relationship between that reality and the data matrix as a representation of that reality so far as the conduct of analysis is concerned and the interpretation of results?

There are a number of steps involved in specifying this relationship (see, e.g., Longley et al., 2001; Mark, 1999). First there is a process of *conceptualizing* the real world. This process extends to the identification of those fundamental properties that are relevant to the application. Such fundamental properties relate both to entities ('things in the real world') and the spatial relationships between entities. In the context of spatial data analysis spatial dependence in attribute values is considered as a fundamental property. Second, a data matrix acquires properties that may distance it from the real world as a consequence of *representational* choices. These are the decisions made about what to include in the data matrix and in what form, usually for the purpose of storing the data in a computer. Decisions must be taken, for example on spatial scale or level of spatial aggregation and the geometric class (points, lines, areas or surfaces) used to represent geographical entities. Third, when attributes and spatial properties are measured, this introduces another source of uncertainty between the real world and what is contained in the data matrix. Inaccuracies in the *measurement process* is an important source of uncertainty that affects the spatial data matrix.

Section 2.1 considers conceptualization and representation issues as they relate both to geographic space and the attributes captured in a data matrix. There is discussion of the nature of spatial dependence. Section 2.2 defines the data matrix, and makes observations on the relationship between it and the geographic reality it is attempting to represent. Section 2.3 considers different dimensions of data quality. The implications of data quality for the conduct of spatial data analysis are discussed in chapter 4.

The final section 2.4, describes methods for obtaining quantitative measures of spatial dependence. Although spatial dependence is a fundamental property it is important to see it as one of many data properties (some inherited from the chosen representation rather than fundamental) to be thought about in analysing spatial data. Discussions about the importance of spatial dependence and the methods of section 2.4 need to be seen in the context of broader issues that are raised particularly in sections 2.2 and 2.3.

2.1   The spatial data matrix: conceptualization and representation issues

2.1.1   Geographic space: objects, fields and geometric representations

Modelling geographic reality means the process of capturing the complexity of the real world in a finite representation so that digital storage is possible. This abstracting of a 'real, continuous and complex geographic variation' (Goodchild, 1989, p. 108) into a finite number of discrete 'bits' involves processes that include generalization and simplification. *Objects* and *fields* represent two fundamental conceptualizations of the entities that comprise geographic reality (Goodchild, 1989; Salgé, 1995; Longley et al., 2001). The difference is most easily expressed through examples. Variables such as temperature, snow depth or height above sea level are appropriately conceptualized as fields. A house (point), road (line) or political unit (area) are usually conceptualized as objects. Objects refer to things in the world whilst a field refers to a single valued function of location in two-dimensional space (see figure 2.1). Usually one or other of these two accord better with our mental perception of the real world and may also provide a better basis for efficient computation (Mark, 1999).

Four classes of digital objects for representing geographic phenomena are usually identified – *points, lines, areas* and *surfaces* (as contour lines for example). Object space is represented digitally by points, lines or areas. A town may be represented as an area (using its administrative boundary to delimit the area) or at another scale of representation as a point. As an area its representation may be refined using census tract-level data (e.g. wards or enumeration districts). Each enumeration district may be represented as an area object using the administrative boundary or as a point object by identifying the area or population-weighted centroid. The population of the town can be

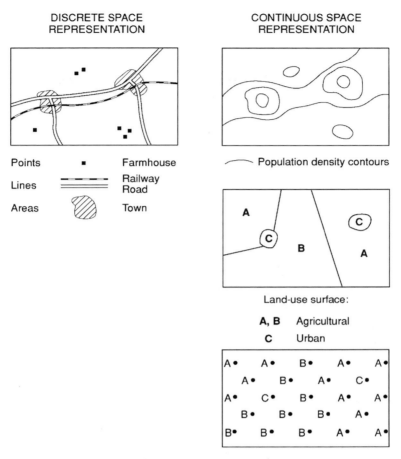

Figure 2.1 Discrete and continuous space representations

represented as address point objects at the scale of individual households. It follows that census tracts represent a *spatial aggregation* of these fundamental entities. A town or a forest are represented as area objects and a boundary line drawn, even though in reality the boundary may be ambiguous and 'fuzzy'.

In the case of a field, data values associated with attributes are possible at each of an infinite number of point locations on the surface. Storing data about a field in a data matrix requires it to be made finite. The field as a surface can be represented by using contour lines. Representing a field using areas often means dividing the region into small regular spatial units called *pixels*. Pixel size specifies the *spatial resolution* of the representation and is the field equivalent of spatial aggregation. To represent a field by points means choosing sample locations. A point measure may be literally sufficient as in the case of a measure of soil or snow depth. In cases like air pollution any measure is a function of the size of block ('support') used to define the quantity.

In the case of area objects and area representations of a field, either the areas are defined independently of data values as in the case of census tracts and image pixels or their boundaries reflect a change in data values. In the first case the areas are said to be *intrinsic*. In the second case the areal partition is imposed *after* analysing data values and the partition defines homogeneous (or quasi-homogeneous) areas or *regions*. Fields can be *segmented* into blocks of pixels with the same or similar values. Census tracts can sometimes be *aggregated* into larger groupings of contiguous tracts that are similar at least with respect to a small number of variables.

There are situations, however, where there is a choice of conceptualization. Population distribution can be conceptualized as object or field. If conceptualized in terms of objects then the representation may be in the form of points (e.g. by residence) or counts by regular areas (pixels) or irregular areas (e.g. census tracts). If conceptualized in terms of a field then the representation may be in the form of a density surface, or by spatially distributing population counts using kernel density smoothing or by interpolation. Bithell (1990) and Kelsall and Diggle (1995) construct relative risk surfaces for disease using kernel density methods to convert population count and disease count data to density surfaces (see chapter 7).

The implications of these representational choices are considered in section 2.3. For further perspectives on spatial representational issues see Raper (1999).

2.1.2   Geographic space: spatial dependence in attribute values

The presence of spatial dependence means that values for the same attribute measured at locations that are near to one another tend to be similar, and tend to be more similar than values separated by larger distances. By 'similar' is meant that if an attribute value is large (small) then nearby values of the same attribute will tend to be large (small). This characteristic has parallels with time series data. Values for the same variable close in time tend to be similar and more similar than values separated by longer time periods. The nature of this similarity may be independent of where (in space) or when (in time) values are measured. For spatial data this implies, no matter where one looks on the map, that the nature of that similarity is the same. The dependency structure in this case is said to be *stationary*. By contrast, if the structure of dependency varies across the map so that any measure of similarity depends on which part of the map is analysed, it is said to be *non-stationary* or that the structure of dependency is *heterogeneous*.

However there are important differences between space and time in respect of dependency which is why different spatial statistical techniques are needed to quantify and analyse spatial as opposed to temporal data (see section 2.4). It is also the reason why different statistical models are needed to describe spatial as opposed to temporal variation (see chapters 5 and 9). First, time has a uni-directional flow. The past may influence the present but the future can only influence the present in the sense of an expectation of that future, not in the sense of an actual realization of that future. Space has no equivalent to the trilogy of past, present and future. Second, spatial dependency is complicated by extending over two dimensions, not one, and because the structure of that dependency need not be the same along the two axes (north/south; east/west). If the dependency structure is the same on both axes it is called an *isotropic* dependency structure, if it is not it is called *non-isotropic* or *anisotropic*. Finally, periodicity which is often encountered in time series data (seasonal effects, business cycle effects; daily and weekly effects) is not often encountered in spatial data.

2.1.3   Variables

Attribute characteristics can refer to the spatial objects themselves or to entities that are associated with or attached to the spatial objects but not directly dependent on them. Attribute characteristics that refer to spatial objects such as the size or spatial extent of an area object raise conceptual and representation issues – what is the length of a coastline or the area of a forest? Attribute characteristics attached to spatial objects such as the number of cases of an offence, the number of plant species are also subject to conceptualization and representation issues. Conceptualization refers to the definition and meaning of the attribute (an offence, a disease, an economic sector, deprivation). Representation refers to how an *attribute* is operationalized into *variables* for the purpose of acquiring and storing data on the attribute (e.g. how deprivation is measured) and to enable analysis to be undertaken. Analysis is undertaken on data collected with respect to one or more variables that measure attributes associated with geographic reality that is typically represented in the form of spatial objects.

Conceptualization and representation issues of attributes are specific to each particular application. Two generic issues that can be discussed here relate to or have implications for attribute representation and data analysis. The first is how variables are classified and the second is the level of measurement of a variable. Classification identifies the place of each variable in an analysis; level of measurement defines what arithmetical operations are permissible.

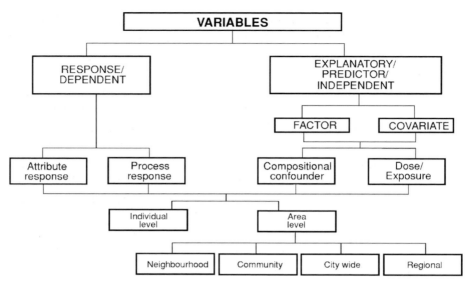

Figure 2.2 Classification of variables

(a)   Classifying variables

Depending on the context, variables are divided into groups and given different labels (see figure 2.2). In the case of data modelling, *Y* represents the variable whose variation is to be explained. In multiple regression modelling where the purpose is to explain variation in *Y* in terms of other variables, *Y* is called the *dependent* or *response* variable and the other variables are called *independent* or *explanatory* or *predictor* variables $(X_1, X_2, \ldots)$. If a predictor variable is measured at the nominal or ordinal level it is often called a *factor* and if measured at the interval or ratio level it is often called a *covariate* (see below for the definition of these level of measurement terms).

When modelling, it is useful to distinguish between variables in a further sense. Some variables measure *individual-level attributes,* such as the age or sex of an individual, or the number of bedrooms or floor space in a house. In some forms of spatial modelling these quantites may be aggregated over all individuals (people, houses) within the area but they can still be referred back to the attributes of individuals located within the spatial unit. These types of variables can be referred to as measuring *aggregated individual-level attributes* – for example the proportion of the resident population in the age group 16–64. These can be used to make an assessment of compositional effects on a response that refers to a spatial aggregate.

Attributes can also be attached to areas to refer to area-level or group-level properties. The Townsend index of material deprivation and the DETR index of local need are area-level measures of deprivation (Townsend et al., 1988; DETR, 2000). An individual, who may or may not be individually deprived as a consequence of a particular combination of personal or household character-istics, might be said to be *exposed to* or in their daily life *experience* the level of deprivation in the area where they live. Social capital, social cohesion and social control or guardianship are attributes defined at the area level to capture eco-nomic, sociological or power structure attributes of areas (e.g. Putnam, 1993). Individuals might be said to be *exposed* to or *experience* the level of social cohe-sion present in the area where they live (Hirschfield and Bowers, 1997). Col-lective efficacy, defined as 'social cohesion among neighbours combined with their willingness to intervene on behalf of the common good' (Sampson et al., 1997) is a neighbourhood-level attribute that in combination with aggregate demographic characteristics of individuals may help to explain geographical variations in certain types of offending. In the context of spatial modelling, variables measuring such attributes are measuring *area-level contextual attributes*. Such attributes may exist at different levels or scales ranging from the neigh-bourhood or ED level, to the level of the ward or some grouping of EDs through to city-wide, regional and higher spatial scales. Typically variables that mea-sure such attributes, whilst they might be constructed from variables measur-ing individual-level attributes (derived from the census), involve combining variables into an index which is meant to quantify (or operationalize) a group-level concept.

In environmental epidemiology, covariates may be included that measure exposure to possible environmental risk factors – such as particulate matter in the incidence of respiratory conditions. These variables are called *exposure* or *dose* variables. Individual-level attributes (e.g. age and sex composition of area populations) must be accounted for in explaining spatial variation in dis-ease incidence because of area variation in demographic groups (compositional effects). There are other types of covariates which must be accounted for. The term *confounder* variable is used to denote a variable that may also influence the level of a response variable (e.g. the rate of some disease) and which if not allowed for may obscure the true nature of a dose–response relationship. Deprivation is a confounder variable in the relationship between exposure to air pollution and the rate of a respiratory condition. This is because deprived populations often live in areas that suffer from higher rates of air pollution and deprived populations may suffer from higher rates of respiratory problems for reasons not directly linked to air quality (such as poor housing or poor diet). Cigarette smoking levels should also be considered as a confounder in such an analysis, and to have a more direct association than deprivation with respira-tory health.

A regression model can be classified as either an *attribute–response* or a *process–response* model. This classification derives from the nature of the model and whether explanatory variables relate to underlying processes or not. For example, a regression model may be used to explain variation in house prices across a region in terms of other individual- and area-level attributes including housing characteristics and area characteristics. In this case the attribute–response model is analysing relationships between attributes and is not constructed in terms of the processes responsible for determining house prices, which presumably should include market processes and the maximizing behaviours of buyers and sellers. In a process–response model the explanatory variables include variables that link directly to the underlying process mechanisms. The previously cited epidemiological models that include dose or exposure variables fall into this category.

(b)   Levels of measurement

Specifying the level of measurement of a variable is important because it specifies the formal properties of the number system underlying the measurement and determines what arithmetic operations are valid and hence what statistical procedures can be employed. In the case of nominal data, cases can only be said to belong to the same class or not, such as land-use type or lifestyle category. At the ordinal level the classes must have an order or ranking as in the case of road status (motorway, A class, B class, minor) or income category (under £5000 p.a; £5000 to under £10 000 p.a.; £10 000 to under £15 000 p.a. etc.). Nominal and ordinal data consist of counts by categories (categorical data) which are called discrete variables. The term 'qualitative variable' is also sometimes used.

At the continuous scale, observations may fall anywhere on a continuum. There are two levels of measurement: interval and ratio. An important difference between the interval and ratio scales is that at the interval scale it is not possible to assert whether, for example, one number is twice as big (or small) as another, because there is no natural origin. In the case of the Townsend index of material deprivation a ward with an index of 4.0 is not twice as deprived as a ward with an index of 2.0. This is because an index value of 0 does not mean an absence of deprivation.

Nominal data can be summarized and compared using modes and frequency distributions. Data at the ordinal level can be summarized and compared using medians and boxplots as well. At interval and ratio levels of measurement, means and standard deviations can also be used. Any variable measured at the ordinal level or higher can be reduced to a lower level of measurement but this results in a loss of information. However there are circumstances when this might be appropriate, as for example when the measurement process is known to contain bias. Interval- and ratio-level data might be degraded to ordinal-level data. The analyst is expressing confidence in the ordering of values, but not the actual data values themselves. This may occur in the analysis of some small-area official crime statistics, for example where the analyst is aware of undercounting but it is not such as to invalidate the real differences that exist between areas. A variable that is not of direct interest (such as a confounder variable in an epidemiological regression model) may have a non-linear effect on the predictor. One way to handle such a variable, for example an interval-valued deprivation index, is to reduce it to ordered classes so that it appears as a set of dummy variables in the model.

Even when data are retained at the ratio level, the analyst may prefer to use statistics that assume a lower level of measurement (the median to measure the centre of a distribution of values rather than the mean). This is because such statistics are robust to possible errors or extreme values that might be present in the data (Hampel et al., 1986).

Table 2.1 gives examples of spatial data classified by the type of spatial object to which a variable refers and the level of measurement of the variable attached to the spatial object. As noted above, attribute values may also be attached to the objects themselves, such as area (in the case of an area object) or length (in the case of a line object).

Variable values are associated with map objects. In order to be able to specify permissible map operations applied to the map objects it is necessary to distinguish between variables that are *spatially extensive* and those that are *spatially intensive*. Quantities such as counts by area are termed spatially extensive and when two areas are merged the corresponding counts can be summed to give the quantity for the newly created map object. Rates, densities and proportions are area dependent. The denominator refers to some attribute of the area (size, population, population at risk) and the variables are called spatially intensive (Goodchild and Lam, 1980). To arrive at the correct value of a spatially intensive variable after aggregation the numerator and denominator must be aggregated separately. This distinction has implications for areal interpolation (see section 4.2.2(b)) and for visualization and statistical analysis (see chapters 6 and 7).

2.1.4   Sample or population?

The spatial objects and the attributes that are present across geographic space may either be conceptualized as comprising the whole population or a single realization from some 'superpopulation'. This is an important aspect of conceptualization that has implications for how spatial data are statistically analysed (the type of sampling theory that is relevant) and the nature of inference.

In some applications the reality that is observed (in terms of objects and attributes) is considered to be the only possible state. If samples are taken then

Table 2.1 *Classification of spatial data by level of measurement and type of spatial object*

Level of measurement	Spatial representation			
	Point (P)	Line (L)	Area (A)	Surface (S)
Nominal (=)	House: burgled/not	Road: under repair/not	Census tracts classified by lifestyle	Land-use type
Ordinal (≥; ≤)	Preference rankings of towns in a region by quality of life	Road classification (Motorway; A, B, . . . class)	Census tracts assigned to income classes	Soil texture (coarse/medium/fine)
Interval (≤; ≥; ±)	Townsend index* for town	Length using Greenwich Meridian as reference	Townsend index* for wards	Ground temperature (°C)
Ratio (≤; ≥; ±; ×; /)	Output from a factory p.a.	Freight tonnage p.a.	Regional per capita income	Rainfall (cm); snow depth (cm)

Notes: *Townsend index of material deprivation (see Townsend et al., 1988).
Permissible operations and relationships between numbers are given in brackets.

the inferences that are made apply to properties of that observed state. If the particular state is observed in its entirety there would be no need to undertake statistical inference except perhaps in relation to, for example, other possible spatial configurations of what has been observed. Statistical inference is used to decide whether a pattern of attribute values can be classified as random or not (the randomization hypothesis). We call this the deterministic case and either the location of the spatial objects or the attribute values or both might be conceptualized as the outcome of a deterministic process.

In other applications the reality that is observed is only one of a theoretically very large (perhaps infinite) number of possible states. In analysing the realization that has occurred (the actual counts of offences or numbers of cases of a disease) the analyst may be more interested in drawing conclusions about the underlying process responsible for it. The analyst is less interested in analysing the realized map of disease incidence or offences and more interested in specifying the generating model, estimating its parameters and testing hypotheses. These parameters, for example the relative risk of a disease or of houses being burgled, are the parameters of real interest which the actual counts of cases in any one period are used to estimate. We call this the stochastic or random case and again either the locations of the spatial objects or the attribute values or both might be conceptualized as the outcome of a stochastic process.

These contrasting conceptualizations of what is observed do not divide neatly into subject areas or types of attributes. In constructing a superpopulation view the analyst may invoke the existence of other regions from the same population, existing at the same point in time, to justify the use of statistical modelling and statistical inference (e.g. geostatistical applications in geology). In other areas the superpopulation view is underpinned by the idea of the process replicating over time in the same location (air pollution maps for different years) making the assumption that the underlying process has not changed in the intervening time period. Patterns of crop yield by small areas in any given year might be considered as one possible realization of values. The process responsible for actual yields in any particular year is complex, involving a very large number of small effects and hence viewed as stochastic in nature. Sometimes the superpopulation view is not justified in any of these ways but used as a conceptual device to avoid the criticism of placing too much emphasis on the particular data set that has been collected (such as a national population census) and on which analysis is then performed. There is often a vaguely defined superpopulation comprising, in the case of census data, the set of other national population assemblages like the one observed at the particular time when the census was taken. Sampling theory is invoked to attach confidence intervals to reported estimates.

The spatial objects and attribute values may both be stochastic or both deterministic. The spatial objects may be stochastic and the attributes deterministic. The spatial objects may be deterministic and the attributes stochastic valued. The situation is further complicated by recalling from section 2.1.3 that the term attribute may refer either to an attribute defined on the object (number or proportion of cases of a particular event attached to the object) or a spatial attribute of the object itself (such as its size or maximum spatial extent in the case of an area; length in the case of a line). This potentially gives rise to a 2 (object location) $\times$ 2 (object attribute) $\times$ 2 (attribute defined on the object) or three-way typology with eight types.

A map showing diseased trees in a region might be conceptualized as stochastically located objects (the trees) with stochastic attributes (diseased or not diseased). The location of vegetation clumps of a particular species might be conceptualized as the outcome of a stochastic process. Although the attribute referring to species type is fixed, the attribute referring to the size of each area might be conceptualized as the outcome of a stochastic process. Census areas are treated as deterministic spatial objects but the number of events occurring within the areas might be treated as the outcome of a stochastic process.

This classification is considered further in the next section.

2.2   The spatial data matrix: its form

Spatial data are classified by the type of spatial object to which variables refer and the level of measurement of these variables. Let $Z_1, Z_2, \ldots, Z_k$ refer to k variables (which may be differentiated in some applications as described in section 2.1.3(a)) and $\mathbf{S}$ to the location of the point or area. The spatial data matrix is represented generically as:

$$
\begin{array}{c}
\text{Data on the } k \text{ variables} \qquad \text{Location}\\
\left[
\begin{array}{ccccc}
z_1(1) & z_2(1) & \cdots & z_k(1) & \mathbf{s}(1) \\
z_1(2) & z_2(2) & \cdots & z_k(2) & \mathbf{s}(2) \\
\vdots & \vdots & \vdots & \vdots & \vdots \\
z_1(n) & z_2(n) & \cdots & z_k(n) & \mathbf{s}(n)
\end{array}
\right]
\begin{array}{l}
\text{Case 1}\\
\text{Case 2}\\
\vdots\\
\text{Case } n
\end{array}
\end{array}
$$

which can be shortened to:

$$\{z_1(i), z_2(i), \ldots, z_k(i) \mid \mathbf{s}(i)\}_{i=1,\ldots,n} \tag{2.1}$$

The use of the lower case symbol on z and $\mathbf{s}$ denotes an actual data value whilst the symbol inside the brackets references the particular case. Attached to a case

(i) is a location $\mathbf{s}(i)$ which represents the location of the spatial object. The bold font on $\mathbf{s}(i)$ identifies this as a vector. Time is implicit in the specification because all observations should be compatible in the time period they refer to. There may be several data matrices each referring to a different time period.

In the case of data referring to point objects the location of the ith point is given by a pair of co-ordinates as illustrated in figure 2.3(a). The axes of the co-ordinate system will usually have been constructed for the particular data set but a national or global referencing system may be used. For some modelling applications the axes are scaled to the unit square. This system of referencing is appropriate whether the locations refer to the points of a discrete space or point samples on a continuous surface.

In the case of data referring to areas the 'location' of each object needs to satisfy an agreed convention. If the areas are irregular shapes then one option

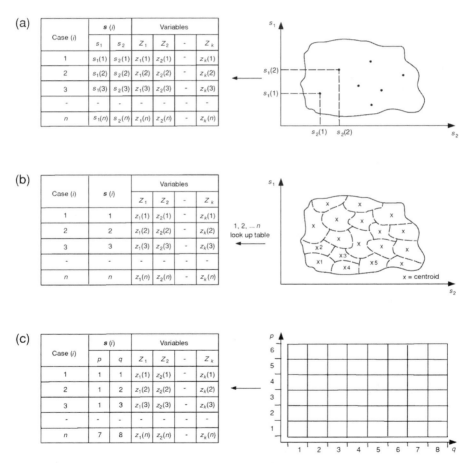

Figure 2.3 Assigning locations to spatial objects

Table 2.2 *Typologies of spatial data*

Types of data[+]		Model or 'scheme'		Example	
GISc	Cressie (1991, p. 8)	Variable value	Spatial index	Variable	Space
Point or area object data	Lattice	Variables (discrete or continuous valued) are random variables	Point or area objects to which the variables are attached are fixed	Crime rates Land use Disease rates Prices	County Urban tracts Census tracts Retail sites
Continuous-valued field data	Geostatistical	Variable is continuous valued function of location	Variable is defined everywhere in the (two-dimensional) study region	Soil pH Surface Temp ($^{\circ}$C)	Watershed Area of water
Randomly located point-object data	Point patterns	(i) Given attribute (ii) Variable is a random variable	Randomly located point objects in the study region	(i) Trees (ii) Trees: diseased or not (i) Hill forts (ii) Hill forts: classified by type	Forest area Forest area Archaeological research area Archaeological research area
Random area-object data	Objects	Spatial extent of each area object is a random variable	Location of area objects (e.g. their centre or origin point) in the study region is a random variable	Lichen patches Vegetation clumps	Moorland Field

Note: [+] Types of data as suggested by the Geographical Information Science literature (GISc) and Cressie (1991).

is to select a representative point such as the area or population-weighted centroid and then use the same procedure as for a point object to provide $s(i)$. Alternatively, each area is labelled and a look-up table provided so that rows of the data matrix can be matched to areas on the map (figure 2.3(b)). If the areas are square pixels as in the case of a remotely sensed image they may be labelled as in figure 2.3(c).

It will be necessary to have a method for keeping track of spatial relationships, particularly adjacencies in the case of area data and this will be discussed in section 2.4.

The classification of spatial data by type of object and level of measurement is a *necessary* first step in specifying the appropriate statistical technique to use to answer a question. That the classification is not *sufficient* is because the same spatial object may be representing quite different geographical spaces (points are also used to represent areas for example). In addition spatial objects and the attribute values attached to the objects may be the outcome of deterministic or stochastic processes as described in section 2.1.4. Table 2.2 provides a typology of spatial data. The table uses the terminology of this chapter but also links to the terminology used by Cressie (1991) for classifying different spatial statistical models.

In describing the nature of spatial data it is important to distinguish between the discreteness or continuity of the space on which variables are measured and the discreteness or continuity of the variable values themselves. If the space is continuous (a field), variable values must be continuous valued since continuity of the field could not be preserved under discrete-valued variables. If the space is discrete (object space) or if a continuous space has been made discrete (e.g. by segmenting it, see section 2.1.1), variable values may be continuous valued or discrete valued (nominal or ordinal valued).

2.3   The spatial data matrix: its quality

The relationship between the real world and the data matrix, including the inheritance of fundamental properties such as spatial dependence, is influenced by the two phases of a mapping from reality to any specific data matrix. These are, first decisions taken on the choice of representation (in terms of both the representation of geographic space and the attributes to be included and how they are to be measured) and second by the accuracy of measurements (on both geographic co-ordinates and attribute values) *given* the chosen representation. Figure 2.4 depicts this relationship and the terms often used to characterize it (see, e.g., Longley et al., 1999). The chosen representation constitutes the *model* of the real world that is employed. Any data matrix can be assessed in terms of the quality of this model so that the first stage of assessment of a

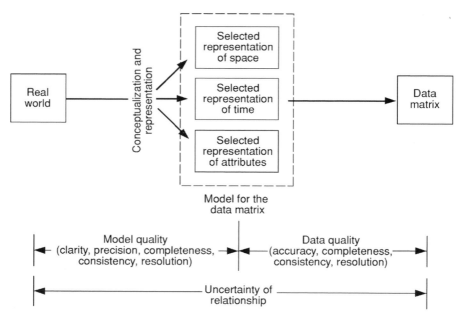

Figure 2.4 From geographical reality to the data matrix

data matrix can be in terms of *model quality*. Model quality may be assessed in terms of the precision (as opposed to vagueness) of a representation, its clarity (as opposed to ambiguity), its completeness (in terms of what is included) and its consistency (in terms of how objects are represented). The level of resolution or spatial aggregation is also a representational issue.

The second stage of assessment is in terms of *data quality* given the model. Importance attaches here to the accuracy of (or lack of error in) the data and completeness in the sense of coverage for example. The overall relationship between the space-(time-) attribute data matrix and the real world it is meant to capture (arguably the most important relationship) is sometimes specified in terms of the *uncertainty* of the mapping. The form of uncertainty associated with any matrix is therefore a complex combination of these two stages, model definition and data acquisition through measurement, that are associated with moving from geographic reality to the spatial data matrix. We now consider the two dimensions of model and data quality.

2.3.1   Model quality

Model quality refers to the quality of the representation by which a complex reality is captured. As described in the previous sections this involves discretizing reality in terms of a finite collection of spatial objects, spatial relationships and variables. Assessment of model quality includes whether, for example, data on the necessary spatial objects and variables, measured appropriately and sufficiently current, are available and have been correctly encoded according to the set of representation rules (Salgé, 1995). Model quality involves assessment of the appropriateness of the spatial representation of an object and the level of detail provided (including spatial detail in terms of resolution or aggregation of fields or objects). Buttenfield and Beard (1994) suggest the use of the term accuracy to reflect the correspondence between a representation or conceptualization and what the analyst wishes to measure. The term error they suggest should be used in the context of measurement processes (see below).

(a)   Attribute representation

Consider first the quality of the representation of an attribute and consider a particular example. The specification for the database may include the need to have a measure of deprivation. Measurement of deprivation starts with a conceptualization of what deprivation is from which may follow a specific operationalization in the form of an index. It may be difficult to speak of error in measuring deprivation when there is no unambiguous (true) value and many ways of operationalizing the measurement of the concept both in terms of what variables to include and which types of arithmetic or logical operations to use in the construction of the index (Lee, 1999a and b; DETR, 2000). Now, suppose it is concluded that the Townsend index should be obtained for each of many spatial units. If the study area includes rural and urban areas, is an index based on the four dimensions of overcrowding, unemployment, housing tenure and car ownership as appropriate to rural as urban deprivation? In rural areas, deprived populations do not own their own house and live in overcrowded conditions but they do tend to own a car and historically at least have suffered less from unemployment.

Model quality also calls for consideration of the extent to which surrogate data represent valid proxies for what the analyst would like to include in the database. Road traffic-count data or even numbers of vehicles per unit area provide estimates of relative levels of exposure to road traffic pollution and NO_2. Wikström (1991) not only defined 'best' denominators but also what he considered the 'best practicable' for computing area offence rates in Stockholm. Swerdlow (1992) lists the minimum set of variables required to undertake an analysis of small-area cancer incidence data (see also Wakefield and Elliott, 1999). These studies help to ensure model completeness in specifying the contents of a data matrix (Brassel, 1995).

(b)   Spatial representation: general considerations

Decisions on how to represent attributes in geographic space are influenced by the specific application and the spatial scale of the analysis. The choice may be made on pragmatic grounds (data are only available on a certain spatial framework or there is a need to find a common spatial framework for integrating different data sets); methodological grounds (more powerful analytical methods are available for some data representations than others); or theoretical grounds (a population density surface may better reflect the mobile nature of population in studying an infectious disease). Martin (1999, p. 75) suggests disease incidence can be represented in terms of either points, lines, areas or surfaces. Choice of representation has implications not only for the type of digital object used to capture the phenomena but also for the choice of analytical techniques and the form of visualization subsequently used. Martin (1999) notes there is no single 'right way' to represent socio-economic phenomena with the implication that 'considerable onus is placed on users to fully understand the implications of the representation strategy which they choose to adopt' (p. 78). The choice of representation should not be made uncritically for in any given circumstance the choice may raise 'technical, conceptual and ethical difficulties' (p. 79).

If analysis can disregard internal differentiation and if the geographical scale of the analysis means that detailed estimates of spatial relationships such as distance or areal configuration are not needed then the representation of an areal object such as a town by a point is appropriate. Individuals are often given fixed-point locations according to place of residence. This may raise conceptual issues. Fine grained geo-coded data provide precise information on location but it needs to be confirmed that the precision of the locational datum is appropriate to the study. In the study of non-infectious diseases, precisely locating an individual (by their address) may not be relevant in identifying exposures to certain types of environmental factors. Precisely locating individuals and then obtaining proximity or adjacency measures may not be relevant when analysing an infectious disease where social interaction rather than spatial proximity may hold the key to understanding the spread of the disease. Individual-level data (for the purpose of research or because of government or commercial interests) also raise ethical issues.

For the types of analyses described in this book, exactness of definition is required: points must have specific locations and areas must have both a precise location and spatial extent. Ambiguity or fuzziness in these definitions is not permitted except as part of a sensitivity analysis when the implications of other representations might be explored. So, the analyst must be able to justify any choice and be explicit on the criteria used to assign such a precise geometry in those cases when this is not a feature of the object. Examples of this include mapping the boundary of a forest, the use of a point to represent the 'location' of a census tract, or locating an individual by their place of residence.

(c)   Spatial representation: resolution and aggregation

Pixel resolution or the level of aggregation determines the amount of spatial detail that is present in the data matrix. Decisions taken on the scale of any partition or the specific location of boundaries (particularly in the case of aggregating point objects such as households into census tracts) introduce uncertainty into the relationship between the contents of the data matrix and reality. This uncertainty arises because geographic space does not form natural units. The extent to which these discrete representations adequately reflect underlying properties of the real world depend for example on such aspects of the representation as the size of the areas or the density of the samples in relation to the spatial variability on the field. A field which varies considerably over short distances will need a denser sample of points to represent it or a more complex set of line contours than one which shows little variation. Areas used to represent a field act as filters, smoothing out variation up to the scale of the areal unit. Objects aggregated arbitrarily and modifiably into areas, fields partitioned into pixels of a given size, lose any variability that is present up to the scale of the spatial unit. Wards (five to six thousand households on average) and enumeration districts (comprising on average 150 households) conceal socio-economic heterogeneity because real variation in such attributes is usually at a smaller scale. Aggregations that differ in terms of scale or partition (at a given scale) will affect how a fundamental property such as spatial dependence is captured in the data matrix.

The level of aggregation will also affect how 'noisy' a mapped data set is. Maps of rates constructed from aggregating small rather than large populations are more affected by counting errors so that statistics are less reliable as a basis for making area comparisons and likely to be less stable from time period to time period. Ward rates are more reliable than enumeration district rates but less reliable than for the town or city as a whole. If the underlying process is random (e.g. rates of a disease) the error variances are bigger the smaller the population.

2.3.2   Data quality

Data quality refers to the performance of the data set given the specification of the model. Any assessment of data quality from the users or producers perspective is in terms of how closely data values represent reality *given* the chosen model for representing that reality (Salgé, 1995). As far as the user is concerned both model and data quality affect the databases' fitness for purpose, but, whereas model quality assessment is specific to the application, data quality assessment involves generic criteria (Guptill and Morrison, 1995).

Assessment of data quality is of particular importance in a field where there is considerable reliance on secondary data sources. The analyst using secondary data must be satisfied they meet acceptable scientific standards but there may have to be a trade-off between data quality and the cost of acquiring better data. The data arising from some lifestyle surveys for example, may be useful in piecing together a picture of an urban environment when taken with other data but may be too noisy or contain too many errors to justify the use of rigorous analytical techniques and statistical tests.

Any assessment of spatial data quality must include both the variable (attribute) values and the spatial objects. Nor are these two dimensions of data quality independent. The right measurement assigned to the wrong location may lead to errors in counts for areas and distance measures (Griffith, 1989). Positional error in defining the extent of a vegetation region may be the result of attribute error in those cases where regional boundaries are defined in terms of attribute variation (Goodchild, 1995). Since data also have time co-ordinates, errors in recording the timing of events have implications for the quality of a spatial data set. A spatial data set involves the recording of attribute values and their co-ordinates in space and time. All three have implications for spatial data quality and errors in any one can have implications for the quality of the others.

Data quality may be spatially heterogeneous, that is the error structure may vary across the map. Location error in remotely sensed data is not uniform across a map even after geometric rectification (Borgeson, Baston and Keiffer, 1985; Ford and Zanelli, 1985; Welch et al., 1985). Heterogeneity of error can arise from the interaction between the process of measurement and the underlying geography being measured. For example, population census surveys and crime surveys usually provide more accurate counts of the number of people or number of crime events in suburban areas than they do in inner-city areas. The errors on remotely sensed images differ by sensor type and according to the nature of the topography.

Guptill and Morrison (1995) identify seven dimensions of spatial data quality: data lineage (description of the history of a data set); positional and attribute accuracy; completeness; logical consistency; temporal specification; semantic accuracy (the accuracy with which features, relationships and attributes are encoded or described given the rules for representation). Veregin and Hargitai (1995), in the context of digital cartography and geographic information systems, emphasize: data accuracy (the opposite of data error), resolution (or precision as it relates to data measurement), consistency and completeness. These four categories are discussed here.

(a)   Accuracy

According to Taylor (1982), data accuracy is the inverse of data error and is defined as the difference between the value of a variable, as it appears in the database for any case, and the true value of that variable. Taylor remarks that it is 'very convenient to assume that every physical quantity does have a true value . . . We can think of the true value of a quantity as that value to which one approaches closer and closer as one makes more and more measurements, more and more carefully. As such the 'true value' is an idealization, similar to the mathematician's point with no size or line with no width' (p. 109). All measurement must entail some error because it arises from the inevitable imprecision of the process of taking a measurement together with the definitional problem that measurements in the real world are not well-defined quantities (Taylor, 1982). It follows that the type of error described by Taylor does not carry the connotation of a mistake and something that can therefore be corrected. Improved processes of experimentation and taking measurements both in terms of the quality of the instrumentation and the skills of the person taking the measurements can reduce this type of uncertainty, although can never eliminate it.

There are other practical variants to the definition of data accuracy apart from the concept of an idealized 'true value'. In some applications, error is defined as the discrepancy that exists with respect to a more accurate but expensive process of measurement, as in the case of certain types of soil measurements (Heuvelinck, 1999). The most accurate process may be impractical – for example obtaining individual-level exposures to air pollutants for large populations. In some cases reality is unobservable because it refers to historical events.

Within the definition provided by Taylor (1982) and in addition to the types of errors or uncertainties he specifies are what might be termed 'real' or 'gross' errors. By a gross error is meant an error arising for example from a failure associated with the process of measurement or, at a later stage the processes of storing, manipulating, editing or retrieving data in the database. A gross error arises from a failure to utilize the level of precision that is possible by the measurement device or the database storage device. Suppose the accuracy allowed by a measuring device in locating an object on a given map is ± 1.0m when translated on to the ground. Take this to mean that skilled technicians repeatedly using the device arrive at a value within ± 1.0m 95% of the time. Any user of the device who takes a measurement that is found to be, let us say, greater than 1.5m from the true value on the ground might be deemed to have generated a measurement containing a gross error. Defining an error as a 'gross' error will be more convincing if evidence, external to the measurement process, can be assembled to demonstrate that the measurement is wrong or that the experiment has been corrupted in some way – for example that a rainguage has been tampered with. Unlike the first type of error, every effort must be made to detect gross errors, eliminate or revise the corrupted data values and improve processes in the future – whilst not falling into the trap of simply discarding data values because they are out of line with expectations. We now consider some of the main types of measurement error in spatial databases.

Point location error can arise from errors in laying down or geo-coding ground markers such as environmental monitoring points. In the case of taking data from a map there can be digitizing errors linked to operator eyesight, patience and hand movement as well as the technology (Dunn et al., 1990), and there have been trials to estimate these errors (Maffini et al., 1989). There are also mapping errors associated with the construction of the source map itself (particularly in the case of older maps) and its scale (and hence the precision with which objects can be represented given map parameters such as pen size). The quality of data from ground surveys will depend for example on the precision of the theodolite. Source map errors arise from expressing a curved surface on a flat sheet of paper and shrinkage and distortion effects associated with the paper. Drummond (1995) discusses the accuracies attained by different national mapping agencies together with their method of reporting them.

The root mean square errors (RMSEs) in the two directions (north/south (s_1-direction); east/west (s_2-direction)) are used to represent positional error (Drummond, 1995). The RMSE is appropriate because measurements are at the interval or ratio scale. The RMSE for a map can be based on the discrepancy between the measured co-ordinates ($s_1(i)$, $s_2(i)$) for n objects, and their true co-ordinates ($s_1(i; \text{true})$, $s_2(i; \text{true})$) that have been obtained from a higher-quality measurement system. So:

$$RMSE = [(1/n)\Sigma_{i=1,\dots,n}[(s_1(i) - s_1(i; \text{true}))^2 + (s_2(i) - s_2(i; \text{true}))^2]]^{1/2}$$

which can be decomposed into the error in different directions.

Gross errors in point data can be subtle and difficult to spot but in other cases, when they give rise to logical inconsistencies, obvious. Errors in a data set that locates only addresses within a city can be screened by superimposing city boundaries. In an analysis of road traffic accidents in the north-east of England, Raybould and Walsh (1995) found events geo-coded in the North Sea! In some systems for recording car thefts, because of inherent uncertainty in fixing the last location of the car, or because of the method of recording, thefts may get assigned to the nearest main road intersection giving a false picture of the geography of car thefts. Swerdlow (1992) cites the following sources of gross errors in health data: coding the place of cancer treatment as if it were the place of residence; using the addresses of hotels or embassies in the case of foreigners who come for treatment; using the nearest post office or cancer registry when the place of residence is unknown. 'As a result, even though such registrations are few, a small-area analysis might well show very high risk of cancer apparently relating to residence in post offices or in the registry itself!' (p. 57).

A continuous boundary is approximated in a digital database by a set of line segments. The selection of the endpoints of each line segment is the outcome of a sampling process, and the size of the errors associated with the boundary will depend on the sampling scheme and in particular on the density of sample points in relation to the complexity of the line. It has been suggested *line location error* be represented by an epsilon band on either side of the line identifying the error bounds on the line (Chrisman, 1989). 'When the width epsilon of the band is set to the deviation of the uncertainty of the line, the sausage represents some form of mean error in the area' (pp. 25–6). However, error along the line is unlikely to be independent or uniform and there appears to have been little empirical work on the shape of this band.

The errors associated with the position of the line segments of an area have implications for other forms of error. First there is likely to be error associated with derived attribute measurements taken from the measured object, such as its area and the length of its boundary. Second, spatial relationships between areas may be affected, for example whether two areas are adjacent or not. Third, there is likely to be attribute error such as errors in counts because point events are allocated to the wrong area.

Attribute error can arise as a result of the processes of collecting, storing, manipulating, editing or retrieving attribute values and as noted errors associated with the spatial objects can induce error in these measurements. Attribute error can arise from the inherent uncertainties associated with the measurement process and definitional problems, including specifying the point or period of time a measurement refers to (Taylor, 1982; Buttenfield and Beard, 1994). There are some special problems that may introduce errors into attributes associated with spatial objects. The measurement of particulate matter and other forms of atmospheric pollution at a location are subject to effects associated with monitoring sites which are positioned with respect to objects that affect the flow of air. UK census data at the enumeration district level and above are altered by the quasi-random addition of -1, 0 or $+1$, to counts – a process known as barnardization.

There are several ways of quantifying attribute errors. For variables recorded at the interval or ratio level the RMSE is again useful. If $z_1(i)$ is the measurement for variable Z_1 for case i, then if the true value is denoted $z_1(i; true)$:

$$RMSE(Z_1) = [(1/n)\Sigma_{i=1,...,n}(z_1(i) - z_1(i; true))^2]^{(1/2)}$$

In the case of nominal and ordinal data, the misclassification matrix (**M**) is used particularly for remotely sensed data, where each spatial unit is of the same size (Congalton, 1991). The columns of **M** refer to ground truth (perhaps obtained from a field survey) and the rows to the classification assigned using remotely sensed data. The diagonal elements of the matrix $\{m_{i,i}\}_i$ identify the number of correctly classified pixels so that the sum of these diagonal values divided by the total number of pixels $(m_{..})$ is the proportion of correctly classified pixels (PCC):

$$PCC = \Sigma_{i=1,...,n}(m_{i,i}/m_{..})$$

In any application not only will misclassification rates reflect data errors but they will also be a function of how difficult it is to distinguish between classes and so error rates are affected by the degree of disaggregation into different classes. Goodchild (1995, p. 73) provides an illustration.

The Kappa (κ) coefficient (Stehman, 1996) evaluates accuracy as the discrepancy between the actual PCC and that value of PCC which would be expected if there was a random allocation of pixels to classes $(E(PCC))$:

$$\kappa = [PCC - E(PCC)]/[1 - E(PCC)]$$

where:

$$E(PCC) = (\Sigma_{i=1,...,n}m_{i,.}.m_{.,i})/m_{..}^2$$

where $m_{i,.}$ is the sum of values on row i of the matrix **M** and $m_{.,i}$ is the sum of values in column i of the matrix **M**. This coefficient, which formulates accuracy as a sampling problem, allows for the fact that a certain number of correct classifications will occur purely by chance. There are other measures of accuracy. The users measure is the proportion of pixels that appear to be in class i which are correctly classified $(m_{i,i}/m_{i,.})$; the producers measure is the proportion of pixels truly in class i that are correctly classified $(m_{i,i}/m_{.,i})$ (Congalton, 1991; Veregin, 1995). When areas are not of equal size other methods based on calculating proportions are used (see for example Court, 1970; Wang et al., 1997).

Where data recording methods are automated and large volumes of data are collected the risk of undetected gross error increases perhaps to the extent of casting doubt on the wisdom of analysing such data statistically at all (Hampel et al., 1986; Leonard, 1983). Careful screening of the data prior to any analysis is essential. It cannot be assumed that the sheer volume of data will overwhelm any problems of data quality that might exist. Even in carefully assembled databases such as the US Census there are recorded examples of serious, indeed bizarre, errors (Coale and Stephan, 1962; Fuller, 1975). Openshaw (1995, pp. 401–5) reviews data problems associated with the 1991 UK Census. Hampel et al. (1986) suggest that as a matter of routine between 1% and 10% of all values in a data set will contain gross errors. Rosenthal (1978) found error rates of between 0% and 4% (with an average in a very skewed distribution of 1%) in 15 data sets in psychology. Lawson (2001, p. 37) reports findings that suggest a discrepancy rate of between 12% and 18% between cause-of-death certification and necropsy. He notes that this is likely to lead to particular problems when dealing with rare diseases.

The common assumption in error analysis that attribute errors are independent (see for example Taylor, 1982) is likely to hold less often in the case of spatial data. Location error may lead to overcounts in one area and undercounts in adjacent areas because the source of the overcount is the set of nearby areas that have lost cases as a result of the location error. So, count errors in adjacent areas may be negatively correlated. Farm boundaries do not correspond to administrative boundaries so allocating farm land use to the administrative area within which the main farmhouse is situated will produce patterns of over- and undercounting which will get worse over time as farm size increases. Coppock (1955) examines the relationship between farm and parish boundaries in the UK.

In the case of remotely sensed data, the values recorded for any pixel are not in one-to-one relationship with an area of land on the ground because of the effects of light scattering. The form of this error depends on the type and age of the hardware and natural conditions such as sun angle, geographic location and season (Craig and Labovitz, 1980). The point spread function quantifies how adjacent pixel values record overlapping segments of the ground so that the errors in adjacent pixel values will be positively correlated (Forster, 1980). The form of the error is analogous to a weak spatial filter passed over the surface so that the structure of surface variation, in relation to the size of the pixel unit, will influence the spatial structure of error correlation (Haining and Arbia, 1993). Linear error structures also arise in remotely sensed data (Craig, 1979; Labovitz and Masuoka, 1984; Short, 1999).

(b)   Resolution

The most important aspect of resolution as it affects data quality is in terms of the spatial dimension of the data. We consider this first before briefly discussing resolution in terms of temporal aspects of the data and attribute values.

High-resolution data, small spatial units used for small-area mapping, contain high levels of noise so it may be difficult to identify underlying structure. Kennedy (1989) and Wilkinson (1998, p. 181) discuss this in the context of disease mapping. Precision in the spatial sense does not equate with precision in the statistical sense. Small-area disease rates often display high levels of variation that is an artefact of small counts with the smallest areas often showing the most extreme rates. The addition or subtraction of only a few cases has a larger impact on rates for areas with small populations than on rates for areas with large populations.

Spatial resolution has implications for how areas can be represented for the purpose of analysis. Areas may be represented by their area or population-weighted centroids, although these terms are not always used in any exact or consistent sense. In the UK, the population centroids of EDs are determined by eye. The smaller the areal object the more it is possible to make the centroid a meaningful representation of the area. The Ordnance Survey of Great Britain's Code-Point defines the centroid of unit postcodes to a precision of 1 metre, whilst its Address-Point attains a precision of 0.1m (in most cases) in defining the centre of an individual house (Harris and Longley, 2000).

There are particular problems when data sets on different spatial frameworks and at different scales of resolution are linked to a common spatial framework. The unit postcode contains an average of about 12 households, and health data, for example, are aggregated from this level to the enumeration district level in order to attach Census information for an ecological analysis. Since the 1991 UK Census, linkage is via an ED-postcode directory in which a postcode is assigned to an ED (its so called pseudo-ED) on the basis of which ED the majority of postcode households lie within. No problem arises if the unit postcode sits entirely within an ED. However unit postcode boundaries do not nest within ED boundaries so there are many occasions when the whole of the disease count for a postcode is attributed to an ED even though part of the unit postcode lies within another ED.

Collins et al. (1998) compared the allocation of new births in 1996 to Sheffield EDs using address matching and the allocation arising from the use of the ED-postcode directory. Address matching is an expensive but more reliable method of allocation because the Ordnance Survey's Address Point coordinate falls inside the permanent building structure of the address. Of all records 16.3% (532 out of 3264) were allocated to the wrong ED by the ED-postcode directory. In terms of overall counts there is some cancelling out. Figure 2.5 shows the geography of the net under- and overcounting. The geography of this misallocation process will reflect the geography of the mismatch

Figure 2.5 The geography of net over- and undercounting using the ED-postcode directory: new births in Sheffield 1996

between ED and unit postcode boundaries. EDs with overcounts will tend to be adjacent to EDs with undercounts since these are the EDs where the overcount is coming from. Of the 532 mismatched records the correlation between their 'correct' Townsend deprivation score (correct in the sense of the deprivation score of the ED they actually belonged to) and their allocated score was 0.495. Over a third of these records had Townsend scores greater than ± 2 from the value in their correct ED (on a range from -5.4 to 13.4 across the 1057 Sheffield EDs). Such matching is used to impute deprivation scores to individual records in epidemiology and in evaluating health service targetting.

The last example illustrates the effect spatial resolution can have on the accuracy of attribute values. Counts by area contain errors when data sets that are either at different spatial resolutions or at identical but incompatible spatial resolutions have to be linked and it is not possible to return to a smaller spatial scale to ensure an accurate mapping. Craglia et al. (2001) in modelling violent crime areas in English cities in terms of socio-economic characteristics of the areas had to reconcile crime regions drawn on street plans with census areas. Even when the source data are of relatively high quality the process of data integration will result in some degrading of the quality of that data. Wherever possible ancillary data to aid the allocation should be used which is why the ED-postcode directory method (that uses household counts) was able to improve on the former OPCS Central Postcode Directory approach (that used a very crude postcode grid reference) for linking unit postcode events to enumeration districts. Theobald (1989) discusses how the spatial resolution at which elevations are measured relative to actual landform variation introduces errors into grid-based digital elevation models.

Temporal resolution refers to the period of time over which data values are aggregated and raises similar issues to those discussed in the context of spatial resolution. Aggregating counts by lengthening the time period for example may conceal small scales of temporal variation but alternatively will increase area counts making rates more robust, thereby helping to reduce the small-number problems cited previously in connection with health data and other forms of mapping.

Variable resolution refers to the precision with which attributes or other quantities are measured. In categorical data it refers to the fineness of the classification whilst in interval and ratio data it refers to how many significant numbers a value is recorded to. Neither should exceed the precision of the instrument used to collect data. Today, storage devices do not place limits on precision levels for variable values so at that stage the problem is usually to avoid spurious precision.

(c)   Consistency

Consistency is defined as the absence of contradictions in a database. It refers to 'the logical rules of structure and attribute rules for spatial data and describes the compatibility of a datum with other data in the data set' (Kainz, 1995, p. 109). Rules derive from mathematical theory and formal tests can be constructed to check that there is consistency both within and between different layers of the data set. Topological rules for spatial objects must not be violated (e.g. there is no more than one household at the same location, at the same time) and there are no contradictions in attribute values (e.g. there are no cases of a disease in a census block with no population).

Inconsistency can originate from data errors – such as geo-coding traffic accidents in the North Sea – but can be a particular problem when linking together different data sets for the same area (e.g. geological surveys and census data) or making comparisons between two time periods (Guptill and Morrison, 1995; Kainz, 1995). It is possible for each of two or more source maps to be consistent but to show inconsistencies when brought together (Kainz, 1995, p. 134). The USA's TIGER (Topologically Integrated Geographic Encoding and Referencing) system provides a structure for joining up geographic references. It describes the block structure used by the US Census and it ensures consistency in relating street patterns and other edges (rivers, roads, railways) to census and postal geographies. Such consistency is an essential underpinning to linking geographical data sets for purposes of spatial data analysis. Consistency should be retained over time to ensure that the same spatial identifiers at one census point refer to the same areas at the next and that any changes are carefully logged. In England and Wales, 70% of enumeration districts changed between 1981 and 1991. Census data in the UK is collected every ten years but health and crime data are collected on a much more frequent basis. Analysing year 2000 crime data against 1991 Census data is a form of data inconsistency that calls for further information on the geography of urban redevelopment or population migration since 1991 in order to interpret or qualify findings and to help guard against entering impossible attribute values into the database. An analyst who assembles a single national data set by linking together regional data sets needs to be sure that the regional data sets adopt the same method of classification and in the case of surveys are based on asking the same questions.

(d)   Completeness

Model completeness includes whether all the variables which have been specified as necessary in order to undertake the analysis are available within the database and whether the spatial scale and geographic scope are sufficient (see section 2.3.1). The concept of model completeness is important in the context of designing a primary data collection programme and for evaluating secondary data sources. However, a data set can be 'model complete' but not 'data complete' and vice versa. Within any data set constructed to a 'model complete' specification there can be missing values, undercounts and over-counts. This is what is meant by 'data incompleteness' and there is clearly overlap in the last two types with data error. 'Spatially uniform' data incompleteness raises problems for analysis but spatial variation in the level of data incompleteness with, for example, undercounting more serious in some parts of the study area than others can seriously affect comparative work and the interpretation of spatial variation.

Swerdlow (1992) lists some of the reasons for data incompleteness in cancer data: errors in registration, including address errors, duplication and lateness in recording that can vary between the catchments of different hospitals or clinicians; errors arising from patients moving across registry boundaries; public awareness campaigns and differential access to healthcare; differences in the criteria used by pathologists in different areas when making diagnoses. In the UK, until a complete cycle of a call–recall screening programme has been undertaken there will be incompleteness because some invitations may not yet have been issued. In the case of mortality data, Lopez (1992) lists diagnostic 'fads' which lead to an overdiagnosing of certain causes of death such as cerebrovascular diseases, differences in medical training and cultural norms as causes of data incompleteness. Undercounting of certain causes of death such as sexually transmitted disease, suicide and alcoholism are due to 'diagnostic reluctance' and where there are specific sensitivities in certain localities this may lead to spatial variability in the level of undercounting. With people living longer, partly as a result of the decline of infectious diseases, the practice of only recording a single cause of death when there could be multiple pathologies will also lead to undercounting. All these factors can lead to forms of under- or over-counting and give rise to spatial variation that is an artefact of how the data were collected.

In the case of official criminal statistics, Bottoms and Wiles (1997) referring to the work of Farrington and Dowds (1984) note that drawing attention to geographical differences between large counties in England can be dangerous because of differences in police investigative and reporting practices. On the intra-urban scale, Bottoms and Wiles (1997) cite their own work and that of Mawby (1989) and others to conclude that in the case of reactive policing statistics, official crime and offender data often seem to reflect real differences between areas of a city. However they add that in any given case this should not be taken for granted and should be investigated. Separate surveys are occasionally carried out to supplement official statistics to try to estimate levels of underreporting of offences (Bottoms, Mawby and Walker, 1987). There are different levels of public reporting of offences between areas. Burglaries in suburban areas will, on the whole, be well reported for insurance purposes, but in some inner-city areas there may be underreporting either because there is no 'incentive' or because of fear of reprisals. Some crimes are more uniformly underreported: victims of sexual assault may be reluctant to report whilst there is probably substantial underreporting of domestic violence. Overreporting can be a problem with particular types of crime where there are inappropriate incentives (usually financial) to come forward with allegations.

The Census provides essential denominator data for computing small-area rates. However refusals to cooperate can lead to undercounting and the 1991 Census in the UK was thought to have undercounted the population by as much as 2% because of fears that its data would be used to enforce the new local 'poll tax'. Inner-city areas show higher levels of undercounting than suburban areas where populations are easier to track. Although there are ten-year gaps between successive censuses, population in- and out-flows in many areas may be such as to preserve the essential socio-economic and demographic characteristics of the areas. However some areas of the city may experience population mobility and redevelopment which result in marked shifts. For this reason other sources of population data have been investigated like the Family Health Service Authorities (FHSAs) patient register to track inter-census population shifts (Lovett et al., 1998).

In the USA missing data rates for Census questions ranged from 0% to 8% in 1990 but with few exceptions every housing unit reported at least one person. This is achieved by undertaking exhaustive follow-up surveys. These are expensive and were estimated to take up approximately 20% of the US Census' ten-year budget. A shift to a sampling approach to deal with non-response might result in a rate of housing unit non-response as high as 20% in addition to the anticipated 0%–8% non-response for specific questions.

Census (and other data) not only have problems of undercounting, values for areas can be missing. Missing data may be due to suppression for confidentiality reasons as in the case of those enumeration districts where there are very small numbers of households. Historical data may be missing because records have been lost or because of the stage reached in surveying an area. It can be difficult to know if some archaeological data sets are incomplete and if so the nature of the incompleteness (Hodder, 1977). In the case of remotely sensed data, some areas of the image may be obscured because of cloud cover. A distinction should be drawn between data that are 'missing at random' from data that are missing because of some reason linked to the nature of the population or the area. Weather stations may be temporarily out of action because of equipment failure, monthly unemployment records lost because of office closure or industrial dispute. These might be considered cases of data missing at random.

However mountainous areas will tend to suffer from cloud cover more than adjacent plains and there will be systematic differences in land use between such areas. This distinction has implications for how successfully missing values can be estimated and whether the results of data analysis will be biased because some component of spatial variation is unobservable.

Spatial data raise special completeness issues because if there is a geography to the incompleteness this undermines comparative work and the description and analysis of spatial variation. There are other forms of spatial data incompletness. Map objects may refer to homogeneous vegetation types but change over time may result in boundary shifts and new vegetation regions which should be captured in the database by new spatial objects or adaptation or deletion of former map objects (Brassel et al., 1995, pp. 96–7). The geographic extent of the database may be too small to enable the analyst to detect large-scale trends or periodicities in the data (Horton et al., 1964; Burrough et al., 1985). The specification of the boundary of the study region for which data are collected proscribes the extent to which analysis can examine the role of external influences on events within the study region. This may call for the collection of data in a pre-defined zone extending beyond the strict boundary of the region of interest.

2.4   Quantifying spatial dependence

Spatial dependence is an inherent property of an attribute in geographic space because of the underlying continuity of space and the operation of various types of processes (see chapter 1). This property will be inherited by any data collected on the attribute. The form of the inheritance will depend on, for example, spatial resolution or sampling density in the case of a field space or the scale of spatial aggregation in the case of an object space (e.g. Chou, 1991).

The first step to quantifying the structure of spatial dependency in a data set is to define for any set of point or area objects the spatial relationships that exist between them. Many forms of spatial data analysis require this initial step. After this has been done there are several ways of measuring spatial dependence. These measures can be applied to the whole map to arrive at a single average measure of spatial dependence or to geographically defined subsets if heterogeneity is suspected.

(a)   Fields: data from two-dimensional continuous space

Suppose point samples have been taken. Spatial relationships are typically defined on the basis of distance or distance bands. The $s(i)$ in the data

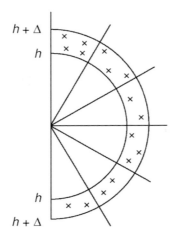

x = sample points in distance band $(h, h+\Delta)$

Figure 2.6 Neighbours of a single point within a distance band and by 30°
segments

matrix provide sufficient information for computing inter-point distances
$d(i, j)$ between any pairs of points i and j. Let the notation $[(i, j) \mid d(i, j) = h]$
denote the condition that j is selected providing it is a distance h from i. This
condition may be relaxed so that j is selected providing it lies within a distance
band $h \pm \Delta$ of i. Thus we write: $[(i, j) \mid d(i,j) = h \pm \Delta]$. Banding may be important
to allow for uncertainty in data point locations and to ensure sufficient num-
bers of pairs from which to compute reliable statistics. However there may be
many pairs in some bands and few in others so estimator precision will vary
and ought to be allowed for in making comparisons between different bands.
In the case of regularly distributed datapoints there may be no pairs in some
distance bands, many in others. Where there are sufficient data, pairing can also
be made to depend on direction so that j is selected providing it lies within distance band $h \pm \Delta$ and in segment k of the half circle to the east of i. Thus we write
$[(i, j) \mid d_k(i, j) = h \pm \Delta]$. This is illustrated in figure 2.6 using 30° segments.

The following discussion is based on the simple case of $[(i, j) \mid d(i, j) = h]$ but
can be easily generalized to the case of banding and/or segmenting. For any
distance h, similarity of values can be assessed graphically using the bivariate
scatterplot $\{(z(i), z(j)) \mid d(i, j) = h\}$. Similarity is indicated by a scatter which is
upward sloping to the right and compact around the 45° line. If the scatter is
widely spread out from the diagonal this is indicative that pairs are not similar
which tends to occur as h increases. Isaaks and Srivastava (1989, p. 52) refer to
plots taken in a specific direction as **h** scatterplots, where **h** is a vector denoting
both distance and direction.

Numerical methods for assessing similarity in the case of variables measured at the ordinal, interval or ratio levels can be based on the squared difference $(z(i) - z(j))^2$ which will tend to be small if $z(i)$ and $z(j)$ are similar and large otherwise. A measure can also be constructed based on the cross-product $(z(i) - \bar{z})(z(j) - \bar{z})$ where $\bar{z}$ denotes the mean value of the $\{z(i)\}$. This quantity will tend to be positive if $z(i)$ and $z(j)$ are similar and either positive or negative otherwise. We now examine numerical descriptors of spatial dependency based on these two quantities.

For any given distance h, the quantity:

$$\hat{\gamma}(h) = (1/2N(h))\Sigma_i \Sigma_j (z(i) - z(j))^2 \qquad (2.2)$$
$$[(i,j)|d(i,j)=h]$$

where $N(h)$ denotes the number of pairs of sites separated by distance h, is the value of the semi-variogram at distance h. It will be small the more alike values separated by distance h are, and will be larger if values are dissimilar. Thus $\hat{\gamma}(h)$ tends to increase as h increases. The semi-variogram function is the plot $\{\hat{\gamma}(h), h\}$ and provides a graphical description of the dependency structure in the data for different distances. The semi-variogram computes half the average squared difference and this is also the basis of the Geary test for spatial autocorrelation described in chapter 7 (Geary, 1954). Figure 2.7(a) is an example of a typical semi-variogram for the case where spatial dependence is strong at short distances and then progressively weakens as h increases until beyond a certain distance (the range) spatial dependence levels off (the sill) close to 0.

We now turn to the second quantity. For any given distance h, define:

$$\hat{C}(h) = (1/N(h))\Sigma_i \Sigma_j (\bar{z}(i) - \bar{z}(i))(z(j) - \bar{z}(j)) \qquad (2.3)$$
$$[(i,j)|d(i,j)=h]$$

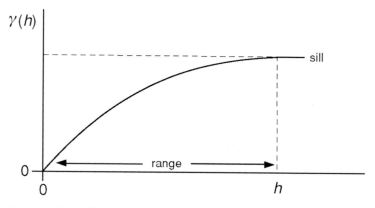

Figure 2.7(a) Model for a semi-variogram $\gamma(h)$

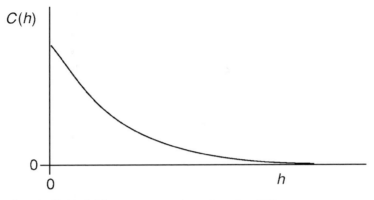

Figure 2.7(b) Model for an autocovariance function $C(h)$

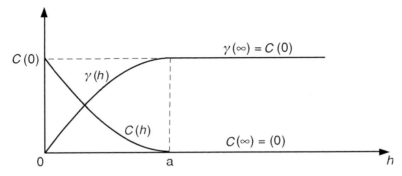

Figure 2.7(c) Relationship between $\gamma(h)$ and $C(h)$

The term $\bar{z}(i)$ denotes the mean of all the values that are included in the first bracket (the $\{z(i)\}$) whilst $\bar{z}(j)$ denotes the mean of all the values that are included in the second bracket (the $\{z(j)\}$). There will, of course, be many values that contribute to both means but these two means will not generally be equal. For ease of computation (2.3) can be written:

$$\hat{C}(h) = [(1/N(h))\Sigma_i \Sigma_j(z(i)\,z(j))] - \bar{z}(i)\bar{z}(j) \tag{2.4}$$
$$\scriptstyle [(i,j)|d(i,j)=h]$$

$\hat{C}(h)$ is the estimate of the autocovariance (or spatial covariance) at distance h. It is the average cross-product and it is large when values are similar (both will tend to be positive or negative in the cross-product term) and close to 0 (because positive and negative values will tend to offset one another) when values are dissimilar. Computing cross-products is also the basis of the Moran test for spatial autocorrelation described in chapter 7 (Moran, 1948). Unlike $\hat{\gamma}(h)$, the plot of the corresponding autocovariance function, $\{\hat{C}(h), h\}$, tends to decrease as h

increases as spatial dependency weakens over increasing distance (see figures 2.7(b) and (c)).

When $h = 0$, it follows from (2.3) that $\hat{C}(0)$ is the variance of the $\{z(i)\}$. If $\hat{\sigma}(i)$ and $\hat{\sigma}(j)$ are the standard deviations for the two subsets of data corresponding to $\{z(i)\}$ and $\{z(j)\}$ in (2.3) then:

$$\hat{R}(h) = \hat{C}(h)/\hat{\sigma}(i)\,\hat{\sigma}(j) \tag{2.5}$$

is the estimate of the autocorrelation (or spatial correlation) at distance h. The plot of the autocorrelation function or correlogram, $\{\hat{R}(h), h\}$, has the same behaviour as the autocovariance function but is standardized in the sense that $\hat{R}(0) = 1.0$. It can be shown that apart from boundary effects $\hat{R}(h) = \hat{R}(-h)$ and similarly $\hat{C}(h) = \hat{C}(-h)$ and $\hat{\gamma}(h) = \hat{\gamma}(-h)$ (Isaaks and Srivastava, 1989, pp. 59–60). In the case of data from a stationary process there is a close relationship between $\hat{C}(h)$, $\hat{R}(h)$ and $\hat{\gamma}(h)$:

$$\hat{\gamma}(h) = \hat{C}(0) - \hat{C}(h) = \hat{C}(0)[1.0 - \hat{R}(h)] \tag{2.6}$$

Where a representation of the field has been obtained using pixels rather than point samples, spatial relationships are defined by looking at the pixels like a set of stepping stones and defining spatial relationships by the number of *steps* required to get from any given pixel (p, q) to any other without backtracking. Define a pixel's *lag one* or *first-order neighbour* as any other pixel that can be reached by a single step that crosses their common edge ('Rook's move'). All the pixels that are *lag two* or *second-order neighbours* of any pixel (p, q) are those that can be reached by crossing two common edges without any backtracking. *Lag three* or *third-order neighbours* are those that can be reached by crossing three common edges without back tracking, and so on for fourth and higher orders of neighbours (figure 2.8). Neighbours can be differentitated by whether they can be reached by taking north/south or east/west steps and how many of each and can also be differentiated by the number of paths that can be followed. So whilst pixels $(p + 1, q + 1)$ and $(p, q + 2)$ are both two steps from pixel (p, q), pixel $(p + 1, q + 1)$ can be reached by two paths both involving one northward and one eastward step, whilst pixel $(p, q + 2)$ can only be reached by taking one path involving two eastward steps. The numbers in brackets in the cells of figure 2.8 denote the number of pathways from pixel (p, q) to the given pixel in those cases where there is more than one. These steps are not distances but because of the regular nature of the partition and since pixels are generally small they approximate distance bands and allow spatial relationships to be classified, as in the case of point samples, not only in terms of distance but direction as well. The same set of graphical and numerical methods as were described above can be adapted to this situation. However it is now necessary to distinguish between

			4(1)					
		4(4)	3(1)	4(4)				
	4(6)	3(3)	2(1)	3(3)	4(6)			
	4(4)	3(3)	2(2)	1	2(2)	3(3)	4(4)	
4(1)	3(1)	2(1)	1	×	1	2(1)	3(1)	4(1)
	4(4)	3(3)	2(2)	1	2(2)	3(3)	4(4)	
	4(6)	3(3)	2(1)	3(3)	4(6)			
		4(4)	3(1)	4(4)				
			4(1)					

Figure 2.8 Steps and numbers of paths from x to any other cell

pairs of the same lag order where there are different numbers of pathways. This means, in the case of a non-isotropic process, computing separate estimates not only for $\hat{\gamma}(0, 2)$ and $\hat{\gamma}(2, 0)$ but also for $\hat{\gamma}(1, 1)$. The full set of semi-variogram estimates are given by:

$$\{\hat{\gamma}(i, j)\}_{i=0,1,....;\ j=...-1,0,1,...}.$$

Pixel data values may be classified into two or more nominal level categories. For example, an image may be classified into vegetation types or land-use classes. It is then possible to count the numbers of adjacent pairs whose categories are the same or the number whose categories are different. Figure 2.9 shows how these counts vary for three different types of map pattern for the case of two categories. In the case of positive dependence the number of joins of the same category will tend to be 'large' and there will tend to be relatively few joins where the categories are different. In the case of negative dependence the opposite will apply. The case of independence (or randomness) lies in between these two cases. This approach to quantifying spatial dependence underlies the join-count test discussed in chapter 7 (see Moran, 1948; Krishna Iyer, 1949; Cliff and Ord, 1981, pp. 11–13).

(b)   Objects: data from two-dimensional discrete space
       Scatterplots using measures based on average squared differences and average cross products and join-counts can again be used for quantifying spatial dependency between pairs of measurements taken on spatial objects in

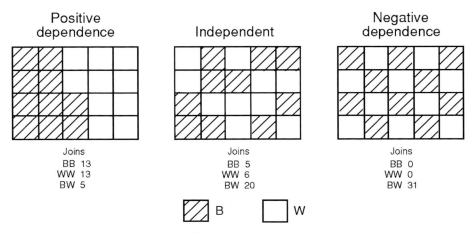

Figure 2.9 Join-counts for different map patterns

discrete space. However in this case the information provided by $\{s(i)\}$ has to be supplemented with neighbourhood information. Neighbourhood information defines not only which object pairs are adjacent to each other but may also quantify the 'closeness' of that adjacency. The information provided by $\{s(i)\}$ may be used in this process of defining neighbourhood information but other data may be employed and assumptions (usually untestable) made. This is needed because in discrete space and for many of the types of processes defined in discrete space there is no single or natural definition of spatial relationships (Gatrell, 1983).

The criteria used for defining neighbourhoods include:

> *Straight line distance*: each point is linked to all other points that are within a specified distance.
>
> *Nearest neighbours*: each point is linked to its k ($k = 1, 2, 3, \ldots$) nearest neighbours. (Note that if point A is one of the k nearest neighbours of B, this does not imply that B is one of the k nearest neighbours of A.)
>
> *Gabriel graphs*: any two points A and B are linked if and only if all other points are outside the circle on whose circumference A and B lie at opposite points (Matula and Sokal, 1980).
>
> *Delaunay triangulation*: all points with a shared edge in a Dirichlet partitioning of the area are linked. A Dirichlet partition, constructed on the points, ensures that the area surrounding any point A contains all the locations which are closer to point A than to any other point on the map (Ripley, 1981; Griffith, 1982). Figure 2.10 shows a set of points from which a Dirichlet partition has been constructed and points joined on the basis of whether they share a common border in this partition.

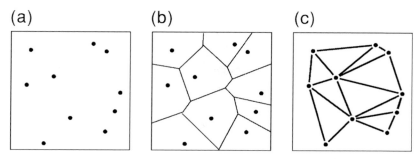

(a)   (b)   (c)

Figure 2.10 Neighbours defined using a Dirichlet partition

In the case of a set of pre-defined areas like administrative units, if they partition the study area they will share common borders so it may be appropriate to define linkages directly in terms of whether the areas share common borders or not. If the areas do not partition the study area one option is to define their location by a point (such as the area or population weighted centroid) and then apply one of the three methods for point objects described above.

Rather than defining linkages between objects in purely geometrical or spatial terms, ancillary data may be used. For example Haining (1987) modelled variation in income levels across a group of urban places in Pennsylvania whose spatial relationships were specified by drawing on central place theory. Cities were classified into orders in the urban hierarchy on the basis of population thresholds. This introduces a strong directionality into the way spatial relationhips are defined in the system of cities. Directionality was also used by Pace et al. (1998) in the analysis of house price data in Fairfax County, Virginia. 'It seems eminently reasonable to assume that the sales price of a neighbouring property will influence the subject property only if the neighbouring sale is earlier in time' (p. 17). Information on which plants are parents and which are offspring, which factories are the main sites and which are branch sites may also allow an ordering to be introduced which goes beyond purely spatial criteria.

Where the analyst wishes linkage to reflect the level of social or economic interaction between two areas then flow data on numbers of journeys or trade data may provide a sounder basis than distance or geometric properties of the set of areas. Linkage may be allowed providing interaction exceeds a certain threshold level (Holmes and Haggett, 1977). Linkages can be constructed to reflect different assumptions about the routes by which effects might be relayed across space as in the case of different pathways for the spread of an infectious disease (Cliff et al., 1985, pp. 182–5).

Geometric or spatial criteria are appropriate for defining relationships between objects if the analyst has no external criteria on which to make a judgement or where physical proximity is the main determinant of similarity. In the case of a continuous surface, distance is a natural criterion. In some circumstances there may be no strong reason to prefer any one of the geometric criteria over another but the analyst should be aware of the implications of different choices. If the purpose of analysis is spatial interpolation (see section 4.4) the analyst might want to define relationships to reduce the effects of clustering of data values on the interpolator (such as a nearest neighbour criterion by segment). However if the aim is to fit a model which is consistent with what is understood about underlying process, a pure distance-based criterion might be appropriate in the case of environmental processes; an interaction criterion in the case of social processes.

Before introducing the versions of equations (2.2) and (2.3) used to describe spatial dependency in discrete space we detour slightly to show how spatial relationships can be described using matrix methods.

Spatial relationships can be represented in the form of a binary contiguity or *connectivity matrix* (C). If there are n objects (points or areas), define a matrix with as many rows and columns as there are objects ($n \times n$). Each area is assigned a unique row and column. If two objects i and j are to be defined as mutually linked then:

$$c(i, j) = c(j, i) = 1$$

where $c(i, j)$ denotes the entry on row i, column j of C. Otherwise any cell has the value 0. Any point or area j where $c(i, j) = 1.0$ will be called a 'neighbour' of i and be denoted $N(i)$. An object cannot be connected to itself (cannot be a neighbour of itself) so $c(i, i) = 0$ for all i. However sometimes a matrix is needed where spatial operations are performed that accumulate all values for a group of areas that include i and other areas connected to i. In this case we use $C^+ = C + I$ (where I is the identity matrix with ones down the diagonal and zeros elsewhere). So it is understood that $c^+(i, i) = 1.0$.

The matrix shown in figure 2.11 is the C matrix corresponding to the definition of adjacency based on two areas sharing a common border. It is a matrix that is symmetric about its diagonal. In the case where object j is a neighbour of i but i is not a neighbour of j (as can arise for example with the nearest neighbour proximity criterion) then whilst $c(i, j) = 1$, $c(j, i) = 0$ and the matrix will not be symmetric about its diagonal.

If C is multiplied with itself, $C^2 = C \times C$, then the non-zero cell entries identify all pairs of areas that can reach each other in two steps – second-order adjacencies. The values in the cells in the resultant matrix $\{c^2(i, j)\}_{i,j}$ identify the number of pathways. This count includes backtracking routes. If i is adjacent to four other areas then $c^2(i, i) = 4$ since there will be four ways of

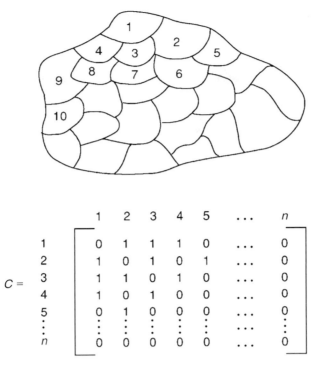

$$C = \begin{array}{c} \\ 1 \\ 2 \\ 3 \\ 4 \\ 5 \\ \vdots \\ n \end{array}
\begin{array}{ccccccc}
1 & 2 & 3 & 4 & 5 & \cdots & n \\
\end{array}
\left[\begin{array}{ccccccc}
0 & 1 & 1 & 1 & 0 & \cdots & 0 \\
1 & 0 & 1 & 0 & 1 & \cdots & 0 \\
1 & 1 & 0 & 1 & 0 & \cdots & 0 \\
1 & 0 & 1 & 0 & 0 & \cdots & 0 \\
0 & 1 & 0 & 0 & 0 & \cdots & 0 \\
\vdots & \vdots & \vdots & \vdots & \vdots & \cdots & \vdots \\
0 & 0 & 0 & 0 & 0 & \cdots & 0
\end{array} \right]$$

Figure 2.11 Binary connectivity matrix based on area adjacency

travelling from i back to itself in two steps. The matrix C^3 provides the same information for third-order adjacencies and so on. These relationships, together with methods for sweeping out the paths that involve backtracking which are deemed redundant, are important for the purpose of writing software for spatial analysis of regional data where the areas are irregular (Anselin and Smirnov, 1996). In the case of analysing pixel data, this matrix representation is unnecessary and, since the matrices are sparse, wasteful of computer storage and possibly computation time. With the exception of pixels on the boundary of the study region all pixels have four neighbours that share a common edge.

There is no requirement to describe relationships between objects as simply present (1) or absent (0). Spatial relationships or those defined using interaction criteria can be defined using a more general *weights matrix*, **W**. Below are examples of different types of weights matrices (if these are applied to areas, distances may be defined by reference to area centroids):

(a) *Distance:* $w(i, j) = d_{i,j}^{-\delta}$ where $d_{i,j}$ denotes the distance between i and j and the parameter $\delta \geq 0$. Distance can be defined in many different metrics (see Gatrell, 1983, pp. 23–34).

(b) *Exponential function of distance*: $w(i, j) = \exp(d_{i,j}{}^{-\delta})$ where $\exp(\ )$ denotes the exponential function.

(c) *Common border*: $w(i, j) = (l_{i,j}/l_i)^{\tau}$ where $l_{i,j}$ is the length of the common border between i and j, and l_i is the length of the border of i (excluding any segment which is on the boundary of the study area). The parameter $\tau \geq 0$.

(d) *Combined border and distance weighting*: $w(i, j) = (l_{i,j}/l_i)^{\tau} d_{i,j}{}^{-\delta}$.

(e) *Interaction weights* (Bavaud, 1998). Export weight: $w(i, j) = n(i, j)/n(i,.)$. Import weight: $w(j, i) = n(j, i)/n(., i)$. $n(i, j)$ is the spatial interaction from i to j; $n(i,.)$ is the total interaction leaving i; $n(., i)$ is the total interaction entering i.

In cases (a) and (b) the weighting is non-zero for all pairs of points or areas but gets smaller as distance increases. The larger the parameter δ the steeper the fall with distance. In case (c) the weighting is only non-zero in the case of areas sharing a common border and decreases as j's share of the border of i decreases, the decrease greater for larger values of τ. In case (d) the weighting is only non-zero in the case of areas sharing a common border and gets smaller as j's share of the border gets smaller and the further away it is. Case (e) is based on shares of export and import totals.

The use of the **W** notation will signify a general weights matrix so it will include the possibility of a binary connectivity matrix **C** as a special case. Note that for all areas i and j, the elements $w(i, j) \geq 0$ and usually $w(i, i) = 0.0$. If the **W** matrix has been row standardized, so that row sums equal 1 as a result of dividing each entry on a row by the sum of the row values, then this will be identified as **W***. So:

$$w^*(i, j) = (w(i, j)/\Sigma_{j=1,...,n} w(i, j))$$

Bavaud (1998, pp. 154–7) describes properties of general weights matrices and shows how they imply properties of the spatial system such as the prominence of any region or place within the total area. Several authors have examined the eigenvalues and eigenvectors of connectivity and weights matrices and identified how they characterize spatial structure (see for example Tinkler, 1972; Boots, 1982, 1984; Griffith, 1996). The principal eigenvalue of matrix **C** provides an index of the connectivity of the set of areas whilst the individual elements of the corresponding eigenvector indicates the centrality of each site within the overall configuration. This work illustrates the assumptions latent in any choice of **C** or **W**. There are further implications of the choice of neighbour that only become apparent in the context of the particular model chosen for representing spatial variation. These implications apply for example to the mean and variance properties of the model (Haining, 1990, pp. 110–13). This will be considered in section 9.1.2.

We are now in a position to define a new group of variables that will be referred to generically as spatially averaged variables and which are obtained as functions of the original set $Z_1, Z_2, \ldots, Z_k$ by performing spatial operations on the data. The general notation for these derived variables will be $WZ_1, \ldots, WZ_k$. The W prefix is simply intended to signal some spatial operation on the original variable. They are obtained as matrix products, so for example:

$$WZ_1(i) = \Sigma_{j=1,\ldots,n} w(i, j) z_1(j) \quad i = 1, \ldots, n \tag{2.7}$$

If W is a row standardized binary connectivity matrix then (2.7) is just the mean of the values in the adjacent regions. This sort of operation is useful for representing neighbourhood conditions around an area i. If W is a row standardized weights matrix based on a distance function and with $w(i, i) \neq 0$, then this operation is useful for some forms of data smoothing (see chapter 7). If W is the unstandardized binary connectivity matrix and $w(i, i) = 1$, then (2.7) is the sum of values in region i and its neighbours. This operation is used in some cluster detection methods (see chapter 7).

In summary, the data used in the spatial analysis of discrete space comprise the original data matrix with data values and an identifier for the location of the spatial object:

$$\{z_1(i), z_2(i), \ldots, z_k(i) \mid s(i)\}_{i=1,\ldots,n}$$

However, in addition at least one weights matrix (W) is needed to capture spatial relationships. These spatial relationships define for each $s(i)$ all the neighbours, $N(s(i))$. The collection of pairs $\{s(i), N(s(i))\}$, or $\{i, N(i)\}$ defines a graph. From this matrix new variables $\{WZ_i\}$ may be constructed.

The general expressions that correspond to (2.2) and (2.3) and which can be used to quantify spatial dependence at different scales in discrete space are:

$$\hat{\gamma}(C^1) = (1/2 \mid N(C^1)\mid) \Sigma_i \Sigma_j c(i, j)(z(i) - z(j))^2 \tag{2.8}$$

and:

$$\hat{C}(C^1) = (1/ \mid N(C^1)\mid) \Sigma_i \Sigma_j c(i, j)(z(i) - \bar{z}(i))(z(j) - \bar{z}(j)) \tag{2.9}$$

where the C^1 simply denotes the use of the connectivity matrix, C. $\mid N(C^1)\mid$ denotes the number of pairs used in the computation. The summation terms used in (2.2) and (2.3) have been simplified by specifying the restriction on

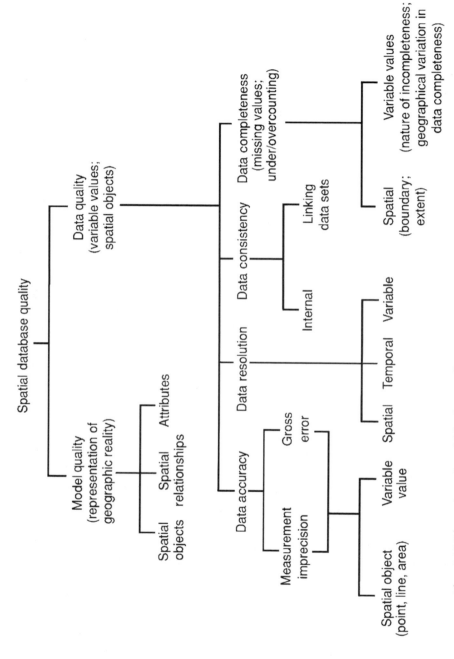

Figure 2.12 Dimensions of spatial database quality

which pairs of data values to use through the elements of the **C** matrix. Equivalently (2.8) and (2.9) can be interpreted using the pixel stepping stone analogy in section 2.4(a). Orders of neighbours or lags are specified by steps that are defined by the connectivity matrix (**C**) and powers of that matrix $\mathbf{C}^2$ ($\mathbf{C} \times \mathbf{C}$), $\mathbf{C}^3$ ($\mathbf{C} \times \mathbf{C} \times \mathbf{C}$), ... suitably swept to remove the redundant steps. In this way semi-variogram function plots and autocovariance function plots can be obtained for irregularly distributed spatial objects. More general versions of (2.8) and (2.9) can be computed using **W** rather than **C**. These will be encountered in chapter 7 as generalized autocorrelation tests for data measured at the ordinal level and above as well as generalized join-count tests for nominal data (Cliff and Ord, 1981; Hubert et al., 1981).

The usual expectation is that values at adjacent locations tend to be similar. Spatial dependence is positive and it is common to refer to the presence of *positive* spatial dependence or autocorrelation in a data set. In the case of continuous space, in the limit (as the distance between any two points goes to zero), it is difficult to visualize any other form of spatial dependence. In the case of continuous data, where point samples are separated or pixels are of sufficient size, it is at least possible that if $z(i)$ is large (small) then $z(j)$ could be small (large). This is called *negative* spatial dependence or autocorrelation. For continuous space the presence of negative autocorrelation can only occur at a distance – between hill top and valley say. In the case of discrete space, a competition process might induce negative spatial autocorrelation between adjacent, but discrete objects. For example, adjacent plants might compete for soil nutrients which might induce negative autocorrelation in plant size.

2.5   Concluding remarks

This chapter has described the framework for spatial data analysis to be used in this book and has brought together a number of model and data quality issues that arise in working with spatial data (see figure 2.12). Any spatial data set provides an abstraction of a complex reality. This chapter has outlined the generic characteristics of data quality in terms of accuracy, precision, consistency and completeness with respect to variables, spatial objects and time. The principal message is that the analyst needs to be alert not only to the problems but how they may impact differentially across a study area or in making comparisons through time or between different study areas. It is for the user to make an assessment of the quality of the data set and to establish fitness for purpose. It may not be feasible to examine the entire data set since this will greatly increase the costs of data capture but a representative sample of the data should be examined in order to evaluate its quality. It will also be necessary to decide which errors are critical (and must be addressed) and which errors are unlikely to lead to serious consequences.

Chapter 4 will examine some of the implications of these data quality issues for spatial data analysis. Different aspects of data quality influence different stages of spatial data analysis. Some concerns, like the presence of data errors, need to be considered at the stage of collecting, preparing and finalizing the data to be analysed. Others, including the resolution or precision of the data, need to be considered in relation to the form and conduct of analysis and in relation to the interpretation of findings. However the next chapter considers sources of spatial data and the special considerations that may arise in obtaining data through spatial sampling.

Spatial data: obtaining data and quality issues

3

Obtaining spatial data through sampling

Spatial data are acquired in many different ways. In section 3.1 primary, secondary and other types of data sources are briefly reviewed. Section 3.2 considers the problem of obtaining data through spatial sampling. Section 3.3 briefly reviews simulation as a means of obtaining spatial 'data'.

Most of the chapter is concerned with obtaining data through sampling. Three types of attribute properties, for which spatial sampling is often needed, are considered. The first is sampling to estimate global properties of an attribute in an area such as its mean value. This arises in estimating crop yield or the proportion of an area under a particular type of land use. The second is sampling for the purpose of constructing maps and making predictions of attribute values at specific locations. This arises when designing a sampling system for rainfall (e.g. rainfall by area) or predicting yields from mining. The third is sampling to detect areas with critically high (or low) values of an attribute ('hot' and 'cold' spots). This arises when seeking to detect land parcels that have been seriously contaminated.

The chapter deals only with attribute sampling. There are situations where a sample design is needed to represent some object such as the location of a line or the boundary of a lake. In geographic information system databases the representation and resolution of boundary lines between polygons is the product of a sampling process. The detail with which any spatial object is represented is a function of the density of sample points (Griffith, 1989; Longley et al., 2001). This aspect of spatial sampling is not considered here.

3.1 Sources of spatial data

Primary data are collected by a researcher to meet the specific objectives of a project. In observational science primary data originate from fieldwork and sample surveys. If hypotheses are to be tested that have a spatial or geographical dimension then surveys should ensure accurate and careful geo-referencing of each observation – as precise as confidentiality will allow. This will help with later stages that may require linkage with data from other surveys. If local-area and contextual influences are to be examined, then focused sampling in contrasting areas is needed. The results of national surveys when applied to local areas may not produce estimates with sufficient precision because the sample size in the local area may turn out to be small. Stratification needs to be built into the sampling strategy if local-area estimates are needed.

Secondary spatial data sources include maps, national and regional social, economic and demographic census data, data generated by public bodies such as health, police and local authorities as well as commercial data sets generated by private institutions in for example the retail and financial sectors. Even when such data (e.g. national censuses) ostensibly represent complete enumerations of the population it is sometimes safer to view them as samples. One of the benefits of this is that the analyst can invoke sampling theory and avoid the criticism of having placed too much emphasis on any particular data set when it is known that had the data been collected for even a slightly different time period, counts would almost certainly have been different (Craglia et al., 2002). The huge growth in certain types of secondary data, generated by public agencies, has frequently been remarked upon.

Satellites are an important source of environmental data and are useful in conjunction with socio-economic, topographic and other ancillary data in constructing descriptions of, for example, urban areas (Harris and Longley, 2000). These developments owe much to modern devlopments in hardware and the creation of geographic information systems that allow the handling, including linkage, of large geographically referenced data sets. Decker (2001) provides a review of data sources for GIS and Lawson and Williams (2001) for spatial epidemiology. With data integration, the process of assigning different spatial data sets to a common spatial framework, comes a range of technical issues about how such integration should be implemented and the reliability of such integrated data sets (Brusegard and Menger, 1989). Data sets may be of varying qualities, have different lineages, different frequencies of collection as well as be on different spatial frameworks that change over time.

Some data are generated by inputing sample data into a model to generate a spatial surface of data values. This may be done if the variable of interest is difficult or expensive to collect. Air pollution monitoring is expensive. Air pollution maps for an area are constructed by combining data on known point, line and area sources of air pollution with climatological data and assumptions about how pollutants disperse. After calibrating and validating model output against such sample data as are available, model output is then used to provide maps of air pollution (Collins, 1998). If a probability model is specified for a variable of interest then not only will the average surface be of interest but also variability about the average. Simulation methods are used to display this variability (see section 3.3).

3.2 Spatial sampling

3.2.1 The purpose and conduct of spatial sampling

The purpose behind spatial sampling is to make inferences about a population where each member has a geographical reference or geo-coding, on the basis of a subset of individuals drawn from that population. Sampling is used rather than say undertaking a complete census for various reasons. The population may be so large a complete census would be physically impossible or impractical (e.g. estimating the average length of pebbles on a beach). There may be an (uncountable) infinity of locations where measurements could be taken as in the case of ground-level air quality, or soil depth in an area with a continuous covering of soil. The cost of acquiring information on each individual may rule out a complete census. The 1991 UK Census only provides data on household employment on the basis of a 10% sample of all the returns partly for confidentiality reasons but also because of the costs of manual coding. Remotely sensed data provides a complete census (at a given resolution) but for cost reasons the data are interpreted by ground truthing based on a sample of sites.

In other situations it is not the size of the population or the cost of acquiring the data that calls for sampling but the level of precision, on the quantity of interest, required by the application. Sampling introduces error in the sense that the property of interest is estimated to within some level of precision – the inverse of the error variance or sampling error associated with the estimator. This error variance can be held to pre-determined limits by the choice of sample size. Taking a complete enumeration (or even a very large sample) may be wasteful of effort if such accuracy is not necessary. Further, the accuracy of a census may be illusory if measurement error is present, and, in a reversal of what might be considered the normal relationship between census and sample, sampling may be needed to improve the quality of 'census' information. In the case of crime data, counts based on police records are known to produce undercounts so household sampling is undertaken in order to improve estimates. Population censuses miss certain groups (such as the homeless) and sampling may be undertaken to improve data on them.

Drawing inferences about a geographical population through sampling calls for a series of decisions to be taken on the sample design. These decisons are taken in relation to the following questions: (i) What is to be estimated? (ii) What sample size (n) is required in order to achieve the desired level of precision? (iii) Since the sampling is spatial, what n locations should be selected for the sample? (iv) What estimator should be used to compute the quantity of interest? (v) What measure of distance (between the estimate and the attribute of interest in the population) should the sample design seek to minimize? The answers to these questions are not necessarily independent of each other. The choice of sampling plan, (iii), is dependent on the selected estimator, (iv), and whether it is a spatial or non-spatial property of the population that is of primary interest, (i). An example of a non-spatial property of a population is the mean level of an attribute in an area or the proportion of the population that exceeds a certain threshold value. They are termed non-spatial properties because the analyst is only interested in 'how much' not 'where' (Brus and de Gruijter, 1997). However spatial properties of a population involve 'where' questions: identifying where in the population threshold values of an attribute are exceeded or where the extreme values are located, and being able to make optimal predictions of attribute values at unsampled locations. 'Where' questions extend to constructing maps of population variability or providing quantitative summaries of that variability in terms of autocorrelation or semi-variogram functions (see chapter 2).

There will be other issues that influence the design of a spatial sample. Increasing estimator precision, by increasing the sample size, always raises sampling costs but the extra costs of collecting larger volumes of data may be particularly important in spatial sampling. There may be problems of accessibility to certain sites and transport costs associated with sampling a geographically dispersed population. It is usually necessary to make a trade-off between economic and statistical criteria in designing a sample. Cressie (1991, pp. 321–2), briefly reviews attempts to formalize such decision making.

Sampling designed to give acceptable levels of sampling error for a regional survey will not achieve the same level of sampling error at subregional levels (because of fewer observations) and unless stratification is incorporated into the sample design some areas are likely to have very few, in some cases no, samples. A city-wide survey can be designed to deliver acceptable levels of precision for the city as a whole, but unless there is stratification say at the ward scale then intra-city comparisons across wards may be undermined by a lack of sample size at the ward level or highly varying numbers of samples between the wards so that estimator precision varies greatly from ward to ward. It is at the sample design stage when attention must be paid to deciding whether intra-area comparisons are important and what aspects of spatial variation are important. If inter-area comparisons are important, the sample must be stratified according to the geographic units to be compared and sample sizes selected for each area to achieve the desired level of precision for the estimator. This will increase the costs of the survey and may lead to levels of precision for the area-wide estimates that are higher than really necessary. Research, particularly at small spatial scales, may want to consider the effects of other nearby ecological systems. If inter-area comparisons in terms of preventative health behaviours are between deprived and affluent neighbourhoods, the researcher may want to differentitate between deprived neighbourhoods that are spatially embedded within other deprived neighbourhoods and deprived neighbourhoods that are spatially embedded within more affluent neighbourhoods. This calls for further levels of spatial stratification.

Even if an estimator of a city-wide attribute, based on a large sample, has high precision, if it is used as the estimator for the value of the same attribute at the ward level it is likely to be a biased estimator. The estimator that uses just the subset of data taken from a particular ward and is an unbiased estimator of the ward-level attribute value is likely to have low precision. Low precision in ward-level estimators is likely to be problematic for the analyst who wants to examine differences at the ward scale. Part of the solution may lie in stratifying the sampling plan and setting sample sizes by strata to ensure appropriate levels of precision. Further improvements can be made by combining the evidence of two scales of sampling (local area and regional) using estimators that combine information or 'borrow strength' for the purpose of local- or small-area estimation (Ghosh and Rao, 1994; Longford, 1999). The city-wide estimator and the ward-level estimator can be combined in a new estimator for the ward-level quantity of interest that weights the two estimators in a way that reflects their relative precision. In this sense the ward-level estimator which has lower precision (larger error variance), 'borrows strength' from the higher precision city-wide estimator. An influential paper by James and Stein (1960) provides an early development of this approach. Gelman et al., 1995, pp. 42–4 provide a Bayesian treatment of this problem. Either the ward-level estimator can be viewed as having been 'shrunk' towards the city-wide estimator (the prior mean) or it can be viewed as the city-wide estimator 'adjusted' towards the observed value of the ward-level estimator. There is further discussion of this in sections 9.1.4 and 10.3.

Designing a sample, particularly a sample for monitoring purposes keeping records through time as well as across a region, may not always start from a 'blank sheet of paper'. There may be a set of sample sites already in position. The problem may be to cut back an existing sampling plan because of costs or because the network is unnecessarily dense. The converse problem is how to add to an existing network of sample points (Arbia and Lafratta, 1997). If the objective is to provide good coverage of the area then kriging theory can be used to identify areas on the map where the most serious gaps exist or where there is overprovision. The starting point is to analyse spatial variation in prediction error given the existing network and then from this to identify which areas need more sample sites (because prediction errors are unacceptably high) or which areas can shed sites because the data for some sites can be predicted from the data generated at other sites (see below, section 3.2.4(c)). There are good examples of these types of problem in the area of rainfall monitoring, particularly for water quality assessment and flood control (Rodriguez-Iturbe and Mejia, 1974; O'Connell et al., 1979; Hughes and Lettenmaier, 1981). The picture is complicated however if the monitoring stations are required to monitor a range of attributes that have different spatial variablity or if the events generating the attributes occur with differing spatial properties. Monitoring rainfall events is complicated by the fact that different types of rainfall events have different spatial properties. Frontal rainfall creates large-scale patterns that could presumably be monitored by a relatively sparse network. Convection rainfall is often associated with highly localized but very intense patterns of rainfall (Bras and Rodriguez-Iturbe, 1976).

Pilot surveys play an important role in survey research, particularly for purposes of evaluating questionnaires. In the case of spatial sampling, the identification of an optimal sampling plan often depends on the pattern of spatial variability in the population. This is usually unknown. In the case of spatial sampling a pilot survey may be needed in order to estimate the pattern of spatial variability in the population in order to make a judgement as to the most appropriate sampling plan. This is likely to be of particular importance where the intention is to put in place a medium- to long-term system for monitoring.

3.2.2   Design- and model-based approaches to spatial sampling

(a)   Design-based approach to sampling

The *design-based approach* or classical sampling theory approach to spatial sampling views the population of values in the region as a set of unknown values which are, apart from any measurement error, fixed in value. Randomness enters through the process for selecting the locations to sample. In the case of a discrete population, the target of inference is some global property such as:

$$(1/N)\Sigma_{k=1,...,N}z(k) \tag{3.1}$$

where N is the number of members of the population, so (3.1) is the population mean. If $z(k)$ is binary depending on whether the kth member of the population is of a certain category or not, then (3.1) is the population proportion of some specified attribute. In the case of a continuous population in region A of area $|A|$ then (3.1) would be replaced by the integral:

$$(1/|A|) \int_A z(x)\,dx \tag{3.2}$$

Design-based *estimators* of quantities (3.1) or (3.2) weight the sample observations by their probabilities of being included in the sample. The merits of different sampling plans in a design-based sampling strategy depend on the structure of spatial variation in the population as will be discussed in section 3.2.4.

The design-based approach is principally used for tackling 'how much' questions such as estimating (3.1) or (3.2). In principal, individual $z(k)$ could be targets of inference but, because design-based estimators disregard most of the information that is available on where the samples are located in the study area, in practice this is either not possible or gives rise to estimators with poor properties (see figure 3.1).

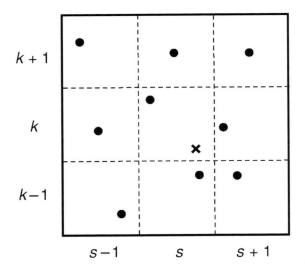

Design-based sample with one observation per strata. In the absence of spatial information the point **✗** in strata (k,s) would have to be estimated using the other point in the strata (k,s) even though in fact the samples in two other strata are closer and may well provide better estimates.

Figure 3.1 Using spatial information for estimation from a sample

(b)   Model-based approach to sampling

The model-based approach or superpopulation approach to spatial sampling views the population of values in the study region as but one realization of some stochastic model. The source of randomness that is present in a sample derives from a stochastic model. Again, the target of inference could be (3.1) or (3.2). Under the superpopulation approach, (3.1) for example now represents the mean of just one realization. Were other realizations to be generated, (3.1) would differ across realizations. Under this strategy, since (3.1) is a sum of random variables, (3.1) is itself a random variable and it is usual to speak of *predicting* its value.

A model-based sampling strategy provides predictors that depend on model properties and are optimal with respect to the selected model. Results may be dismissed if the model is subsequently rejected or disputed. Matern (1986, p. 69) remarks that model-based predictors of (3.1) should only be used when detailed knowledge is available about the structure of the underlying population – information that is rarely available. Hansen et al. (1983, p. 792) further suggest that design-based methods lose relatively little efficiency as compared with model-based methods even when the models are perfect descriptors of the data.

In the model-based approach it is the mean (μ) of the stochastic model assumed to have generated the realized population that is the usual target of inference rather than a quantity such as (3.1). This model mean can be considered the underlying signal of which (3.1) is a 'noisy' reflection. Since μ is a (fixed) parameter of the underlying stochastic model, if it is the target of inference, it is usual to speak of *estimating* its value. In spatial epidemiology for example it is the true underlying relative risk for an area rather than the observed or realized relative risk revealed by the specific data set that is of interest. Another important target of inference within the model-based strategy is often $z(i)$ – the value of Z at the location i. Since Z is a random variable it is usual to speak of *predicting* the value $z(i)$.

There is now no need for randomness in the sample plan. Whereas in the design-based approach the surface never changes and the evaluation of a sampling strategy must consider taking repeated probability samples, in the model-based approach each realization produces a new surface of values and the same sampling plan could be adopted in each case (Brus and de Gruijter, 1997). In fact even in cases where a complete 'census' has been taken – for example, a count by area of all new cases of a disease in a given period of time – this may still be viewed as a sample from the underlying model and the data used to estimate the parameters of that model. As implied above, the model-based approach is also of importance for tackling 'where' questions. Model-based estimators utilize much of the spatial information available in the sample. They weight the sample observations using both the model that is presumed to be generating the data and the configuration of the sample locations. This is the reason they are particularly important for spatial prediction.

(c)   Comparative comments

As Brus and de Gruijter (1997) note there has been considerable discussion about appropriate methods of spatial sampling. In the discussion to their paper, Laslett provides historical context within geostatistics and soil science. Model-based spatial sampling has its origins in mining and the development of geostatistics in order to cope with the effects of 'convenience' sampling when the purpose is spatial interpolation. Classical sampling theory applied in this context tends to overestimate large values and underestimate small values. Adopting a model-based approach provides the required regression to the mean. In areas where sampling is more uniform, such as in soil sampling for example, the benefits of the geostatistical approach to spatial interpolation may not be so evident (see section 4.4.2(v)).

Other discussions, particularly in a social science context, have drawn attention to the conceptual meaning of basing inference on a superpopulation view (Galtung, 1967). The mathematical model assumes a 'hypothetical universe' of possible realizations that in practice are never observed and a sample of size one that permits valid inference about model parameters only under certain conditions (see section 9.1.2). However Bernard in the discussion to Godambe and Thompson (1971) remarks: 'one is rarely ... concerned with the finite de facto population of the UK at a given instant of time: one is more concerned with a conceptual population of people like those at present living in the UK'. Notwithstanding this comment the important issue is again whether a model-based strategy provides sounder inference on the properties of interest. In those areas of research where there is inherent randomness in the underlying processes the model-based approach is the appropriate choice.

In summary a model-based sampling strategy should be used for predicting values at particular locations, mapping and for estimating the parameters of the underlying stochastic model (such as the model mean μ) but not quantities like (3.1) or (3.2) unless the model is known. Design-based sampling should be used for estimating global properties of the (realized) population of values such as the population mean (3.1) or (3.2) but not for estimating individual values or mapping. In the next chapter, in the context of data completeness and estimating missing values, the problem of spatial prediction and estimation will be revisited (see section 4.4).

3.2.3   Sampling plans

The main classes of sampling plan for estimating a map property are random, stratified random and systematic and these will be considered first (Ripley, 1981). Sampling may take the form of points or quadrats. For a discussion of quadrat sampling see for example Kershaw (1973).

Under random sampling, n sites are selected so that each member of the population has an equal and independent chance of selection (figure 3.2(a)). Under stratified random sampling the population to be sampled is partitioned into areal strata and, within each of these, sites are selected according to the method of random sampling. Figure 3.2(b) shows a stratified random plan based on nine strata with one sample taken from each stratum. Figure 3.2(c) shows one type of systematic sampling, centric systematic sampling, with square strata

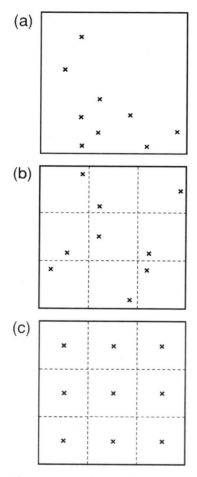

Figure 3.2 Spatial sampling: (a) random, (b) stratified random, (c) systematic

and one point taken from the centre of each stratum. Randomization can be introduced by selecting the site in the first stratum at random. The remaining $(n-1)$ sample points occupy the same relative positions in their respective strata as the first. There are many variants of this form of sampling (Koop, 1990). In the case of both stratified random and systematic sampling, other strata shapes may be adopted, such as hexagonal or triangular strata. In social and economic applications the strata may be chosen to capture different regions, such as administrative areas, so that comparisons can be made between them.

Random sampling does not ensure even coverage of the area to be sampled and in fact often gives rise to relatively large unsampled areas and to groups of sample sites that appear geographically clustered. Stratification reduces both these problems although there can still be some evidence of gaps and clusters in the spatial distribution. Systematic sampling is often easy to implement and provides well-defined directional classes and large numbers of samples separated by specified distances.

Other sampling plans may be appropriate in particular circumstances, although with the proviso that there can be problems with computing sampling variances. In social science research, cluster sampling is often used. If the population is grouped into clusters, rather than trying to sample by one of the above methods the clusters are first sampled. The sample of individuals is then drawn, often at random, from the individuals in the selected clusters. This is similar to stratified random sampling except that there are two stages of randomization and at the first stage some of the strata are randomly removed from the sample. Kahn and Sempos (1989) give the example of a health survey involving a large number of economically similar but geographically dispersed villages. Instead of drawing a sample of size n from the whole population that could involve costly and time-consuming travel to visit all the villages, a sub-sample of villages is drawn and the sample of size n is drawn at random from within this subset. This method of sampling will give good population estimates if the villages are just random assemblages of members of the population. It will not give good population estimates if the villages are economically more homogeneous than the population as a whole – that is, if there is strong positive correlation between the members of each cluster.

Variation in the population may be associated with different spatial scales or spatial hierarchies and the contribution from each of these scales may be the target of inference. In plant ecology areas are exhaustively divided into small blocks (the smallest scale the analyst is interested in) and then these blocks can be combined to provide a sequence of blocks of increasing size (area). Analysis to detect, for example, vegetation patch size proceeds using this census. For a discussion of the methodology see for example Cressie, 1991, pp. 591–7.

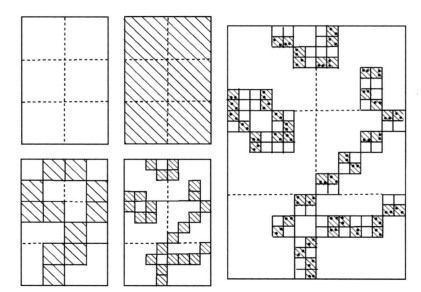

Figure 3.3 A nested sampling scheme

However, if a large range of spatial scales are to be examined necessitating a fine partition of the area then considerable data collection could be involved. One solution is to follow a nested sampling approach.

Nested sampling requires that the population is divided into blocks (level 1) which are then subdivided (level 2) and then level 2 blocks are subdivided into level 3 blocks and so on. Blocks at any level nest within blocks at the higher level. Figure 3.3 shows a form of nested spatial sampling. Each randomly selected datapoint contains variation that derives from each of the levels and the contribution from each level can be estimated by hierarchical analysis of variance. Data requirements are reduced whilst preserving the ability to analyse different spatial scales.

A problem with the scheme in figure 3.3 is that the sampling plan does not control for distance between the samples so scale effects may be masked or at least confounded by the variation in inter-sample distances that exist even at the same scale. Youden and Mehlich (1937) proposed a sampling plan where the distance between pairs is fixed. Primary sampling points are fixed at a specified distance apart. From each of these, further sample sites are randomly chosen at a fixed distance (randomly chosen direction) from the primary sites and from these other sites are selected a fixed distance apart as shown in figure 3.4. This analysis can be used to show at what spatial scale most variation occurs (Webster and Oliver, 2001, pp. 93–4). Miesch (1975) shows how accumulating

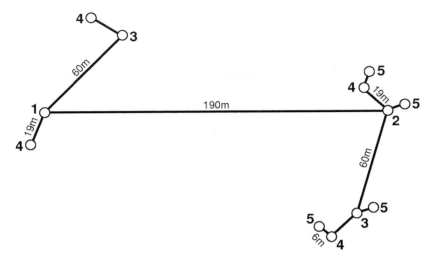

Figure 3.4 Fixed interval sampling

the different components of variance can provide rough estimates of the variogram at different distances. This can then be used to plan a survey to obtain a more careful estimate of the variogram, concentrating sampling effort where the variation occurs (Oliver and Webster, 1986).

Spatial variability in the levels of explanatory variables can be used to explore relationships between explanatory and response variables (see section 1.1). One of the contributions of geographic variation to furthering scientific understanding is to provide the researcher with a natural laboratory through which to look for possible relationships between variables. It will be important therefore to select sampling sites that whilst controlling for other factors achieves good variability on the explanatory variables of interest. This is a form of stratified sampling where the stratification is in terms of the variablity of explantory variables. The Harvard 'six cities' study, exploring the relationship between air pollution and mortality, selected cities that differed markedly in terms of the average levels of air pollution their populations were exposed to (Dockery et al., 1993).

3.2.4   Selected sampling problems

(a)   Design-based estimation of the population mean

We consider the problem of estimating (3.1) which we now denote $\bar{Z}$. Results given here are for a single area or region that may be subdivided into strata. These results can be extended to the case where separate samples have been taken in different areas or regions and the analyst wants to draw comparisons between the areas (Griffith et al., 1994). Significant differences between areas might be identified, for example where confidence intervals do not overlap.

An intuitively simple and robust estimator is given by:

$$(1/n)\Sigma_{k=1,...,n} z(k) = \bar{z} \tag{3.3}$$

This is in fact the design-based estimator under random sampling where the probability of inclusion for an individual in the sample is n/N (Brus and de Gruijter, 1997).

Under random sampling and assuming the individuals are independent, from classical sampling theory, (3.3) is an unbiased estimator of $\bar{Z}$. The error variance of $\bar{z}$ as an estimator of $\bar{Z}$ is $E[(\bar{z} - \bar{Z})^2] = \sigma^2/n$ where σ^2 is the population variance. It can be shown, again from classical sampling theory, that s^2/n, where:

$$s^2 = (1/(n-1))\Sigma_{k=1,...,n}(z(k) - \bar{z})^2 \tag{3.4}$$

provides an unbiased estimator of this error variance. These results can be used to place confidence intervals on the estimates, since, for large n, $\bar{z}$ is normally distributed (see, e.g., Freund, 1992). As sample size (n) increases estimator precision increases.

If the individuals are not independent as would be the usual expectation for a spatial population, then the error variance of $\bar{z}$ as an estimator of $\bar{Z}$ with finite population correction $f = n/N$ is (Dunn and Harrison, 1993, p. 595):

$$((1-f)/n)(N/(N-1))[\sigma^2 - (2/N(N-1))\Sigma_j \Sigma_{k(j<k)} \text{Cov}(z(j), z(k))] \tag{3.5}$$

where the second term inside the square brackets measures the average covariance between all pairs of individuals in the population. For large N, the term in front of the square brackets in (3.5) simplifies to $(1/n)$ and again s^2/n provides an unbiased estimator of this error variance (Haining, 1988, p. 579). For the continuous space analogue of (3.5) where $\bar{z}$ is the estimator of (3.2) see Ripley (1981, p. 32).

The result (3.5) is based on taking expectations both over the positioning of the randomized sample points and over the distribution of the values $\{Z(k)\}$ which have a constant mean (μ) and spatial covariance that depends only on distance separation and possibly direction. Error variances for $\bar{z}$ as an estimator of $\bar{Z}$ can be derived by the same methods also under stratified random and systematic sampling. There are reasons to anticipate that these two sampling methods should outperform random sampling. As noted in section 3.2.3, random sampling leaves unsampled 'holes' in the sample plan whilst also having clusters of sample points in others (see figure 3.2(a)). This seems inefficient because given the dependency between near neighbour individuals this means that random sampling produces some information redundancy. Adjacent sampled individuals carry 'overlapping' amounts of information about the population. Stratified random sampling goes some way to ensuring a more uniform coverage of the population and reducing information redundancy. Systematic sampling by fixing the interval between sample points in adjacent strata carries this process further.

Results corresponding to (3.5) for stratified and systematic sampling are given by Dunn and Harrison (1993, p. 595) and for the continuous space analogue by Ripley (1981, p. 23). In the case of stratified random sampling the second term inside the square brackets in (3.5) is replaced by the average across all strata of the average covariance between all possible pairs of individuals *within a stratum*. In the case of systematic sampling the error variance is the average covariance between individuals in the systematic sample minus the second term inside the square brackets in (3.5).

From these results it follows that the design-based error variance (using the estimator $\bar{z}$) is minimized in the case of stratified random sampling by taking small strata – in order to maximize the average within-stratum covariance. In the case of systematic sampling the design-based error variance is minimized when samples are taken far enough apart so that the covariance between elements of the same systematic sample are as small as possible. The conclusion is that stratification ensures that both stratified random and systematic sampling will outperform random sampling. Ripley (1981, p. 25) expects systematic sampling to be best in the presence of strong local positive spatial correlation unless there is spatial periodicity in the population. Empirical evidence in Dunn and Harrison (1993) confirm the benefits of these two methods over purely random sampling but show that their relative efficiencies, with respect to one another, are far less clear cut. They suggest that their relative performance 'appear to reflect the complex and varied autocorrelation functions of the data' (p. 600).

Estimates of error variances (s^2/n) for stratified random and systematic sampling can be obtained by applying (3.4) to each stratum and averaging over each of these strata. There must be at least two sample points in each stratum if (3.4) is to be computable. If there is only one sample per strata then one option is to impose larger strata (post hoc stratification) of size two for the purpose of estimating s^2 (Ripley, 1981, pp. 26–7). These methods do better than treating the stratified sample as if it were a random sample and computing (3.4) once on the whole data set. A 95% confidence interval is obtained by taking the square root of the estimated error variance $(s/n^{1/2})$ and multiplying by 1.96, the value from the standard normal tables.

The estimator (3.3) can be used for estimating the proportion of the population which has some property, that is when $z(k)$ is 0 or 1. In this case s^2 defined by (3.4) simplifies to:

$$(n/(n-1))\,[p(1-p)] \tag{3.6}$$

where p is the proportion of 1s in the sample of size n so that $\Sigma_{k=1,\dots,n}z(k) = \Sigma_{k=1,\dots,n}z(k)^2 = np$. The study by Dunn and Harrison (1993) evaluates post hoc stratification (of size 2) for binary data. Equation (3.6) is computed for each stratum – so that p is computed separately for each stratum – and then averaged over the strata. This quantity is then divided by n to obtain the estimated error variance. They show that this method overestimates the true sampling error. However this estimator does better than treating the sample as if it were a random sample of size n and computing $p(1-p)/(n-1)$ as the estimated error variance. They conclude however that further evaluation of methods is needed.

(b)   Model-based estimation of means

There are now two kinds of means to consider – the mean of the realized values $(\bar{Z})$ and the mean of the underlying model (μ).

The best linear unbiased predictor (BLUP) of Z under the model-based approach to spatial sampling is given by kriging theory (Cressie, 1991, p. 173). It is a predictor of the form:

$$\Sigma_{k=1,\dots,n}a(k)z(k) \tag{3.7}$$

where the $\{a(k)\}$ depend on the *known* spatial stochastic model and hence on the spatial properties of the underlying model. Another way to look at (3.7) is as follows. The population of realized values is $z(1), \dots, z(N)$ and suppose the sites have been indexed so that $z(1), \dots, z(n)$ denote the values in the sample. The BLUP of $\bar{Z}$ can be expressed in the form:

$$(1/N)\,[\Sigma_{k=1,\dots,n}\,z(k) + \Sigma_{k=n+1,\dots,N}\,\tilde{z}(k)] \tag{3.8}$$

where $\tilde{z}(k)$ is the BLUP of an unsampled $z(k)$. The BLUP of $z(k)$, $\tilde{z}(k)$ and its prediction error will be discussed further in section 3.2.4(c). However, there may be two reasons to prefer (3.3) and the sampling theory developed in the previous section to an estimator such as (3.7). First the prediction errors associated with (3.7) will be too large if the idea of other realizations is controversial in the application. This is because the prediction errors are calculated over all possible realizations and not just with respect to the one outcome that has in fact occurred (Isaaks and Srivastava, 1989, pp. 506–13). Second, in practice the model which determines the $\{a(k)\}$ in (3.7) is rarely known. By contrast in the design-based strategy $a(k) = 1/n$ for all k and follows from the probability of inclusion in the sample which *is* known.

The estimator for μ in the case of a normal population will be discussed in chapter 8. At this point, note that (3.3) is an unbiased estimator of μ. The problem with this estimator is that s^2/n underestimates the sampling error if observations are spatially correlated. If a large amount of data are available, one solution is to sample an independent subset using stratification and use (3.3) together with s^2/n. This will avoid the need to identify a model for the spatial dependence, be easier to implement, and if n is chosen large enough provide an estimator with the desired precision for the application.

(c)   Spatial prediction

The problem considered here is the selection of a sampling plan for an attribute of interest so that good predictions can be made for unsampled sites and a good map of the attribute can be drawn. By a 'good' map is meant that 'best' predictors are chosen (they have the property that they minimize the mean squared prediction error over the class of linear unbiased estimators). Sampling plans are selected so that the prediction error at any point on the map that has not been directly observed is less than some specified amount. The sampling intensity and the position of sample points are critical to obtaining good predictions and for constructing maps from sample data. Such problems arise in many areas of application including mining, hydrology and precision agriculture and provide the context for the development of geostatistics (Webster and Oliver, 2001).

The selection of a sampling plan that will meet these objectives is based on the application of the theory of kriging (optimal spatial prediction). The methodology depends on being able to specify a model (superpopulation) for spatial variation in the attribute to be mapped. This then allows the identification of optimal weights for local prediction, even if for the application it is difficult to defend the idea of multiple realizations. The predictor is the expected value of the attribute Z at a given site, o, given the available data $z(1), \ldots, z(n)$: $E[Z(o) \mid z(1), \ldots, z(n)]$. The theory of kriging will be discussed in more detail in chapter 4 so the reader may wish to return to the following description at a later stage.

Kriging theory shows that the largest prediction errors will usually be found along the boundary of the study region (if sampling is sparse there) and at the centres of the largest gaps between the sampled points in the interior of the study region. A systematic sampling plan (with sample sites coming close to the boundary) will be better than an irregular sampling plan, particularly one that creates clusters of sampled sites. In the case of a rectangular grid of sample points, the maximum prediction error will be in the centre of any rectangle of sampled points. For a given sampling density, triangular grids give lower prediction errors but for practical reasons rectangular grids are preferred and of these centric aligned systematic grids are best (Burgess and Webster, 1980; Webster and Burgess, 1981). Burgess, Webster and McBratney (1981) show how a plot of the maximum value of the minimized prediction error, taken from the site at the centre of any rectangle of sampled points, plotted against sample size can be used to select sample size to achieve a required level of prediction error.

A map can be constructed to show how prediction errors vary over the sampled area. This information can then be used to suggest where new sampling points might be added. Conversely by experimentation, deleting sampling points or small subsets of sampling points, it is possible to examine the impact of such a deletion process on prediction error across the map (see section 3.2.1 for examples).

These predictors and estimates of prediction error are based on the superpopulation or model-based strategy. The analyst has to decide if the idea of there being 'other realizations' (at other points in time or other locations in space) has any meaning because if it does not then these prediction errors for a predictor of the expected value of a population of possible values of $Z(o)$ may be too large. Prediction error in this case should be with reference to $z(o)$ (since no other values are possible) and based on the idea of re-sampling the single data set with a similar sampling plan.

In the case of an environmental pollutant the superpopulation view is usually reasonable because a single source of pollution might be responsible for a number of polluted sites or it might be reasonable to consider the same process being replicated through time at the same place. Even in a geological context, if the study area is large, there may be several subregions with similar statistical properties so that prediction errors can refer to this collection of areas (Isaaks and Srivastava, 1989, p. 507).

(d)   Sampling to identify extreme values or detect rare events

The researcher has reason to believe that an area contains extreme values of some attribute, that is values above some critical value (z_{crit}). The attribute of interest might be soil contamination arising from industrial processes and the critical value has been set by government. The aim is to identify where such sites are located and the total area above some critical value. If an exhaustive data set has been acquired by small quadrats that partition the area, so there is no sampling problem, the next step is to map the indicator function $(I(.))$:

$$I(z(k)) = 1 \quad \text{if } z(k) > z_{\text{crit}}$$
$$\qquad\quad = 0 \quad \text{if } z(k) \leq z_{\text{crit}}$$

Quadrats (k), above z_{crit} can be coloured black, those less than or equal to z_{crit}, white. The total area above the critical threshold can then be calculated by summing the areas of the quadrats coloured black. Such information may be useful in assessing the costs of any clean-up operation or deciding whether the area is safe on health grounds for a particular change of use. Critical values may also be defined in terms of the distribution of observed attribute values. The indicator function might be used to identify all areas that have an attribute value more than (say) three standard deviations above the mean, or in the top 10% of values.

When no data are available a sampling strategy is needed. If a sample has been taken the problem may be to convert the sample data into useful information on the two questions of 'where' and 'how much'. We consider first the situation where there are no data.

Selecting regions to sample   If no sample data are available then the problem is to devise a sampling plan. Neither random, stratified random nor systematic sampling, described in section 3.2.3, are generally used in these circumstances. They tend to discard too much information the analyst may have about where the critical values are likely to be found. 'Judgement' sampling, in which the sampler draws on experience and knowledge about where the most contaminated sites are likely to be found, is more commonly used (Brus and de Gruijter, 1997). This question of where to look (and when to look elsewhere) also arises in archaeological searches for rare artefacts (Switzer, 2000).

Even using judgement sampling the sampler may be confronted by a number of possible areas that experience suggests could have rare artefacts or levels of an attribute above z_{crit}. Suppose for simplicity there are two areas where there might be high levels of a contaminate. The analysts prior belief is that area 1 is more likely to be where the high levels are to be found rather than area 2 so that area 1 is the best place to start looking. This prior belief may be based on the types of industrial activities that went on in the two areas. So, judgement suggests to start sampling area 1 first, but if sampling fails to throw up sample values above z_{crit} is there a point (after taking a certain number of samples) at which sampling should switch to area 2 and when is that point reached?

This is a problem which can be formalized in Bayesian terms as a process in which prior beliefs are progressively modified in the light of new data. At some point the posterior distribution (the prior distribution modified by the data) may be such that when combined with a switching rule leads the sampler to move areas in the search for high levels of the attribute.

Area 1, the area where the sampler believes extreme values are most likely to be found, is subdivided into n quadrats and $\theta(1)$ denotes the probability of finding an extreme value in any given quadrat. This probability is based on the samplers judgement in the light of what is known about the areas history – perhaps in relation to experience from elsewhere. For convenience in this example, assume that the prior distribution for $\theta(1)$ is the beta distribution with parameters $a(1)$ and $b(1)$. That is:

$$\theta(1) \sim \text{beta}\,(a(1), b(1)) \tag{3.9}$$

where $\sim$ denotes 'has the probability density'. The beta distribution parameters imply that the expected value and variance of $\theta(1)$ is (Gelman et al., 1995, pp. 476–7):

$$E[\theta(1)] = a(1)/(a(1) + b(1)) \tag{3.10}$$

$$\text{Var}[\theta(1)] = a(1)b(1)/\{[a(1) + b(1)]^2[a(1) + b(1) + 1)]\} \tag{3.11}$$

Suppose the same prior density is specified for $\theta(2)$, for area 2, but with beta distribution parameters, $a(2)$ and $b(2)$. Since the sampler has chosen to start in area 1 rather than 2, these parameters are assumed to satisfy the relationship:

$$a(1)/(a(1) + b(1)) > a(2)/(a(2) + b(2)) \tag{3.12}$$

The sampler draws n samples from area 1 of which y have values that exceed z_{crit}. We assume that the likelihood for this sampling process is the independent binomial model for the number of areas with an attribute value greater than z_{crit}. So:

$$\{y \mid \theta(1)\} \sim \text{binomial}(n, \theta(1)) \tag{3.13}$$

The assumption of independence derives from the sampling process adopted by the sampler. The fact that adjacent quadrats are likely to be spatially correlated if quadrat size is less than the scale of pollution patches is not important here. (If the sampler were to adopt a sampling strategy in which the choice of the kth site to sample was dependent on the location and attribute values in $r(r \geq 1)$ preceding samples, the assumption of independence would be invalid.)

The posterior density for $\theta(1)$ given the data (y extreme values from a sample of size n) can de derived using Bayes rule:

$$\{\theta(1) \mid y\} \sim \text{beta}(a(1) + y, b(1) + n - y) \tag{3.14}$$

(Gelman et al., 1995, pp. 35–7). The posterior mean of (3.14) is:

$$E[\theta(1)\,|\,y] = (a(1) + y)/(a(1) + b(1) + n) \qquad (3.15)$$

A possible sampling rule would be: if the first n samples yield no values above the critical level ($y = 0$) switch the search area from 1 to 2 when:

$$(a(1))/(a(1) + b(1) + n) < a(2)/(a(2) + b(2)) \qquad (3.16)$$

(Switzer, 2000).

Consider another example. The Poisson model arises naturally in counting the number of cases of a rare disease. Suppose the problem is to detect clusters of cases of the rare disease in regions with poor medical records. We assume the same situation as in the previous example with the sampler starting to take samples in the area where cases are expected (area 1) but needing a decision rule to help decide when to look elsewhere. Let $\phi(1)$ denote the disease rate in area 1 and on the basis of previous experience or other information about the area, the prior distribution is assumed to be gamma distributed with parameters $\alpha(1)$ and $\beta(1)$:

$$\phi(1) \sim \text{gamma}(\alpha(1), \beta(1))$$

So, the expected value of $\phi(1)$ is (Gelman et al., 1995, pp. 44–5):

$$E[\phi(1)] = \alpha(1)/\beta(1) \qquad (3.17)$$

The probability of finding y cases in area 1 with a population at risk of $n(1)$ when the underlying rate is $\phi(1)$ is Poisson distributed:

$$\{y\,|\,\phi(1)\} \sim \text{Poisson}(\phi(1)n(1)) \qquad (3.18)$$

The posterior density for $\phi(1)$ given new data (y cases are found in a sample of n individuals) is:

$$\{\phi(1)\,|\,y\} \sim \text{gamma}(\alpha(1) + y, \beta(1) + n) \qquad (3.19)$$

(Gelman et al., 1995, pp. 48–9). The posterior mean of (3.19) is:

$$E[\phi(1)\,|\,y] = (\alpha(1) + y)/(\beta(1) + n) \qquad (3.20)$$

A possible sampling rule would be: if the first n individuals yield no cases of the disease ($y = 0$) switch the search area from 1 to 2 if and when:

$$(\alpha(1))/(\beta(1) + n) < \alpha(2)/\beta(2) \qquad (3.21)$$

Mapping areas with extreme values   We now turn to the case of a continuous surface of values, where sample data have been collected according to a sampling plan. The analyst now wishes to draw a map showing where the attribute value exceeds a critical value and compute areas. The approach to be discussed assumes a superpopulation or model-based view of the data.

Kriging (section 3.2.4(c) and 4.4.2) provides a means of predicting values on a surface given sample data. Confidence intervals can be attached to each prediction using the prediction standard errors. By supplementing sample data with predicted values particularly in areas where sampling is sparse, the surface can be quantified at various sites and contour lines drawn to give a continuous representation of the surface (Ripley, 1981, pp. 75–7). The extent to which the resultant maps show spatial detail will depend on sample intensity as well as sample positions and detail is lost as sampling becomes less intensive.

But for identifying areas with extreme values there are two problems with adopting the methodology of section 3.2.4(c). First, prediction standard errors derived from sampling cannot be used to assess the variability of non-linear functions of the predicted surface such as line lengths or areas (Ripley, 1981, p. 64; Chilès and Delfiner, 1999, pp. 449–51). Second, describing a distribution using the results of a limited sample is not as informative as examining model properties and in particular looking at the distribution that derives from the model (Switzer, 2000, p. 629). Kriging predictors are *expected values* of the random variable $Z(o)$ given the sample data. Kriging smooths the variation in the map as a whole and this smoothing is not uniform across the map. There is overestimation of small values and underestimation of large. This is an undesirable property if interest focuses on the variability in the surface or if the analyst is particularly interested in identifying areas with extreme values. In order to get a better picture of the distribution it is necessary to examine not just the central tendency of the distribution but also its variability and especially its tail properties. *Conditional* simulation can be used to achieve this. Conditional simulation produces representations consistent with known data values and 'aims to retain the overall texture of the variation of the statistics of the original data in the simulated values' (Frogbrook and Oliver, 2000, p. 226). The next section describes unconditional and conditional simulation.

Once multiple conditional realizations of a probability model have been obtained, there are several options for presenting results. Sample simulations can be shown (e.g., Bloom and Kentwell, 1998; Frogbrook and Oliver, 2000). To provide a summary, areas of the map that exceed the threshold for an extreme value in more than say 90% or 95% of simulations could be highlighted or maps that show the percentage of simulations where a specified extreme value

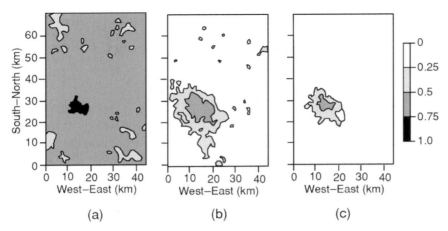

Figure 3.5 Gaussian simulations of Strontium 90: maps of probability of exceeding (a) 0.2, (b) 0.5, (c) 0.7 Ci/km (Savelieva et al., 1998, p. 463)

is exceeded can also be generated (e.g., Savelieva et al., 1998). Figure 3.5 shows simulated maps of the probability of extreme levels of Strontium 90 contamination arising from the Chernobyl fallout. Showing maps of the average across the set of independent simulations is not recommended as it will be similar to the map that would be obtained by kriging and so for the purposes described here will be too smooth. The variance of these simulated maps will tend to the kriging variance.

3.3   Maps through simulation

If a probability model can be specified for data then a better understanding of spatial variability can be obtained by generating multiple realizations from the model. *Multiple* realizations are needed because any simulation is but one of a large number of representations of the specified probability model. The following procedures can be used to generate multiple realizations of a normal random field with a specified mean and covariance or semi-variogram structure. The discussion is in two parts: first *unconditional* and then *conditional* simulation. Because this section uses results that will not be discussed until later in the text (chapters 4 and 9) the reader may prefer to return to this later.

The aim of *unconditional* simulation is to obtain $\iota = 1, \ldots, M$ realizations of a spatial model at n locations where the attribute Z is a normal random variable with a specified mean $\boldsymbol{\mu}$ (of length n) and specified $n \times n$ variance–covariance matrix $\boldsymbol{\Sigma}$. Valid models for spatial data will be discussed in chapter 9.

Unconditional simulation yields realizations that are consistent with the probability model but simulated data values are not required to correspond with known data values.

A valid simulation procedure is as follows. First, the Cholesky decomposition of the $n \times n$ matrix Σ is obtained. The Cholesky decomposition states there is a lower triangular n by n matrix $\mathbf{L}$ such that $\mathbf{L}\mathbf{L}^T = \Sigma$. This decomposition should be obtained analytically if possible, otherwise numerically. A numerically efficient method is to decompose Σ into the matrix product $\mathbf{Q}\Lambda\mathbf{Q}^T$ where the columns of $\mathbf{Q}$ are the n eigenvectors of Σ and Λ is a n by n diagonal matrix of corresponding eigenvalues. It follows that $\mathbf{L} = \mathbf{Q}\Lambda^{1/2}\mathbf{Q}^T$. Note that although this aspect of the simulation is a 'one-off' operation it can be a computationally demanding operation if n is large (>1000). If n is large it may be necessary to partition the area into large overlapping neighbourhoods and keep watch for spurious discontinuities on the final simulated surface (Chilès and Delfiner, 1999, p. 468).

Next, for each of the M simulations that are required, obtain a vector of length n of uncorrelated normal random variables with mean zero and unit variance, $\xi(\iota)$ $(\iota = 1, \ldots, M)$. Third, compute $\mathbf{L}\xi(\iota)$ $(\iota = 1, \ldots, M)$ to obtain the M realizations of the variance–covariance part of the field. Each vector can now be added to the mean vector to give M *unconditional* realizations of the field:

$$\mathbf{z}^*(\iota) = \boldsymbol{\mu} + \mathbf{L}\xi(\iota) \quad \iota = 1, \ldots, M \tag{3.22}$$

In the case of *conditional* simulation, realizations are from a normal probability model where $\boldsymbol{\mu}$ and Σ are specified but at the locations where data values are known the realization matches these. This is achieved by transforming an unconditional simulation of the probability model. The principle underlying conditional simulation is due to Matheron (1976) and is as follows. Let $z(\mathbf{s})$ denote the true but unknown value at location $\mathbf{s}$. Let $\hat{z}(\mathbf{s})$ denote the kriging predictor of $z(\mathbf{s})$ based on data at a set of sample points (see (4.37), where $\mathbf{y}$ is used instead of $\mathbf{z}$). Now, clearly:

$$z(\mathbf{s}) = \hat{z}(\mathbf{s}) + [z(\mathbf{s}) - \hat{z}(\mathbf{s})] \tag{3.23}$$

The second term on the right-hand side of (3.23) is the kriging error however this is also unknown. An estimator of the kriging error is required. This is obtained as follows. The same expression as (3.23) for the output of an unconditional simulation is:

$$z^*(\mathbf{s}) = \hat{z}^*(\mathbf{s}) + [z^*(\mathbf{s}) - \hat{z}^*(\mathbf{s})] \tag{3.24}$$

where $z^*(\mathbf{s})$ is the unconditional simulation value at $\mathbf{s}$ (3.22). Now $\hat{z}^*(\mathbf{s})$ is the kriging predictor obtained using the data taken from the unconditional simulation at the same locations used to calculate $\hat{z}(\mathbf{s})$. Now if $z^+(\mathbf{s})$ denotes the conditional simulation value at $\mathbf{s}$, then:

$$z^+(\mathbf{s}) = \hat{z}(\mathbf{s}) + [z^*(\mathbf{s}) - \hat{z}^*(\mathbf{s})] \tag{3.25}$$

The data value at $\mathbf{s}$ is honoured because kriging is an exact interpolator. There is further discussion together with a demonstration that (3.25) yields map output with the required properties providing there is no systematic measurement error (normal $\boldsymbol{\mu}$, $\boldsymbol{\Sigma}$ model; $z^+(\mathbf{s}) = z(\mathbf{s})$ at those $\mathbf{s}$ where there are sample data values) in Chilès and Delfiner (1999, pp. 465–8).

The argument can be summarized as follows. If $z^+(\mathbf{s}; \iota)$ denotes the realized value at location $\mathbf{s}$ in conditional simulation ι then (Cressie, 1991, p. 208):

$$\begin{aligned}
z^+(\mathbf{s}; \iota) &= \hat{z}(\mathbf{s}) + (z^*(\mathbf{s}; \iota) - \hat{z}^*(\mathbf{s}; \iota)) \\
&= z^*(\mathbf{s}; \iota) + \mathbf{c}^T \boldsymbol{\Sigma}^{-1}(\mathbf{z} - \mathbf{z}^*(\iota))
\end{aligned} \tag{3.26}$$

where $z^*(\mathbf{s}; \iota)$ is the realized value at location $\mathbf{s}$ in simulation ι from an unconditional simulation (3.22). $\hat{z}^*(\mathbf{s}; \iota)$ is the simple kriging predictor as defined by (4.37) except the data vector $\mathbf{z}$ ($\mathbf{y}$ in (4.37)) is replaced with $\mathbf{z}^*(\iota)$ given by (3.22). The unconditional simulations are only involved through the estimation of the kriging errors (the second term on the right-hand side of (3.26)). Chilès and Delfiner (1999, p. 468) say, of conditional simulation, that it '"vibrates" in between the datapoints within an envelope defined by the kriging standard error'. Maps obtained by conditional simulation are useful qualitatively because they provide realistic pictures of the spatial variability based on the evidence in the data; they are useful quantitatively because they allow the analyst to assess the impact of spatial uncertainty on outcomes (Chilès and Delfiner, 1999, p. 453). The methodology can be extended using co-kriging to the multivariate case (see, e.g., Savelieva et al., 1998).

There is discussion of simulation methods in Ripley (1981, pp. 16–18, 64–72), Cross and Jain (1983), Haining et al. (1983), Cressie (1991, pp. 200–9), Goovaerts (1997) and Chilès and Delfiner (1999, chapter 7). Simulation methods, particularly conditional simulation methods, along with interpolation methods (see chapter 4) are used for downscaling data, that is transfering data from larger to smaller scales (Bierkens et al., 2000, pp. 111–44). Switzer (2000) describes a method for efficiently sampling from the model distribution rather than adopting unrestricted random sampling as a result of which many realizations might be generated that are similar to one another whilst leaving unsampled other areas of the space of realizations.

4

Data quality: implications for spatial data analysis

This chapter is concerned with examining the implications of different aspects of data quality for the conduct of spatial data analysis. It was noted at the end of chapter 2 how particular aspects of data quality may have an impact on particular stages of spatial data analysis. Whilst some quality issues impact on the data collection and data preparation stages prior to undertaking analysis, other quality issues impact more on the form and conduct of the statistical analysis or on how results can be interpreted.

The first section deals with error models and the implications of different types of error for data analysis. Section 4.2 considers various problems associated with the spatial resolution of data. The problems discussed include: the impact of varying levels of precision across a map divided into areas; the change of support problem (moving from one spatial framework to another); the problems associated with ecological analyses including aggregation bias and the modifiable areal units problem. Sections 4.3 and 4.4 deal with consistency and completeness problems which include the missing data problem. Some of the results in these later sections use data models which are discussed in chapter 9.

4.1 Errors in data and spatial data analysis

4.1.1 Models for measurement error

All data contain error as a consequence of the inaccuracies inherent in the process of taking measurements. Error models are important in data analysis. An error model allows quantification of the probability that the true value lies within a given range of the measured value. Valid error models allow exploration of the effects of error propagation where arithmetic or other operations are performed on one or more variables that individually contain error. Specification of an appropriate model for the errors is an important element of regression modelling.

(a)   Independent error models

The independent and identically distributed (iid) normal model, $N(\mu, \sigma^2)$, where μ denotes the mean and σ^2 the variance of the distribution, is widely used as a model for measurement error. This is because in many situations the errors that occur are a compounding of many small independent sources of random error (Mikhail, 1976). Suppose a quantity is measured that has a true value X. Suppose also that each source of error deflects the measurement up or down by a quantity ε and that these two possibilities occur with equal probability. If there are n sources of error then the measured value could range from $(X + n\varepsilon)$ to $(X - n\varepsilon)$. If v of the n sources give positive errors and $(n-v)$ give negative errors then the measured value will be: $(X + (2v - n)\varepsilon)$. Let the probability of v positive errors out of n sources be given by the binomial probability distribution with parameters n and $p = 1/2$. As n increases and $\varepsilon \to 0$ this distribution converges to a normal distribution about the true value X. Providing $\varepsilon \downarrow 0$ and $n \uparrow \infty$ such that $(\varepsilon\sqrt{n})$ remains fixed, then the standard deviation of the normal distribution is given by $\sigma = \varepsilon\sqrt{n}$ (Taylor, 1982, pp. 197–9).

The mean (μ) represents any systematic bias in the measurement process and the variance (σ^2) measures the dispersion around the mean. The square of the RMSE provides an estimate of σ^2. The model can be used to compute the probability that the error will lie between two values (a, b) of X.

The $N(\mu, \sigma^2)$ model may be appropriate for the errors associated with the measurement of a point location in one dimension. For a point location on a two-dimensional surface the bivariate normal density will allow for error in both dimensions.

The independent normal model may be plausible for the errors associated with the individual sample points along a boundary but not for the errors associated with the measurement of the boundary itself. Nor is it likely to be plausible for the attributes of the area feature (such as its size or boundary length) that are derived from the representation. This is because these errors also depend on the positioning and density of sample points in relation to the shape and overall complexity of the boundary.

The assumption of normality may be reasonable for errors in measuring attribute values although in the case of synthetic aperture radar data, theoretical arguments have suggested a Rayleigh distribution (Besag, 1986). The independence assumption may also be justifiable for attribute measurement error.

However, as illustrated in section 2.4, there are circumstances where models are needed that allow for spatial dependence amongst the errors – particularly for errors at neighbouring locations.

(b)   Spatially correlated error models

A general model for dependent normal measurement error on a variable X across a set of n locations (points or areas) is provided by the multivariate normal distribution, MVN($\boldsymbol{\mu}$, $\boldsymbol{\Sigma}$) $\boldsymbol{\mu}^T = (\mu(1), \ldots, \mu(n))$ is the vector of means representing the bias in the measurements at each location and:

$$\Sigma = \begin{bmatrix} \sigma_1^2 & \sigma_{1,2} & \sigma_{1,3} & \cdots & \sigma_{1,n} \\ \sigma_{2,1} & \sigma_2^2 & \sigma_{2,3} & \cdots & \sigma_{2,n} \\ \vdots & \vdots & \vdots & \cdots & \vdots \\ \sigma_{n,1} & \sigma_{n,2} & \sigma_{n,3} & \cdots & \sigma_n^2 \end{bmatrix}$$

The term σ_1^2 is the variance for the measurement error at i and $\sigma_{i,j}$ is the covariance in the errors between locations i and j.

When there is spatial correlation in attribute errors it may be reasonable to assume that it is local in scale (see 2.3.2). So, for any location i let:

$$\sigma_{i,j} \begin{cases} \neq 0 & \text{if } j \in N(i) \\ = 0 & \text{if } j \notin N(i) \end{cases}$$

where $N(i)$ denotes a low-order neighbour of location i, that is area i and j share a common boundary or are perhaps at most one step removed (see section 2.4). There are a number of possible models that meet this requirement whilst also satisfying the condition of positive definiteness for Σ which is a requirement for a valid covariance function (Morrison, 1967, p. 60).

A possible model with a local covariance structure is given by the moving average model (Haining, 1978) where:

$$u(i) = \nu \Sigma_{j \in N(i)} w(i, j) e(j) + e(i) \tag{4.1}$$

where ν is a constant, $\{e(i)\}$ denote independent drawings from a $N(0, \sigma^2)$ distribution. It follows that the variance of $u(i)$ is:

$$\text{Var}(u(i)) = \sigma_i^2 = (1 + \nu^2 \Sigma_{j \in N(i)} w(i, j)^2) \sigma^2 \tag{4.2}$$

The covariance between $u(i)$ and $u(j)$ where $j \in N(i)$ is given by:

$$\text{Cov}(u(i), u(j)) = \sigma_{i,j} = (\nu(w(i, j) + w(j, i))$$
$$+ \nu^2 \Sigma_{k \in \Im(N(i), N(j))} w(i, k) w(j, k)) \sigma^2 \tag{4.3}$$

where $\Im(N(i), N(j))$ denotes the set of sites that are neighbours of both i and j. The covariance between $u(i)$ and $u(k)$ where $j \in N(i)$, $k \in N(j)$, but $k \notin N(i)$ is given by:

$$\text{Cov}(u(i), u(k)) = \sigma_{i,k} = \left(v^2 \Sigma_{j \in N(i)} \, w(i, j) \, w(k, j)\right) \sigma^2 \qquad (4.4)$$

All other covariances are zero. Cliff and Ord (1981, p. 150) define a moving average model where covariances are zero after the border adjacency, as do Kiefer and Wynn (1981). These are not the only local error models. Ripley (1981, p. 55) describes a process due to Zubrzycki that can generate a covariance function for a spatial surface that decays steeply to 0.

Correct model specification includes specifying a valid model for the errors. To illustrate, take the case of a simple bivariate regression model where the response variable is Y and the explanatory variable is X and:

$$Y(i) = \beta_0 + \beta_1 X(i) + e(i)$$

where β_0 and β_1 are the intercept and slope or regression parameters respectively. The model is to be fit to data values $\{y(i), x(i)\}$ associated with n spatial objects. The term $e(i)$ is the error term which in a correctly specified model (in terms of Y and X) and with no error in measuring values of X would account for measurement error on Y. Ordinary least squares fitting assumes the $\{e(i)\}$ are independent and identically distributed with a mean of 0 and variance σ^2. However there will be situations arising in spatial data analysis where this assumption may not be supported by the data and a more general model needed that allows for spatial dependence amongst the errors. This has implications for model fitting.

4.1.2   Gross errors

(a)   Distributional outliers

Data values which are extreme with respect to the overall distribution of values, may not be wrong but amongst such *distributional outliers* is one area to start looking to detect gross errors. The detection of extreme values on a variable is important because their presence is likely to give rise to a number of consequences. Particularly in small samples, non-resistant descriptive statistics like the mean or standard deviation are affected by the presence of such data values. Where the presence of such gross error is suspected, or if the distribution of values is known to contain some extreme but valid data values, statistics like the median and the inter-quartile range (the difference between the upper and lower quartile) provide descriptions of the centre and spread of a set of data values that are resistant to their presence. In the case of regression modelling, extreme data values in the X or explanatory variable give rise to large leverage effects whilst in the Y or response variable give rise to large residuals or outliers (see figure 4.1). The presence of large leverages and/or outliers can have a disproportionate influence on regression parameter estimates and model predictions. Regression diagnostics are provided in statistical packages like MINITAB and SPSS for example to assess the influence of individual observations by deleting cases (Belsley, Kuh and Welsch, 1980).

Methods for identifying outliers in a data set usually draw on a combination of graphical and numerical techniques. Some extreme forms of location error can be detected just by overlaying the data cases on a map of the study area and using local knowledge. Location error may also come to light by examining attribute values by area. Extreme attribute values may be caused by location error in the data.

Attribute error in the set of values on a variable Z may be suspected from a histogram plot in which data intervals on the horizontal axis are specified in terms of fractions of standard deviations. Cases lying more than three standard deviations above or below the mean may be considered extreme and flagged for closer investigation. The above method using the mean and standard deviation is equivalent to basing the identification of an outlier on the mean shift outlier model (Weisberg, 1985, p. 114). The model specification is:

$$z(i) = \beta_0 + \delta u(i) + e(i) \tag{4.5}$$

where β_0 and δ are parameters, the $\{e(i)\}$ are independent and identically distributed errors with a mean of 0 and variance σ^2 and $u(i) = 0, (i \neq j)$ and $u(j) = 1$. The jth case is under suspicion and is deemed an extreme value if $\delta \neq 0$ because its mean value is $\beta_0 + \delta$ whilst for all other values $(i \neq j)$ the mean is β_0. Gross measurement error on the jth case may be responsible for $\delta \neq 0$. The above test is equivalent to regressing Z on U and testing $\delta = 0$ against the two-sided alternative $(\delta \neq 0)$. The t-test with $n - 2$ degrees of freedom can be used if the $\{e(i)\}$ are iid $N(0, \sigma^2)$.

The D statistic is a resistant technique for checking for the presence of extreme values. It is defined:

$$D = n^{1/2}(\bar{z} - \text{med}(z))/(0.7555\hat{\sigma}) \tag{4.6}$$

where $\hat{\sigma} = (U - L)/1.349$ and U and L are the upper and lower quartiles respectively. $(U - L)$ is called the inter-quartile range, $\text{med}(z)$ is the median value, $\bar{z}$ is the mean and n is the number of cases. This criterion can be used on any symmetric distribution of values (e.g. a normal distribution) to flag the presence of atypical values. Their presence is suspected if $|D| > 3.0$. A resistant approach to isolating particular cases is associated with the boxplot. An extreme value

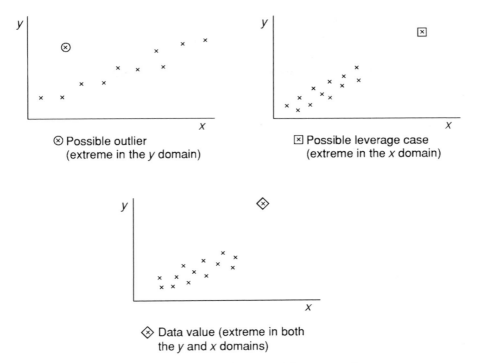

Figure 4.1 Graphical identification of cases with extreme values

criterion for any individual value, $z(i)$, is given by:

$$z(i) > U + \phi(U - L) \quad \text{or} \quad z(i) < L - \phi(U - L) \tag{4.7}$$

where $(U - L)$ denotes the inter-quartile range and ϕ is usually set to 1.5. Since these quartile statistics are resistant measures they will not be much affected if there are some extreme values, unlike the mean shift outlier model.

An error may only be apparent when examined in the context of other variables. For example an error may be made in recording a disease rate for an area, $z(i)$, that is not apparent from looking at the distribution of $\{z(i)\}$. The given value is not necessarily a distributional outlier. In figure 4.1 the outlier is not a distributional outlier on the set of y values. An extreme value may only be suspected when it is analysed in the context of another characteristic, X, for example deprivation. The evidence for a possible error may only be apparent when a scatterplot of $\{z(i)\}$ against $\{x(i)\}$ is constructed or after fitting a regression model of disease rate on deprivation and examining outlier diagnostics (Weisberg, 1985). The test is again based on Weisberg's mean shift outlier model (4.5), although Z is now regressed on both U and X:

$$z(i) = \beta_0 + \delta u(i) + \beta x(i) + e(i) \tag{4.8}$$

The jth case is an extreme value if $\delta \neq 0$ because the mean of $z(j)$ is $\beta_0 + \delta + \beta x(j)$. This test can be generalized to more than one explanatory variable.

(b)   Spatial outliers

Attribute values may be extreme given their position on the map. Such attribute values are termed 'spatial outliers' because their values are extreme relative to the set of neighbouring values on the map. It is possible for a data value to be a spatial outlier without being extreme in the distributional sense which is why methods other than those described above are needed. A map of the data may help flag up possible spatial outliers. Further visual evidence can be obtained from a scatterplot of $\{z(i)\}$ on the vertical axis against their corresponding values in $\{\mathbf{W}^*\mathbf{Z}(i)\}$ on the horizontal axis where the row sums of $\mathbf{W}^*$ equal 1 and $w(i, i) = 0$. $\mathbf{W}^*\mathbf{Z}(i)$ is the value in the ith entry of the product of the matrix $\mathbf{W}^*$ with the vector of values on Z. So, this plot displays the attribute value against the average of its neighbours.

The plot may highlight individual cases if they are sufficiently different from their neighbours, however it would be useful to have criteria for deciding when to flag particular cases for further investigation. If the data are on a square lattice, Cressie (1984, 1991, pp. 38–40) suggests computing the D statistic (4.6) for each row and column. These tests can be adapted to non-lattice data by assigning data values to row and column *bands*. Cressie (1991, pp. 396–8) assigned the 100 counties of North Carolina to the nodes of a 9 by 24 square lattice. In non-lattice cases, however, results could be sensitive to the chosen allocation procedure.

Fitting the model:

$$z(i) = \delta u(i) + \beta_1 \mathbf{W}^*\mathbf{Z}(i) + e(i) \tag{4.9}$$

where $u(i) = 0 \, (i \neq j), u(j) = 1$ offers a method for testing whether the jth value is extreme given its neighbouring values. The test can be used if there is no spatial trend or after removing the trend. The method is based on Weisberg's (1985, pp. 115–16) mean shift outlier test described above (4.8).

Figure 4.2 shows a plot of $z(i)$ against $\mathbf{W}^*\mathbf{Z}(i)$ where $z(i)$ is the standardized mortality rate for accidents in Glasgow in community medicine area i (Haining, 1990, pp. 199–200). The circled case is an outlier at the 1% level based on fitting the model:

$$z(i) = \beta_0 + \beta_1 \mathbf{W}^*\mathbf{Z}(i) + e(i) \tag{4.10}$$

by ordinary least squares and flagging cases with a standardized residual exceeding $|3.0|$. This case is probably not an error. It refers to the city centre and the extreme value (relative to the average of the neighbouring rates) is

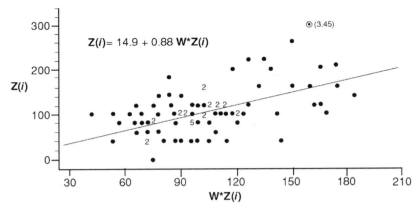

Figure 4.2 Model-based identification of a spatial outlier

probably a consequence of using population as the choice of denominator. Such areas have low resident populations but high daytime and early evening populations and high levels of traffic.

It should be noted though that ordinary least squares fitting of either (4.9) or (4.10) is likely to overestimate the number and seriousness of the outlier problem. This is because ordinary least squares fitting in the case where $\beta_1 \neq 0$ underestimates the residual standard error. However the purpose here is simply to flag cases for closer investigation.

(c)   Testing for outliers in large data sets

The previous tests seem reasonable if the analyst is using them to check a particular value that has been suspected in advance of reviewing the data. In the case of very large data sets it may be important to draw on automated methods that scan through the entire data set. But automated methods that pass through the entire data set in search of possible errors are performing n (the number of cases) significance tests. The probability of finding at least one outlier as a result of performing n independent significance tests approaches 1 as n increases. Any testing procedure therefore should be modified to reflect this multiple testing. Weisberg (1985, p. 116) proposes resetting the critical value based on the Bonferroni inequality. For an overall α significance level of 0.05, a level of $0.05/n$ is chosen for each test. Choosing the $(\alpha/n) \times 100\%$ point of the t-distribution for each of the n tests gives an overall significance level of α. However, if the data set is very large dividing by n will probably result in a very conservative test (underestimating the number of possible extreme values). The application of this adjustment in the case of the model defined by (4.9) raises further problems however because the n tests are not independent. Not only do the n tests use overlapping subsets of the data but if $\beta_1 \neq 0$ then the underlying data values are spatially correlated. If the test is not to be conservative even for moderate sized data sets n needs to be reduced to reflect the 'effective' or 'equivalent' number of independent tests. In the light of these problems a method based on simply ranking the extreme values and identifying breaks in the distribution would seem easiest and most practical.

The methods described above are appropriate for detecting single extreme values but when there are many it is possible that they could mask one another, making their detection difficult. Spatial clusters of errors can arise too, for example through error propagation effects as described in section 4.1.3. One approach to the detection of multiple extreme values is to develop methods that search all subsets of cases for outlying subsets (Hawkins et al., 1984). These methods are not discussed further except to note that techniques based on the principal of analysing all possible *spatial* subsets have been developed to test for spatial clusters of disease and these will be considered in chapter 7.

It is important to emphasize that extreme data values are not evidence in themselves of gross errors and in some geographical problems extreme values arise because of the top heavy or 'primate' structure of the underlying system (Cox and Jones, 1981). Capital cities and economic enclaves in developing countries and core areas of developed economies may appear as outliers or leverage cases in plots and analyses of regional socio-economic data. In other circumstances finding extreme values is the objective of the analysis as in the case of finding disease or crime hot spots or areas with special environmental or geological characteristics not found elsewhere. Thus in some cases extreme values are providing evidence of what is sometimes termed 'spatial heterogeneity'. If, however, an extreme value is the result of error the data value may be correctable or it may have to be discarded. The latter course of action can lead to other problems associated with the consequences of data incompleteness (see section 4.4).

4.1.3   Error propagation

Error propagation results from carrying out arithmetic operations on interval and ratio data that contain error. Errors can also propagate as a result of performing logical operations – for example, identifying areas or point sites that simultaneously satisfy conditions associated with two or more variable values $(x(i)>x.\text{AND}.y(i)<y)$; at least one of the variables satisfy a specified condition $(x(i)>x.\text{OR}.y(i)<y)$ and so on (see Arbia et al., 1998, p. 150 for examples.) Error propagation in this case has implications for identifying regions on a map and for the statistical analysis of nominal- and ordinal-level data.

In analysing error propagation effects, the size and geography of the errors is important. Where errors possess spatial continuity, blend in with map structure, and appear visually plausible this has implications both for the detection of errors and for the interpretation of the results of statistical analysis. 'Small-scale, visually plausible patterns of error are perhaps more likely to escape detection' (Haining and Arbia, 1993, p. 294). Carter (1992) and Lee et al. (1992) describe how spatial correlation in elevation estimates combined with known levels of precision in digital elevation models impacts on slope estimates and drainage basin estimates.

The Geman and Geman (1984) 'corruption model' has been used to explore error propagation in raster data sets. A special case of this model assumes that n ground truth values $\{t(i)\}$ are corrupted as a result of location error and spatially correlated measurement error into a set of n observed values $\{z(i)\}$. In particular:

$$z(i) = \Sigma_{j \in N(i)} \, w^+(i, j) \, t(j) + u(i) \tag{4.11}$$

where again $N(i)$ denotes the set of pixels that are 'adjacent' to pixel i. For all i and j, $w^+(i, j) \geq 0$ and $\Sigma_{j \in N(i)} \, w^+(i, j) = 1.0$. Total error $(z(i) - t(i))$ can be partitioned:

$$z(i) - t(i) = (w^+(i, i) - 1) \, t(i) + \Sigma_{j \neq i} \, w^+(i, j) \, t(j) + u(i) \tag{4.12}$$

The first two terms on the right-hand side represent a form of location error arising from the fact that if $w^+(i, i) < 1.0$, what is recorded for pixel i and which is supposed to represent ground truth at the corresponding ith parcel of land in fact represents ground truth at an average of one or more nearby parcels of land. Attribute error includes a further component arising from reflectance picked up by the sensor which acts as a filter, observing ground truth for any parcel of land i as a weighted function of ground truth in the adjacent areas (Forster, 1980). The second term, $\{u(i)\}$, can represent this additional source of error and can be modelled as a sample from a MVN$(\mathbf{0}, \Sigma)$ distribution. Spatial correlation in the measurement process is specified using the matrix Σ. This component of error is treated as spatially localized so that only the off-diagonal values in Σ representing the near neighbours of each pixel i, are non-zero.

A model similar to (4.11) was analysed using Taylor series methods by Heuvelink et al. (1989) and Heuvelink (1993). In a series of papers, Haining and Arbia (1993), and Arbia, Griffith and Haining (1998, 1999) analyse and visualize in map form error propagation arising from various arithmetic and logical operations using Geman and Geman's model. Analytical as well as Monte Carlo simulation methods are employed. The covariance structure in Σ was based on the filter identified by Forster (1980). Monte Carlo methods allow properties to be identified by examining multiple simulated realizations of a process. Properties of the propagated error are given under different map operations as well as the contribution of different sources of error to both aspatial and spatial error properties. The effect of location error is least when there are high levels of spatial correlation in ground truth values (e.g. the landscape is broadly uniform in character). It increases as ground truth shows less and less spatial correlation relative to the scale of the location error (e.g. a highly mountainous area). The formation of 'error regions' – that is contiguous pixels on the map with particularly high levels of error – is noted in the case of the ratio operation. This arises particularly where ground truth is spatially correlated, there is location error and there are high levels of spatial correlation in the attribute measurement process (as described by the matrix Σ). Similar but less-striking spatial structure seems to emerge in the case of the other arithmetic and logical operations.

Much of this work in the geographical and environmental sciences has been motivated by the need to understand the reliability of the output from geographic information systems which bring together data sets with different origins, generated at different scales often of varying quality and which are then subject to various map overlay operations (Unwin, 1995). One approach to this problem is to try to derive properties of the propagated error starting from model-based assumptions about the errors in the source maps, as in the examples above. An alternative approach is to perturb the final maps to see by how much they must be changed before associations and relationships identified from the observed map start to break down (Moran and Bui, 2000). If the perturbations need to be severe before any marked changes in conclusions are uncovered, then the analyst may feel confident in the results based on the observed map; where small perturbations of the data lead to marked changes, caution is called for. Even following this empirical approach some limits and some structure need to be imposed on the perturbation process so as to be consistent with what is known about the uncertainty in the data – and for this a model is useful.

The methodology of Monte Carlo simulation is widely used for analysing error propagation – see, for example, Goodchild et al. (1992), Veregin (1994, 1995). Models are often relevant to quite specific circumstances and assume spatial continuity and spatial homogeneity of error, whereas in reality error is likely to vary across the map. In some types of surveys there may be discontinuities arising from parcelling up an area into blocks and assigning different data collectors to different blocks (Milne, 1959). Goodchild (1995) expresses the view that 'as we learn more about the nature of errors, and why they occur, it will be possible to produce more refined models of error that take . . . spatial heterogeneity into account' (p. 74). However, the generality of findings arising from this area of research has often been questioned as it is difficult to disentangle general findings from the specificity associated with particular case studies. At the present time the practical use of much of this work for spatial data analysts is probably limited to raising awareness of the way error can corrupt statistics. It helps to qualify the results of statistical analysis and in some cases helps to decide which of several competing statistics might be the more robust to likely errors in the data. This latter concern applies to maps of vegetation indices where there are several to choose from, including those based on image differencing, orthogonalizing transformations such as Gram-Schmidt (Kauth and Thomas, 1976) and principal components analysis (Richards, 1986). The paper by Arbia et al. (2003) offers some suggestions. Health (Jarman, 1993; Townsend et al., 1988) and mortality (Friedman, 1994) indices also manipulate spatial data values through arithmetic operations. Error propagation effects on such maps again need to be recognized.

4.2   Data resolution and spatial data analysis

For a given study area, the results of any analysis of aggregated spatial data will depend on the choice of scale and partition in the case of vector data, and the scale, orientation and origin of the grid in the case of raster data. Space has no natural origin or partitioning. Even re-orienting a grid or taking a new origin will result in a different aggregation, although where spatial correlation is strong in all directions in the variable of interest up to the scale of the pixel size neither of these should raise serious problems. Land-use categories that occupy small scattered areas may be lost if the scale of resolution is large so that their presence is underestimated, whilst that of land-use types appearing in large contiguous areas is overestimated. Arbia et al. (1996) analyse the effect of grid resolution on image misclassification and have shown how error increases as a consequence of moving from fine to coarse resolutions, although with an effect that is moderated by the level of spatial correlation in the underlying surface.

The scale of a grid in the case of remotely sensed data, or the size of the areal units in the case of recording population characteristics, could be altered. For a given study region partitioned into subareas, terms such as 'size of the data set' or 'number of cases' are scale-dependent quantities. As sample size increases as a result of using a finer and finer resolution, any null hypothesis is likely to be rejected if a sufficiently fine resolution is selected. The usefulness of undertaking classical hypothesis testing, particularly when data volumes are large so that any simple null hypothesis will almost certainly be rejected in favour of a general alternative, continues to provoke debate that ranges from adjusting significance levels to reflect the volume of data (Leamer, 1978) to abandonment of classical hypothesis testing altogether (Nester, 1996).

Where the scale of variation of the phenomenon is greater than the resolution of the spatial unit through which its attributes are recorded then data values for adjacent spatial units will be spatially dependent. If there is a further reduction in the size of the areal unit leading to, say, a doubling of the number of 'cases' this is similar to augmenting the data file by simply duplicating some of the observations already in the file. The 'effective' sample size, in the sense of the amount of independent information carried in the data file about the process or the spatial surface, has not increased by the same amount as the increase in the number of cases. The 'effective' sample size for the purpose of statistical inference in the case of spatially dependent data is less than the number of cases because neighbouring data cases carry overlapping amounts of information (Clifford et al., 1985). The concept of the 'effective sample size' is one way to quantify the effect of spatial dependence in data values for the purpose of carrying out statistical testing (see section 3.2.4(a) and chapter 8).

4.2.1   Variable precision and tests of significance

The number of crimes committed or cases of a particular disease in an area are often viewed as a sample from a random process (see section 2.1.4). What is then computed for each area, such as the rate of the disease (number of cases divided by the population at risk) provides only an *estimate* of the true underlying risk. Suppose the underlying process is homogeneous, that is the same random process is operating across all members of the population at risk. Intuitively we would expect the precision of the estimator to be higher (the variance of the estimator smaller) the larger the number of individuals used in any aggregation and the more common the event that is being recorded.

These remarks lead to the conclusion that any map of rare disease rates, for example, needs to be interpreted cautiously if it is used to infer variation in risk across the map. The variance of the estimator for any area is greater the smaller the underlying population – the denominator used in the calculation of the rate. It follows that maps of rates where the areal framework has produced large differences in the denominator is likely to show evidence of rate variation that may be a statistical artefact of the spatial framework rather than any intrinsic variation across the region in the true underlying disease rate. This gives rise to two specific consequences (see, e.g., Mollie, 1996). First, most extreme rates are found in administrative units with small populations.

Second, most of the standardized rates, where the observed rate is significantly greater than the expected rate under the null hypothesis of a random allocation of cases across the map, are found in administrative units with large populations. This latter remark follows because as the denominator increases estimator variance decreases so that it only takes relatively small departures of the observed count from the expected count to generate a statistically significant difference. The methodology for constructing 'reliable' maps, will be discussed in more detail in chapter 7 as part of exploratory spatial data analysis and chapter 10 as part of univariate modelling.

Size–variance relationships arise in other contexts (Haining, 1990, p. 49). For example, suppose $Y(i)$ is any continuous valued variable that measures the average of $n(i)$ equally variable observations (with variance σ^2) in spatial unit i. It follows that the variance of $Y(i)$, $\mathrm{Var}(Y(i)) = \sigma^2/n(i)$. If $Y(i)$ is the response variable in a regression model then this property would violate one of the assumptions of ordinary least squares regression that error variances must be constant (homoscedasticity). Violation of this assumption would be checked by constructing a scatterplot of the square of each residual obtained from the ordinary least squares fit of Y on the set of explanatory variables against the corresponding $1/n(i)$. If the scatterplot is upward sloping or 'wedge-shaped' to the right this provides evidence of a size–variance relationship (heteroscedasticity). The remedial action is to refit using weighted least squares estimation with weights given by $n(i)$, thereby downweighting the contribution of datapoints with large variances. Note that if $Y(i)$ is a total of $n(i)$ observations then $\mathrm{Var}(Y(i)) = n(i)\sigma^2$ and a similar argument can be constructed for this situation (Weisberg, 1985, p. 83).

4.2.2   The change of support problem

This section looks at two different change of support problems, that is situations where data are collected with respect to one spatial framework but need changing or transforming to a new framework. In geostatistics a commonly encountered change of support problem arises when the analyst wants to make inferences about the arithmetic average of a variable for an area on the basis of point data, that is the transfer of data from smaller to larger scales or 'upscaling' (Bierkens et al., 2000). In the case of regional data, data are collected on one spatial framework but need transforming to another. In the geographical literature this is called the areal interpolation problem.

(a)   Change of support in geostatistics

Suppose samples have been taken from a continuous surface at locations $s(1), \ldots, s(n)$. These locations are considered here as *points* or very small areas. The n observed values on the measured variable Y, might be the yield of some mineral or a measure of soil contamination or air pollution. These data values are denoted $\mathbf{y} = (y(1), \ldots, y(n))$ and they are assumed to be realizations of a random process. The analyst needs to predict the average value of the variable Y for an area A, $\bar{Y}(A)$. In an agricultural context samples may refer to levels of a nutrient in a small volume of soil and A is the field they are taken from. In mining the purpose might be to assess the commercial viability of mineral exploitation in A; in the case of soil contamination the problem might be to assess whether children, eating a particular volume of soil, are likely to put themselves at risk of poisoning (Heuvelink et al., 1999). Point sample data are used to make predictions for areas. This is known as a change of support problem because the observed data are with respect to point supports $(s(1), \ldots, s(n))$ and the prediction is with respect to an area support (A).

For simplicity assume a constant mean so that for any point $E[Y(i)] = \mu$. It follows that $E[Y(A)] = \mu$. Intuitively it might be expected that the sample mean:

$$\bar{Y} = (1/n)\Sigma_{i=1,\ldots,n}\, y(i) \tag{4.13}$$

would make a good predictor of $Y(A)$. However a predictor that gives equal weighting to each observation is ignoring the geographical distribution of the sites relative to the location where the prediction is needed. In statistical terms this means ignoring the spatial correlation or dependence in Y – both in the sample data and between the sample datapoints and the location where the prediction is required. Furthermore (4.13) does not take into account the difference in the variances of $Y(s(i))$ and $Y(A)$. Computing an average over a large area has the effect of reducing the variance of the point sample data, although the reduction occurs less rapidly the more continuous, the more spatially correlated, the data (see, e.g., Isaaks and Srivastava, 1989, p. 462). It can be shown that $\mathrm{Var}(Y(A)) < \mathrm{Var}(\bar{Y})$ (see, e.g., Cressie, 1996, pp. 163–4). The sample mean is therefore an inefficient predictor (it has a large sampling variance) and this can lead to significant over or under prediction of $Y(A)$. In the context of commercial mining for example, where sampling is used to predict areas or blocks of commercially viable deposits, this could be a costly mistake.

If the surface mean is not known but can be assumed constant, from standard results in geostatistics (Cressie, 1996, p. 163), the best linear unbiased predictor of $Y(A)$, which means that of all linear unbiased predictors it has the minimum mean squared prediction error, is given by:

$$\hat{Y}(A) = [\mathbf{c}^T \Sigma^{-1} + (\mathbf{1m})^T \Sigma^{-1}]\mathbf{y} \tag{4.14}$$

where:

$$\mathbf{m} = (\mathbf{1}^T \mathbf{\Sigma}^{-1} \mathbf{1})^{-1}(1 - \mathbf{1}^T \mathbf{\Sigma}^{-1} \mathbf{c}).$$

$\mathbf{1}$ is a column vector of 1s; $\mathbf{c}^T = [\text{cov}(Y(A), Y(1)), \ldots, \text{cov}(Y(A), Y(n))]$ where cov denotes covariance. $\mathbf{\Sigma}$ is the $n \times n$ symmetric variance–covariance matrix. The (i, j)th element of $\mathbf{\Sigma}$ is $\text{cov}(Y(i), Y(j))$ and the ith diagonal element is the variance of $Y(i)$ which is also denoted $\text{Var}(Y(i))$. The predictor (4.14), called the *ordinary* block kriging predictor, addresses the shortcomings of (4.13). The inclusion of $\mathbf{\Sigma}^{-1}$ in the estimator adjusts for the spatial dependency between the sample data and the inclusion of $\mathbf{c}$ exploits the spatial dependency between the sample points and the area to be predicted.

All these quantities including the elements in $\mathbf{c}$ can be estimated from the point support data (Isaaks and Srivastava, 1989, pp. 324–7). There will be further, more detailed discussion of kriging in section 4.4.2(v). At this point however note that the prediction error for (4.14) is:

$$\text{Cov}(Y(o), Y(o)) - \mathbf{c}^T \mathbf{\Sigma}^{-1} \mathbf{c} + \mathbf{m}[1 - (\mathbf{1}^T \mathbf{\Sigma}^{-1} \mathbf{c})] - \text{Cov}(Y(A), Y(A)) \quad (4.15)$$

where $\text{Cov}(Y(o), Y(o))$ is the average point variance in the sample and $\text{Cov}(Y(A), Y(A))$ is the average within block variance. In due course the reader should compare (4.15) with (4.42). The prediction error measured by (4.15) is reduced relative to (4.42) by the final term, the average within block variance. In general the larger the support A the smaller the variance of Y. Equation (4.15) shows that the prediction error for $Y(A)$ is also smaller than the prediction error for any point prediction, say $Y(o)$. Webster and Oliver (2001, pp. 162–4) provide illustrative examples.

(b)   Areal interpolation

Areal interpolation is undertaken in order to make attribute values recorded on one spatial framework available on another spatial framework. This may be required when attempting to make comparisons over time – for example across two or more censuses when it is likely that some census tract boundaries will have been altered. In another context an analyst needs to link socio-economic, environmental and health data in a single database on a common spatial framework. However, environmental data on air pollution may be recorded by grid squares from a model, land use may be recorded using area classes where boundaries denote changes of land use; socio-economic data may be from the census and based on enumeration districts; health data may be recorded by postcodes. Enumeration districts may be used for the common spatial framework. Health data records may be linked to enumeration districts using an ED/postcode directory (see Collins et al., 1998 for a discussion of the errors this can introduce). Kelsall and Diggle (1995) in an analysis of disease rates attach numerator and denominator data to a common grid framework and construct a relative risk surface. Markoff and Shapiro (1973) provide some examples from historical research.

Mrozinski and Cromley (1999) classify areal interpolation problems into one of four types. However their first type, the missing data problem, will be considered in the context of data incompleteness (see section 4.4). In identifying the three other types, they distinguish between area-class map data and choropleth map data. Area-class data are data where the spatial unit boundaries reflect breaks or changes in the attribute level of a continuous or field variable – such as a land-use map. Attribute values are typically at the nominal or ordinal level of measurement. Choropleth data arise where the spatial unit boundaries are independent of the recorded attribute – such as when socio-economic and demographic data are recorded by census tract. Attribute data associated with choropleth maps are typically at the interval or ratio level. They may refer either to counts such as total population or total income (spatially extensive) or rates or averages such as per capita income or population density (spatially intensive) (Goodchild and Lam, 1980).

Type (1): choropleth data are available on one areal partition but need to be interpolated to another areal partition that may be a spatial framework that derives from an area-class framework or another, different, choropleth framework.

Type (2): data are available on two variables, one a choropleth map and the other an area-class map. Values need to be interpolated to the intersections created by overlaying them.

Type (3): data are available on two variables, both choropleth maps and values need to be interpolated to the intersections created by overlaying them.

In the case of type (1) problems there is a single set of 'source zones' that are to be interpolated to a different set of 'target zones'. In types (2) and (3) there are two sets of source zones which when overlayed create a set of target zones arising from their intersection. It is these new zones ('resels' or resolution elements) created by the intersection for which interpolated values are required. In type (1) and (2) problems the known 'volume' associated with the source zone data must be preserved on the new target zones on just one variable. In type 3 problems this volume preserving (or pycnophylactic) property needs to be honoured for both variables.

The methodology of areal interpolation divides into cartographic methods and statistical methods. Cartographic methods exploit information on which areas overlap and sometimes adjacency information which is used for smoothing to avoid sharp discontinuities on the interpolated surface and/or preserve volume. Statistical methods typically involve fitting models to the source data drawing on a range of ancillary data to explain variation. These methods may be sequential (fit a model to the observed data and then use the model to interpolate) or iterative (fit model then estimate data values then refit the model and so on). We consider cartographic methods first and concentrate on type (1) problems, noting however that solutions to type (1) problems can usually be applied to type (2) problems. There are links between the cartographic methods described here and smoothing methods used in exploratory spatial data analysis which are discussed in section 7.1. The method chosen in any particular case must honour the type of variable, that is whether it is spatially extensive or intensive.

The known values on variable Z for n source zones are denoted $\{z(s(i))\}$ where $s(i)$ now signifies the ith source zone. If the source zone variable is a count (spatially extensive) the true unobserved value for Z in target zone j, $t(j)$, is:

$$z(t(j)) = \Sigma_{s(i)}\, z(s(i) \cap t(j)) \tag{4.16}$$

where $(s(i) \cap t(j))$ is the intersection of the ith source zone with the jth target zone and $z(s(i) \cap t(j))$ denotes the true but unobserved count associated with this area of intersection. If the source zone variable is a ratio (spatially intensive), the true unobserved value for Z in target zone j, $t(j)$ is:

$$z(t(j)) = \Sigma_{s(i)}\, z(s(i) \cap t(j))\, [A(s(i) \cap t(j))] / [A(t(j))] \tag{4.17}$$

where $z(s(i) \cap t(j))$ denotes the true but unobserved rate associated with this area of intersection and where $A(.)$ denotes area.

The problem is to *estimate* values of the variable of interest across the set of k 'target zones' – $\{\hat{z}(t(j))\}$. In the case where the source area $s(i)$ is represented by a single point, such as the centroid of a set of individual point objects that occupy $s(i)$, then $z(s(i) \cap t(j))$ in (4.16) can be estimated by $z(s(i))$ if the representative point falls within $t(j)$, but is 0 otherwise. This is called the point-in-polygon method. This method should work well if the individual objects cluster close to the representative point (Sadahiro, 2000).

A variant of the point-in-polygon method is the kernel method. This distributes the weight according to a probability density function (the kernel) around the representative point (Silverman, 1986; Sadahiro, 1999). Bracken and Martin (1989) used the kernel method to continuously map the spatial distribution of the UK population from enumeration district-level census data that are associated to population weighted centroids. Kelsall and Diggle (1995) use kernel density estimation to assign data from different irregular spatial frameworks to a common framework to compute disease rates. A problem with this is that there is often no basis on which to decide whether the real map meets these conditions. The resulting map of the attribute on the set of target zones tends to be too smooth.

If the source zone variable is spatially extensive and can be assumed to be uniformly distributed within the source zone an estimator for (4.16) is given by:

$$\hat{z}(t(j)) = \Sigma_{s(i)} z(s(i)) [A(s(i) \cap t(j))]/[A(s(i))] \tag{4.18}$$

If $z(s(i))$ is an area dependent ratio or proportion (spatially intensive) an estimator for $z(t(j))$ is given by substituting $z(s(i))$ for $z(s(i) \cap t(j))$ in (4.17) so:

$$\hat{z}(t(j)) = \Sigma_{s(i)} z(s(i)) [A(s(i) \cap t(j))]/[A(t(j))] \tag{4.19}$$

Again (4.19) will be appropriate providing the ratio is uniform across the source zone. With this and other estimators it may not be straightforward to decide on the area of intersection. If the operations are performed in a geographic information system the analyst needs to guard against errors in recording area boundaries which compound when the two sets of zones are overlayed giving rise to false intersections.

These area weighting estimators (4.18) and (4.19) are based on geometric properties of the source and target zones. Where the assumption of a uniform distribution of the variable across source zones is known to be false, dasymetric mapping excludes from the area calculations those areas where the variable of interest is known to be 0. For example, water areas and industrial areas would usually be excluded in calculating population counts and satellite data overlaid on the source and target zone maps may be useful (Langford et al., 1991). Evidence suggests that this modification can substantially improve area-weighting estimates (Fisher and Langford, 1995; Mrozinski and Cromley, 1999). Brindley et al. (2002) allocate modelled pollution data at a fine spatial scale to enumeration districts using data on where the population are located within each ED. This is to obtain an ED-level measure of pollution that attempts to reflect more closely what the people who live there are exposed to. A representative point in the target zone is used to select the appropriate data value from the source zone.

If population varies as a function of some category like land use or rock type the dasymetric method needs to reflect this. Providing it is possible to obtain estimates on how Z varies as a function of category and providing the geographic distribution of the category within each target zone is known then this should improve on binary dasymetric area weighting. These are examples of so-called 'intelligent' interpolation (Fisher and Langford, 1995). Goodchild et al. (1993) use data from a third set of spatial units they refer to as control units and within which the spatially extensive variable Z is uniform. The control unit boundaries need not be congruent with either the target or source zones. Let $c(k)$ denote control zone k then:

$$z(s(i)) = \Sigma_{c(k)}\, z(c(k))\, [A(s(i) \cap c(k))]/A(c(k)) \tag{4.20}$$

Now if $z(c(k))$ can be estimated, $\hat{z}(c(k))$, then:

$$\hat{z}(t(j)) = \Sigma_{c(k)}\, \hat{z}(c(k))\, [A(t(j) \cap c(k))]/A(c(k)) \tag{4.21}$$

Goodchild et al. (1993) discuss various statistical methods of estimating $\{z(c(k))\}$ using $\{z(s(i))\}$ including Poisson regression and constrained least squares.

Mrozinski and Cromley (1999) place areal interpolation for count data within a matrix framework also found in spatial interaction modelling. An intersection matrix can be defined comprising S rows (corresponding to the source zones) and T columns (corresponding to the target zones). The matrix consists of 1s where zones i and j intersect, 0 otherwise. The problem is to estimate the variable values associated with the non-zero cell entries in the intersection matrix. Variable values associated with the other cells in the matrix are known to be 0. In type (1) and (2) problems only row sums are known (a singly constrained model) whilst in type (3) problems both row and column sums are known (a doubly constrained model). Once cell values have been estimated target zone estimates can be obtained by aggregation. Iterative operations are performed on cell entries until convergence takes place. The operations of polygon smoothing and dasymetric polygon smoothing (for type (1) and (2) problems) which Mrozinski and Cromley describe are extensions to Tobler's (1979) original pycnophylactic method. The methodology involves assigning to any intersection zone a density given by the source zone it sits within, averaging this density with values from the contiguous intersection zones and iterating until convergence occurs. A volume-preserving constraint is introduced and counts are obtained at the end by multiplying the density values with the areas of the intersection zones. Polygon smoothing performed a little better than area weighting in their test data but when the dasymetric adaptation was introduced there was no clear winner in terms of overall root mean squared error. These methods raise a number of questions about the existence and uniqueness of solutions under different conditions including the effect of increasing the number of intersections as well as how to represent the interaction between the source and target zones and what types of ancillary data might be useful.

The other main approach to areal interpolation draws on statistical modelling. Flowerdew and Green (1989) propose a statistical modelling approach to areal interpolation which can be used if ancillary data are available in the form of area class or choropleth data. Their approach is an extension of the 'intelligent' method of interpolation for spatially extensive data and is similar to Goodchild et al. (1993). They apply their method to estimate population counts for target zones where population varies according to whether land type is grassland (1) or woodland (2). The full data set $\{z(s(i))\}$ together with data for each source zone on the area under each land-use type, $\{A(s(i); 1)\}$ and $\{A(s(i); 2)\}$, are used to estimate the parameters of a Poisson regression model (with identity link and no constant term). These estimated parameters ($\hat{\lambda}_1$, $\hat{\lambda}_2$) are the expected population counts per unit area for each land-use type. Providing the area under each land-use type in each target zone is known ($\{A(t(j); 1)\}$, $\{A(t(j); 2)\}$) then:

$$\hat{z}(t(j)) = \hat{\lambda}_1 A(t(j); 1) + \hat{\lambda}_2 A(t(j); 2)$$

Estimates for smaller zones formed by the intersection of the population map and the land-use map can also be determined by this method. The counts will need scaling to ensure the target zone total matches the source zone total. Langford et al. (1991) develop a similar approach where the ancillary datum is a land-cover classification derived from remotely sensed data.

Flowerdew, Green and Kehris (1991) generalize the model-based approach using the intersection matrix described by Mrozinski and Cromley (1999). In their generalization, variable values are estimated using the EM algorithm which is a statistical technique for dealing with estimation problems where there are missing data (see section 4.4). The algorithm computes expected (E) values of the 'missing data' (the variable values in the non-zero entries of the intersection matrix corresponding to the target zones) given the model and the observed data. At the first iteration these values could be derived from the areal weighting approach. The completed data set is then used to fit the model by maximum likelihood (M). The algorithm then iterates between the E and M steps until convergence occurs. The example by Flowerdew et al. (1991) again uses the Poisson model but others can be used in the algorithm.

There are problems with their model-based approach as Flowerdew et al. (1991) observe. The method is computationally intensive and processing speed is much slower than the point-in-polygon method. This is an important consideration when processing large spatial databases like national census data (Sadahiro, 2000). Model goodness of fit is no guarantee as to how well the target zone values will be estimated by the model. Parameter estimates in the Poisson regression model are global estimates and do not take into account possible spatial heterogeneity in the association between, say, population and land-use type.

The above factors may help to explain why area weighting and simple dasymetric methods based on localized mapping often perform better than more elaborate methods in comparative trials (Fisher and Langford, 1995). Cockings et al. (1997) compare interpolation methods using Monte Carlo methods and show that, whilst areal weighting method errors are strongly related to geometric properties of the target zones, dasymetric method errors are more correlated with population or attribute parameters. However trying to make links between error properties and characteristics of the zones is made difficult by the complex inter-relationships between the various confounding effects (Cockings et al., 1997, p. 327).

An underlying problem in this area is that the analyst performs these operations in order to transform the data on to a more appropriate spatial framework. No new data are created, rather estimates are obtained based on transforming the original data. If the analyst now treats these estimates as data, these new 'data' now contain *additional* unknown errors. However there are relatively few criteria to help decide which method to use in any particular situation and no diagnostics to assist in evaluating when a chosen method is or is not doing well and no quantitative measure of the precision of the estimates.

Empirically based studies indicate the performance of different methods in particular cases but leave open the question of the generality of findings. This experimental shortcoming can be overcome by Monte Carlo methods (Fisher and Langford, 1995). It can also be handled using the stochastic modelling approach of Sadahiro (1999, 2000). However these analyses at best indicate the relative performance of different methods under different conditions. The presence of spatial heterogeneity means that any one method may perform very differently in different subareas of the map. The situation is analogous to trying to analyse the effects of error propagation (see section 4.1.3).

Areal interpolation is an estimation problem but we are not in a position to attach confidence intervals to the estimates or to map the geography of the likely errors. The success of most estimators appears to depend on satisfying a set of often quite restrictive assumptions. Where there is strong spatial correlation at a scale that exceeds the scale of the areal partition then iterative smoothing methods should do well but perhaps not otherwise. The maps in Mrozinski and Cromley (1999) suggest that it is where there are sudden changes in population density associated with urban places with sharp boundaries that large interpolation errors are found. In these circumstances further ancillary data will be helpful.

This is a very problematic area of spatial analysis and will remain so whilst data are collected on different spatial frameworks. The past focus has been on methods for transforming data from one spatial framework to another but perhaps more attention needs to be given to the *consequences* of having to change spatial frameworks. This means assessing the robustness of findings to induced errors and exploring the variability in model findings that arise from data uncertainty.

4.2.3   Analysing relationships using aggregate data

Ecological inference is the process whereby grouped data (also known as ecological or aggregate data) are analysed and results used to infer individual-level relationships. It is of interest to those working with spatial data because data aggregated by area is one type of grouped data.

Ecological inference is important because there may be a lack of reliable individual-level data. It may be impractical to collect data at an individual level or if it were collected it would only be available with uncontrollably large imprecision, for example exposure rates to environmental risk factors. Small-area averages of natural radiation levels may actually be a more reliable basis for inference than data based on estimating individual exposure and show greater robustness to measurement error. Some data may only be available for areas, for example income data and electoral data.

Gelman et al. (2001) use the following model to represent the nature of ecological inference. Consider $j = 1, \ldots, m(i)$ individual units in area i ($i = 1, \ldots, n$). Let $y(j, i)$ be the response for individual j in area i, let $x(j, i)$ be the corresponding predictor and let the $e(j, i)$ be independent with mean 0 and also independent of the $x(j, i)$. Define:

$$y(j, i) = \alpha(i) + \beta(i) x(j, i) + e(j, i) \tag{4.22}$$

The interest is in estimating $\alpha(i)$ and $\beta(i)$, but no individual level data are available, only averages for each area denoted $\bar{y}(i)$ and $\bar{x}(i)$. Assuming that average errors $\{e(i)\}$ are close to 0 then it follows from (4.22):

$$\bar{y}(i) = \alpha(i) + \beta(i) \bar{x}(i) \tag{4.23}$$

Ecological regression estimates the parameters in (4.23) by regressing $\bar{y}(i)$ on $\bar{x}(i)$ with one datapoint per area, that is fitting:

$$\bar{y}(i) = \alpha + \beta \bar{x}(i) + u(i) \quad i = 1, \ldots, n \tag{4.24}$$

where $u(i)$ is independent with mean zero and also independent of the $\bar{x}(i)$. Ecological inference involves using the estimates of the aggregate parameters α and β in (4.24) in place of the local regression coefficients $\alpha(i)$ and $\beta(i)$.

Ecological (or aggregation) bias is the difference between estimates of relationships obtained using grouped data (4.24) and those estimates obtained using individual-level data (4.22). The analyst who takes the estimate obtained from grouped data and uses it to infer an individual-level relationship, without specifying the conditions under which the estimates are reasonable, would be said to be guilty of committing the ecological fallacy.

Simpson (1951) drew attention to the dangers of only analysing the margins of complex (multi-dimensional) contingency tables and there is much evidence showing how correlation and regression estimates yield different results depending on whether individual- or aggregate-level data are analysed. Early examples can be found in Gehlke and Biehl (1934), Neprash (1934) and Robinson (1950). Robinson's study of race and literacy in the USA showed that whilst at the individual level the tetrachoric correlation for the 2×2 table of counts was only 0.203, at the state level the product moment correlation of the percentage figures was 0.773. Assembled evidence in epidemiology and elsewhere suggests that the correlation coefficient is more seriously affected than regression parameters (Firebaugh, 1978 and Morganstern, 1982).

Other forms of aggregation bias can arise in spatial analysis. There are potentially two sources of aggregation bias in modelling the relationship between say house prices and a set of predictor variables that includes distance from the city centre using aggregate data. First the use of areally grouped data for the response and predictor variables, second the use of the area centroid to represent the location of the houses in any unit rather than the true spatial average of the individual houses in each areal unit (Okabe and Tagashira, 1996).

Inferring individual-level relationships from grouped data is important in many scientific areas because the individual is the object of study and hence the target of inference. However groups can act as effect modifiers and hence so can areas. The overall composition of the constituency in a first-past-the-post electoral system could affect whether and how individuals vote. There is evidence that individual propensities to commit acts of vandalism can be influenced by the density of young males in an area. Individual preventative health behaviours may be influenced by norms and attitudes in the immediate environment. The analyst will want to try to distinguish individual-level effects from area-level effects. The converse to the ecological fallacy is the atomistic fallacy (disaggregation bias) which is the error that can arise from identifying associations from individual-level data whilst ignoring area-level or contextual effects. Multi-level modelling (see section 9.2.3) is used to model these types of problems.

In fields such as geography and areas of environmental science the target of inference is often at the area level rather than at the individual level. How do different forms of spatial aggregation affect statistics like correlation and regression which are known *not* to be invariant to aggregation effects? When carrying out comparative work including checking findings in one area by carrying out a parallel study elsewhere or at a different time in the same place the analyst might be concerned that any differences observed do not reflect real differences. They might, in fact, be an artefact of two different spatial aggregations that have produced different degrees of within-area homogeneity.

In geography the generic name for this is the modifiable areal units problem (MAUP). 'If a statistic is calculated for two different sets of areal units which cover the same population, or sample, a difference will usually be observed even though the same basic data have been used in both analyses. This difference is cited as evidence of the modifiable areal units problem' (Holt, Steel and Tranmer, 1996, p. 181). The term 'modifiable' is used because neither the choice of number of spatial units (the scale of the analysis) nor their particular configuration (the selected partitioning or zoning given the scale of analysis) is fundamental and any one of a number of other choices could have been made. For a brief overview of the extensive work in geography examining the volatility of regression parameters and correlation coefficients see Wong and Amrhein (1996).

If results differ between different scales of analysis this may reflect the operation of scale-dependent processes. Two economic activities might compete for land at a local scale but be found clustered together when their distribution is examined at a larger scale (using larger spatial units) because of input–output linkages or other forms of interaction. If altering the partition whilst holding scale constant produces a different set of results this is because the new partition has introduced a different smoothing of the data and brought about a reconfiguration of the within-area homogeneity.

In fields such as epidemiology (dose–response relationships) and econometrics (quantity–price relationships) ecological inference is undertaken with continuous-valued variables. In political science, ecological inference is used on data where variable values at the individual level are binary (vote–not vote). Ecological inference is undertaken where: (i) there are no reliable individual-level data; (ii) there are individual-level data on the variables of interest, for example, originating from a national survey which has provided some sparse coverage across smaller spatial units; (iii) there is individual-level data available on a set of other variables – so-called 'grouping' variables – that explain the within-area homogeneity that underlies the causes of ecological bias. A distinction can also be made between studies in terms of their target of inference. Is the target of inference the individual-level relationship *within* each of the areas that partition the study region so that there are as many inferences as there are areas in the study region (the conditional approach), or is the target of inference the individual-level relationship *across* all the areas in the study region (the marginal approach)?

We now consider three aspects of the problem of making valid inferences from aggregate data.

(a)   Ecological inference: parameter estimation

We review the problem of ecological inference for discrete valued data (on both the response and predictor variables of (4.23)) and draw on King (1997), Holt et al. (1996), Wrigley et al. (1996) and Tranmer and Steel (1998) to identify what underlies aggregation bias. Four methodologies for pursuing ecological inference, and in particular estimating parameters of interest, due to Goodman (1953), Freedman et al. (1991), King (1997) and Tranmer and Steel (1998) are described.

Let $T(i)$ denote the proportion of the voting-age population who turn out to vote in an election in area i – called the turnout in area i. Let $X(i)$ denote the proportion of the voting-age population who are non-white in area i, so that $1 - X(i)$ is the proportion who are white. $\{T(i)\}$ and $\{X(i)\}$ are the (aggregate) data. It follows that if $\beta^o(i)$ is the proportion of voting-age non-whites who vote and $\beta^w(i)$ is the proportion of voting-age whites who vote in area i then by definition:

$$T(i) = \beta^o(i) X(i) + \beta^w(i)(1 - X(i)) \tag{4.25}$$

The parameters of interest are $\{\beta^o(i)\}$ and $\{\beta^w(i)\}$ and they are unknown. If individual-level data were available the cells of the 2×2 contingency table (vote/not vote; white/non-white) could be filled in and the parameter values read off. In the absence of such data, if there are n areas then there are n equations but twice as many ($2n$) parameters. The parameters of interest in (4.25) cannot therefore be estimated and this is called the indeterminacy problem. ((4.25) is a reparameterization of (4.23) for discrete-valued variables so that the area averages are proportions. In (4.23) the equivalent problem to that described for estimating (4.25) is that in (4.23) there is one equation and two unknowns.)

One approach to tackling this problem is Freedman et al.'s (1991) neighbourhood model which assumes that at the chosen level of aggregation ethnicity has no influence on voting behaviour so that $\beta^o(i) = \beta^w(i) = T(i)$ (see Gelman, 2001, p. 105). Goodman's (1953) regression approach makes the constancy assumption that $\beta^o(i) = B^o$ and $\beta^w(i) = B^w$. Although voting propensity is allowed to vary between ethnic groups it does not vary by area. The parameters can be estimated by weighted least squares in a regression model of $T(i)$ on $X(i)$ and $(1 - X(i))$ where the intercept coefficient is set at 0. The weights are chosen to reflect the argument that the variance in the dependent variable $T(i)$ is expected to be inversely proportional to the population in the ith area (see section 4.2.1). King (1997, pp. 56–73) has an extended discussion of the problems associated with estimating Goodman's regression model. These two methods adopt the marginal approach to individual-level inference and provide no evidence on how voting propensities might vary between voting areas.

Consider now the nature of ecological or aggregation bias arising in the Goodman regression approach to estimation. Following King (1997, p. 45) write: $B = B^o - B^w$. Write $\hat{B}$ for the difference in the weighted least squares estimates of B^o and B^w. That is: $\hat{B} = \hat{B}^o - \hat{B}^w$. Aggregation bias is then given by: $\hat{B} - B$.

Now, let $X(j, i)$ and $T(j, i)$ denote the individual-level data where (j, i) signifies individual j ($j = 1, \ldots, m(i)$) in area i ($i = 1, \ldots, n$). These are binary (0 or 1) variables. Let $\mathbf{X}$ and $\mathbf{T}$ denote the vectors of length N ($= \Sigma_{i=1,\ldots,n} \, m(i)$) that contain the individual-level data. Now let $\bar{\mathbf{T}}$ and $\bar{\mathbf{X}}$ denote the simple average of $\{T(j, i)\}$ and $\{X(j, i)\}$ over all individuals in all areas. Then:

$$[\mathbf{T} - \bar{\mathbf{T}}] = [\mathbf{X} - \bar{\mathbf{X}}]B$$

This is the individual-level expression of the relationship, and follows from the definition (4.25) after introducing Goodman's constancy assumption. The value of B (the individual-level parameter) can then be computed exactly either by least squares or as a cross tabulation:

$$B = [[\mathbf{X} - \bar{\mathbf{X}}]^T[\mathbf{X} - \bar{\mathbf{X}}]]^{-1} [\mathbf{X} - \bar{\mathbf{X}}]^T[\mathbf{T} - \bar{\mathbf{T}}]$$

Now consider the effects of aggregation. Aggregation is introduced via a N by n grouping matrix ($\mathbf{G}$) which assigns individuals to areas so that $g(k, i) = 1$ if individual k ($k = 1, \ldots, N$) is in area i ($i = 1, \ldots, n$). It can be shown (see King, 1996, p. 47) that $\mathbf{G}(\mathbf{G}^T\mathbf{G})^{-1}\mathbf{G}^T = \mathbf{H}$, when multiplied with a vector, replaces each individual observation by its area mean. So, the aggregate version of the model is:

$$\mathbf{H}[\mathbf{T} - \bar{\mathbf{T}}] = \mathbf{H}[\mathbf{X} - \bar{\mathbf{X}}]B$$

and B is obtained by least squares of $\mathbf{H}[\mathbf{T} - \bar{\mathbf{T}}]$ on $\mathbf{H}[\mathbf{X} - \bar{\mathbf{X}}]$ or by weighted least squares of $[\mathbf{T} - \bar{\mathbf{T}}]$ on $[\mathbf{X} - \bar{\mathbf{X}}]$:

$$\hat{B} = [(\mathbf{H}[\mathbf{X} - \bar{\mathbf{X}}])^T\mathbf{H}[\mathbf{X} - \bar{\mathbf{X}}]]^{-1}(\mathbf{H}[\mathbf{X} - \bar{\mathbf{X}}])^T\mathbf{H}[\mathbf{T} - \bar{\mathbf{T}}]$$

Estimation bias induced by the aggregation can now be examined by analysing $(\hat{B} - B)$ but the challenge is to put the difference into a form that sheds light on the source of the discrepancy. King (1997, pp. 46–53) discusses a number of different representations. Following a derivation by Palmquist the discrepancy can be decomposed into the *product* of two components (see King, 1997, pp. 51–3). The first component is what is termed a 'specification shift' which is described as the effect of using individual data but being forced to include the grouping operator **G** in the regression. King (1997, p. 52) describes this problem as 'reverse omitted variable bias'. This enforced specification shift will have most effect on the estimate of B when **G** (the system of aggregation) causally intervenes between T and X. *Hence any form of homogeneous grouping using either of the two variables X and T or any other variable causally associated with X or T, will introduce specification shift.* The method of aggregation that has the least effect is random aggregation.

The second component is called an 'inflation factor' and is given by:

$$\left[\sigma^2_{(X(j,i))}/\sigma^2_{(X(i))}\right] - 1 \tag{4.26}$$

where $\sigma^2_{(X(j,i))}$ is the variance of X over all individuals ($X(j, i)$), and $\sigma^2_{(X(i))}$ is the variance over all the group-level values ($X(i)$). *This quantity (4.26) is least when aggregates that are homogeneous in X are used (and 0 when all aggregates comprise individuals identical in terms of X) and considerably greater when random aggregates are used.*

This analysis identifies two situations when aggregation bias will be small – either when random aggregates are used (because, although the inflation factor is large, the specification shift is zero) or when pure homogeneous aggregates on X are used (because, although the specification shift is large, the inflation factor is 0). This observation follows because the bias is a product of these two components.

Spatially defined groups will display some degree of homogeneity because of spatial correlation. In the case of census data: 'individuals who live in the same area are exposed to common influences and as a result exhibit similarities . . . individuals with similar characteristics choose to live in the same area' (Tranmer and Steel, 1998, p. 818). This means that grouping by geographical proximity will tend to produce aggregates that are homogeneous in terms of the independent variable (X in the above example) and/or the dependent variable (T in the above example) rather than random aggregates. This introduces specification shift but providing the aggregate is homogeneous so that there is no within-area variation the analyst can expect no aggregation bias because the inflation factor associated with homogeneous aggregates is zero. In practice, of course, areal aggregates are rarely if ever purely homogeneous which is why aggregation bias is always present. Where the analyst has the opportunity to construct his or her own areal aggregates (areas or regions) these findings have implications for the criteria that should be employed in region building. Approaches to region building are considered in the context of exploratory spatial data analysis (see section 6.2.2).

Tranmer and Steel (1998) analyse aggregation bias on variance, covariance and correlation statistics using a model for within-area homogeneity in which:

$$y(j, i) = \mu + \alpha(i) + e(j) \tag{4.27}$$

where μ is a regional mean effect; $\alpha(i)$ is a random variable with 0 mean and variance $\sigma^2(i)$ representing the area effect associated with the ith area; $e(j)$ is a random variable with a mean of 0 and variance σ_e^2 representing a pure individual effect. The area-level effect is common to all individuals from the same area. A similar variance–covariance model is specified for a second variable and co-variances between the area effects and the individual effects are defined for the two variables.

This is a multi-level model. Intra-area homogeneity is modelled by the random variable $\alpha(i)$ because all individuals in the same area have the same value of $\alpha(i)$. The $\{\alpha(i)\}$ are independent so area-level effects are independent. In the case of an aggregate such as an enumeration district this means individuals across the street but in different enumeration districts have independent area-level effects but two individuals some distance apart but in the same enumeration district have identical area-level effects. This approach to analysing the effects of intra-area homogeneity has also been used by Arbia (1989). Cliff and Ord (1981, p. 127) constructed a nested hierarchical model for two variables, each with m hierarchical levels together with a common factor at each of the m levels.

In analysing the difference between correlation coefficients computed on individual-level data and on area-level data Tranmer and Steel (1998) identify the difference as a function of the relative sizes of three quantities denoted $\delta_{1,1}$, $\delta_{2,2}$ and $\delta_{1,2}$ (pp. 823–4). Two of these ($\delta_{1,1}$ and $\delta_{2,2}$) are each inversely related to Palmquist's inflation factor and refer to the two variance terms (corresponding to the two variables) in a correlation estimate. The third, $\delta_{1,2}$, is a within area measure of cross-correlation between the two variables and measures the effect of the aggregation on the estimate of the relationship between the two variables. This term corresponds to Palmquists specification shift effect. The size of any aggregation bias in estimating the individual-level correlation coefficient using aggregate data is a function of these three terms and also the average number of individuals per area. This latter term can also have a serious effect on the bias because it multiplies the effects of each term $\delta_{1,1}$, $\delta_{2,2}$ and $\delta_{1,2}$ and is often large – even an enumeration district may have about 500 individuals.

Aggregation bias in estimating regression coefficients follows from their analysis as a function of the same terms except that there is only one variance inflation term which is associated with the independent variable in the regression model (Tranmer and Steel, 1998). Tranmer and Steel note that their results can be used to disaggregate individual-level and area-level correlation and shed light on the MAUP by showing how different zonings by producing different effects on the three data dependent quantities produce different values of correlation and regression statistics.

King (1997) provides conditional estimates of voting propensities that vary by ethnicity and voting area. These area-level estimates can then be treated as fixed values and mapped (see, e.g., p. 25) or their variation modelled in terms of independent explanatory variables. The deterministic method of bounds (King, 1997, pp. 77–90) constrains estimates of the propensities to lie at least within the interval $[0, 1]$ and possibly a narrower band depending on information provided by the data. The voting propensities $\beta^o(i)$ and $\beta^w(i)$ are assumed to be randomly drawn from a truncated bivariate normal probability distribution, conditional on $X(i)$, and specified by mean and variance–covariance parameters that are estimated from the data on $\{X(i)\}$ and $\{T(i)\}$. The truncation is set by the method of bounds and the same truncated distribution is used to draw the propensities for each area, although the conditioning (on $x(i)$) will vary.

King's approach treats the voting propensities that would be obtained even were the full table available as noisy estimates of an underlying true voting propensity. The truncated normal model represents a prior distribution for the voting propensities, and its parameters are hyperparameters. Because these hyperparameters are estimated from the full data set for $\{X(i)\}$ and $\{T(i)\}$, the methodology 'borrows strength' from the whole data set in providing estimates of individual-area propensities. A number of extensions to the methodology have been discussed in order to further reduce aggregation bias, including incorporating covariates into the specification of the mean value hyperparameters (King, 1997, p. 170).

The Bayesian approach proposed by King (1997) has generated considerable comment with criticism focusing on the reliability of the standard errors as indicators of the magnitude of actual estimation errors, the subjectivity of the diagnostics and the generality of the method (Freedman et al., 1998, 1999; King, 1999). Diagnostics provide checks on data requirements and statistical assumptions and, at the time of writing, criticism focuses on whether the available diagnostics provide sufficient information to assess when it is safe, and more importantly when it is not safe, to use the method. This area of ecological inference involves estimating parameters but diagnostics in relation to the underlying assumptions of the method are critical if the user is to be able to make a valid assessment of the reliability of the answers. Gelman (2001, pp. 105–7) discusses model checking for ecological regression and identifies what information is available to suggest when a model should be rejected.

An area of concern when analysing spatial data is the effect on King's methodology of spatial correlation in the parameters of interest. King (1997, pp. 164–8, 2000) concludes from his experience based on Monte Carlo simulation that spatial correlation 'does not appear to have major consequences for the validity of inferences from the basic model' (p. 168). However, the following parallels with Bayesian modelling of area-specific relative risk rates are worth noting. A spatial model rather than a model of spatial independence could be used for the prior distribution as employed in some areas of spatial epidemiology (Mollie, 1996). A specification that includes spatial dependence has the effect of restricting the 'borrowing of strength' to just the adjacent areas and on the evidence from the spatial epidemiology literature can for certain models result in considerable smoothing of the final map – which may or may not be appropriate. The expected presence of area-level contextual effects and the operation of social networks inducing contagion into the process of deciding whether to vote and if so for which party further complicates the problem of specifying an appropriate prior model.

The estimation methods discussed to this point assume that no individual-level data are available. Tranmer and Steel (1998) consider the situation where there are some limited individual-level data on other variables though not the variables for which inferences need to be drawn. The context for this work is the availablity in the UK of the 2% Sample of Anonymized Records (SAR) relating to districts. Their work shows that adjustments to ecological-level correlations and regression estimates can be made where appropriate individual-level data are available on a set of other variables called 'grouping variables'. Whereas in (4.27) within-area homogeneity appears as an unobserved area-level effect (as also in King's basic model), the grouping variables are introduced because they are observable and are associated with the within-area homogeneity. King's specification shift or 'reverse omitted variable bias' is addressed (in the context of their model) by specifying the variables that the enforced aggregation has introduced into the individual-level regression. It is these variables which need to be controlled for if a reliable estimate of the relationship (between the two variables of interest) is to be obtained.

It is not necessary for the individual-level data to have a locational reference nor even to come from the same data source, providing they refer to the same population. Their approach, which is also discussed in a different context in Cressie (1996) involves the inclusion of data on these variables, as 'grouping variables', into the multi-level model. Thus the original model (4.27) now becomes (Tranmer and Steel, 1998, p. 827):

$$y(j; i) = \tilde{\mu} + \boldsymbol{\beta}^T \mathbf{z}(j) + \tilde{\alpha}(i) + \tilde{e}(j)$$

where the tilde over the original terms denotes that they are now conditional on the grouping variables $\mathbf{Z}$; $\mathbf{z}(j)$ denotes the (column) vector of values of the grouping variables for the ith individual and $\boldsymbol{\beta}$ is the (column) vector of coefficients that relate $y(j;i)$ to the grouping variables $\mathbf{Z}$.

Tranmer and Steel (1998, pp. 829–30) derive adjusted correlation and regression coefficients that they show yield a considerable improvement in the estimation of the individual-level relationships. The estimation bias introduced by the aggregation into the covariance between two variables $Y(1)$ and $Y(2)$, and which is captured by the grouping variables is removed by a term:

$$\bar{\boldsymbol{\beta}}_{Y(1),\mathbf{z}}{}^{T}(\bar{\mathbf{S}}_{\mathbf{z},\mathbf{z}} - \hat{\mathbf{S}}_{\mathbf{z},\mathbf{z}})\bar{\boldsymbol{\beta}}_{Y(2),z}$$

where $\bar{\boldsymbol{\beta}}_{Y(1),\mathbf{z}}$ is a (column) vector of estimates of the regression coefficients of $Y(1)$ on the grouping variables at the aggregate level and similarly for $\bar{\boldsymbol{\beta}}_{Y(2),\mathbf{z}}$. $\bar{\mathbf{S}}_{\mathbf{z},\mathbf{z}}$ is the aggregate level covariance matrix of the grouping variables and $\hat{\mathbf{S}}_{\mathbf{z},\mathbf{z}}$ is the estimate of the individual-level variance–covariance matrix (on the grouping variables). The adjustment for the variances follow and from these quantities the adjusted correlation and regression estimates can be computed. Tranmer and Steel (1998) suggest that individual-level data from the SAR on variables such as age, housing and ethnic group structure may function effectively as grouping variables. Figure 4.3 illustrates the improvement obtained through this method.

(b)   Ecological inference in environmental epidemiology: identifying valid hypotheses

In environmental epidemiology ecological inference arises in the analysis of certain types of dose–response relationships. For example 'response' is the rate of a disease by area and 'dose' refers to exposure to an environmental risk factor by area. Unlike the earlier problem, the data may not be discrete valued. An individual may either have the disease or not, but their exposure to the risk factor may be continuous valued. This raises additional problems because more information is lost as a consequence of aggregation than in the discrete-valued case. With discrete data only cell values in a table are lost and in the 2×2 case where marginal sums are available it is sufficient to estimate only one value in order to be able to complete the table. In the case of continous data the mean provides information only on the centre of the distribution of values, nothing on the spread of values is available. Further, in the voting example linearity was a property of the relationship (it did not have to be assumed), but in epidemiology there is no reason to expect dose–response relationships to be linear.

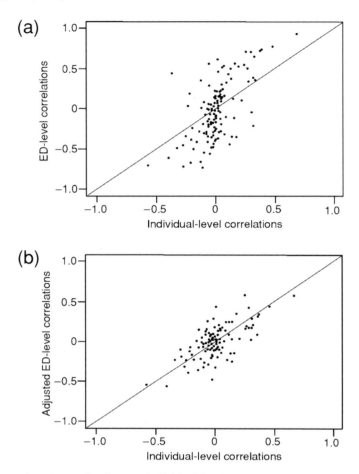

Figure 4.3 ED-level versus individual-level correlations for the Reigate data
(a) before and (b) after adjustment (Tranmer and Steel, 1998, pp. 827 and 830)

Richardson (1992) identifies aggregation problems of interest to epidemiologists. (I) What is the difference between individual dose–response relationships and the relationship obtained after aggregating over all individuals in
a group? If all individuals have identical parameters in a linear relationship
then these parameters can be estimated from the aggregate data – but not if
the relationship is non-linear. (II) Assuming linearity of the dose–response relationship within each aggregate, the difference between the estimated slope
coefficient from the N aggregate values and the average of the N within aggregate slope coefficients is the ecological bias. This ecological bias can be decomposed into two elements. The first component is bias due to inter-aggregate
variation in the disease rate amongst those *not* exposed to the risk factor (differences in the intercept parameter of the linear relationship across aggregates).

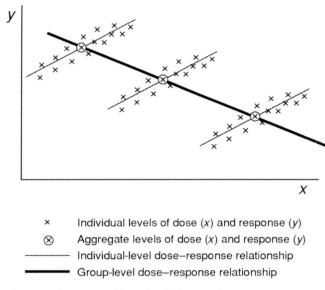

Figure 4.4 Sign reversal in ecological regression

The second is bias due to groups acting as effect modifiers in the dose–response relationship (differences in the slope parameter across aggregates). A further source of bias is the effect due to confounding variables, such as socio-economic variables not allowed for in an analysis (Jolley, Jarman and Elliott, 1992). These biases can be arbitrarily large even resulting in sign reversals as shown in figure 4.4. If the source of the bias can be identified it may be possible to adjust using multiple regression providing the assumptions of the model (e.g. additive joint effects) are satisfied (Greenland and Morganstern, 1989).

The problems associated with ecological inference mean that the analysis of aggregate data in environmental epidemiology is viewed as the weakest form of analysis for establishing exposure–disease relationships (although see Armstrong, 2001, for a defence of ecological analysis). Such analyses are often undertaken in order to generate scientifically valid hypotheses about the existence of an association rather than estimating its strength (Richardson, 1992). These hypotheses may then be followed up by other methods such as cohort studies, particularly if there is supporting evidence from other independent studies. In environmental epidemiology, the emphasis is on design construction to try to counter the problems of ecological analysis that include specification bias, confounding exposure misclassification and effect modification. Richardson (1992, pp. 199–200) lists good practice in the design of geographical studies: allow for heterogeneity of exposure; use well-defined population groups; employ survey data to help identify relevant exposure data; allow for latency times in the disease; allow for migration effects. The last considerations suggest that time series data on population movements will be important in helping to identify associations. Particular care needs to be taken to try to identify all relevant confounders especially socio-economic and lifestyle factors and other environmental factors – although in practice it is impossible to be sure that all relevant confounders have been allowed for.

(c)   The modifiable areal units problem (MAUP)

Tobler (1989) identifies two forms of the MAUP. The first of these is observing the effects of different spatial aggregations and the interpretation of the patterns revealed by different aggregations. Analysis which attempts to measure area-level effects (either because areas are the object of interest or because they are effect modifiers) will need to pay particular attention to the choice of the areal framework to ensure it is meaningful in terms of the underlying processes. The second of Tobler's forms of the MAUP is observing the effects of different aggregations on the behaviour of statistics. Underlying this is the concern that the analyst may be using a statistic that is not appropriate to the problem posed.

The MAUP consists of two distinct effects on the properties of estimators: those associated with the *scale* of the analysis and those associated with the particular *partition* (given the scale of the analysis). By 'scale of analysis' is usually meant the number of subareas a study area is partitioned into because this determines the size of each spatial unit (the areal filter) through which events are observed. Holt et al. (1996) point out that this is imprecise terminology and that it is necessary to specify which property of an estimator is affected – whether it is for example its expected value or its variance. Research in this area, particularly in geography, has tended to focus on the former.

Consider first the issue of scale and in particular the effect of analysing the same set of variables for the same region but repeating the analysis using different sized areal partitions. Measures of association will vary if scale-dependent processes are influencing outcomes. It was noted in the case of ecological inference that measures of association at any given scale of aggregation confound different scales of association. The scale 'problem' is to derive measures of association at any given scale that are pure measures of the association at that scale and not confounded with effects from smaller or larger scales. This calls for multi-level modelling.

In situations where all relationships between variables are a consequence of individual-scale processes so that there are no area-level (contextual) processes, one solution to the partition problem is to select statistics that are invariant to aggregation. This implies discarding statistical techniques like correlation and regression, indeed all statistics based on computing variances, since they are not invariant to aggregation. Where area-level effects are present, this strategy is inappropriate since such a statistic would by construction be ignoring a potentially important element in the relationship. Another solution, as discussed above, is to develop adjustments to the way the statistics are computed in order to separate area-level from individual-level effects (Holt et al., 1996). Grouping variables may be used as discussed by Cressie (1996) and Tranmer and Steel (1998). Green and Flowerdew (1996) observe that just as estimates of relationships at the individual level are affected by area-level influences, so the area level is likely to be affected by higher, regional-level effects. They suggest adding the regional values of the independent variable into the regression model. These regional values for the ith area are averages for a defined area (typically the adjacent neighbours) surrounding the ith area. Unlike multi-level modelling which constructs a strict hierarchy these higher-level spatial units overlap (see section 9.2.1 and equation (9.31)).

Where data for small areas are available regionalizations that seek to control for partition effects might be constructed. If analysis involves making comparisons between two regions, partitions might be constructed for the two regions that are similar in terms of their levels of within-area homogeneity for each of the variables (Openshaw, 1996). This is likely to be difficult to achieve in practice and in fact may not give the desired result. This is because the aggregation effect also depends on covariances (not just variances) and because the ecological correlation (for example) compounds individual-level and area-level influences in measuring the relationship. Holt et al. (1996) and Tranmer and Steel (1998, p. 824) are more specific and suggest forming regionalizations that maximize the pure area-level correlation since, they argue, it is this quantity that is of interest geographically rather than obtaining estimates of individual-level relationships. Their work has highlighted the underlying complexity of the relationships responsible for the MAUP and why it is 'not amenable to simple attempts to unravel it' (Holt et al., 1996, p. 198). It is important to analyse the MAUP in the context of a well-defined model that identifies the different scale components (from the individual-level upwards) underlying the behaviour of each of the variables.

4.3   Data consistency and spatial data analysis

Consistency checks are essential to ensure that data values do not fall outside permitted ranges, such as percentages that must lie in the 0% to 100% range or measures of dispersion or distances which must be positive valued. Problems can arise when merging spatial units or moving to a common spatial framework using interpolation methods. Counts can be summed but in computing new percentages or averages then the analyst should return to the original data. Consistency checks are needed to ensure that error is not introduced into a database as a consequence of undertaking inappropriate or inaccurate manipulations of the data. GIS software for example does not necessarily provide warnings on when inappropriate spatial operations have been performed (Mrozinski and Cromley, 1999, p. 288). Many forms of statistical analysis have to be performed outside a GIS and inconsistencies can enter a database as a result of errors in transferring files, or from creating several copies of a file that may then, inadvertantly, undergo different revisions and updating.

The most important context for carrying out consistency checks is when different databases have to be merged or synchronized particularly if those databases have been collected by different agencies. The problems are likely to be especially acute when the data sets have not been collected at exactly the same scale. When merging data sets that refer to different time periods the analyst needs to be aware of this, report the differences in time period, and consider the possible implications for interpreting findings. Population data derived from a census may provide a poor measure of the appropriate denominator for computing an incidence rate when the health data refer to an intercensus period. In addition to ensuring consistency in data attributes, it is also necessary to ensure consistency when merging spatial objects so that, for example, houses are not located in the middle of bodies of water (Kainz, 1995).

Data inconsistency is a form of data error but is considered apart from data error. Inconsistency errors may be subtle or severe, but in theory, at least, can be avoided by carrying out the appropriate checks on the database both during and after carrying out data operations.

4.4   Data completeness and spatial data analysis

The distinction was drawn in section 2.3 between model and data completeness. If a database is not *model* complete then not all the important variables needed for statistical analysis are available in the database. This can lead to model misspecification resulting in biased estimates of the parameters of those variables that are measured accurately and correctly included in the model. Failure to include a significant variable (say, X_k) means that the estimate of the regression parameter for a variable of interest (say, X_1) on the response variable (Y) measures more than the *direct* effect of X_1 on Y. It also measures all the *indirect* effects associated with the influence of X_1 on Y through X_1's relationship to X_k (see figure 4.5). In practice it is impossible to be sure that all relevant

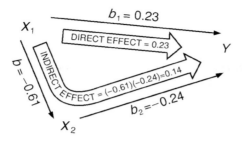

Figure 4.5 Effect of an omitted variable on parameter estimates

factors have been identified but the most important ones should be identified and included in the model. The alternative, not often feasible in observational science, is to randomize the level of X_1 over all cases since this removes the bias of missing variables.

Sections 4.4.1 and 4.4.2 examine methods for handling two aspects of spatial data incompleteness. Section 4.4.1 examines the missing-data problem. The term 'missing data' refers to those situations where there is a true but unrecorded value for a case. Typically the term 'missing data' and the methods of section 4.4.1 are applied to data sets that refer to areas but may be adapted to the case of data that refer to point locations on a spatially continuous surface. The missing-data problem is distinguished here from spatial interpolation and spatial prediction problems. The latter problems arise where data have been recorded at a number of locations on a spatially continuous surface and the analyst now wants to provide an estimate or prediction at one or more *other* locations on the same surface. Methods for spatial interpolation and spatial prediction are the subject matter of section 4.4.2. There is however overlap between the methods of sections 4.4.1 and 4.4.2.

Throughout these sections we refer to either 'predicting' or 'estimating' values. The term estimation is used in statistics when deciding on the value of a parameter which is assumed to have some unknown single value. If it is assumed that there is a single true data value then we speak of *estimating* the value. However the process of measurement or the nature of the underlying mechanism generating the data may suggest thinking of a data value as one of a possible distribution of values from some probability model (section 2.1.4). In this case it is usual to speak of *predicting* a data value and attaching a measure of the prediction error that reflects (i) the inherent variability arising from sampling the *probability* model together with (ii) any *statistical* uncertainty that derives from estimating the unknown parameters of the probability model. There are parallels here with design-based and model-based approaches to spatial sampling (see chapter 3 and in particular section 3.2).

The emphasis throughout the rest of this section is on using local information to make estimates or predictions. However there are circumstances where criteria other than spatial nearness may be introduced to weight observations. There may be a substantive basis for discriminating between near values. In soil surveys the unknown value for a site might be taken to be the mean value of the relevant attribute computed from those nearest values that are in the same soil class as the unknown value.

4.4.1   The missing-data problem

Interest in missing data may originate because the analyst wishes to complete the record and therefore would like to obtain plausible values for the missing data. Data missing at one level of recording leads to an undercount when the datum forms part of a figure for a larger unit of which it is a part. To handle these problems the analyst might want good estimates or predictions of specific missing values on a variable, and an important consideration is whether there are other data that might be used to help provide these values. A second area of interest stems from the need to carry out a statistical analysis such as fitting a regression model and there are gaps in the multivariate data set. Discarding cases with one or more missing values could lead to throwing away a great deal of useful data particularly if missing values are scattered across the data matrix. The analyst is not interested in predicting the missing values per se but rather wants to be able to make valid inferences about the entire target population not just a subset for which the data happens to be complete, making full use of the data that are available.

The mechanism or process responsible for data being missing is important. A missing-data *mechanism* is said to be ignorable in the case of a single variable if the missing data are missing at random. This means the missing observations are a random sample of the sampled units. If the probability of a value being observed on a variable depends on its value (for example all the largest or smallest values have been suppressed) then the missing-data mechanism is not ignorable. Any analysis that does not allow for this will be subject to bias (Little and Rubin, 1987, p. 10). Where there are two variables (Y and X), but only one (Y) is subject to missing values, then the mechanism is ignorable for likelihood- or model-based methods. This holds even if the observed data on Y are not a random sample of the sampled units, providing they are independent of Y and they are random samples of the sampled values within subclasses defined by values of X. If the observed values are independent of both X and Y then the missing-data mechanism is ignorable for sampling-based methods as well. Little and Rubin (1987, pp. 14–18) discuss these issues and give examples.

Most methods for dealing with data sets with missing values assume the missing-data mechanism is ignorable. Some methods are available for non-ignorable missing-data mechanisms, the most direct involving follow-up surveys to try to get some of the information needed (Little and Rubin, 1987, pp. 259–64). In the case of completing Census data at the small-area level this involves follow-up work in each tract and this is expensive.

Spatial data raise some additional issues and we consider three. First, 'missing at random' does not necessarily imply that missing values must be geographically distributed 'at random' but it is an additional consideration. If observations have been deleted at regular intervals this could raise problems if an important scale of spatial variation coincides with that interval. If data values are missing from one area of a study region resulting in a sparse coverage there, or if there are clusters of missing data then the remaining data values will have a large influence on the fit of any model used to describe surface variation. Furthermore, prediction errors will vary over the map. Unwin and Wrigley (1987) illustrate this point with respect to trend surface modelling using the leverage measure (Belsley et al., 1980). Small errors in the remaining data values in the sparsely covered subregion may have a disproportionate influence on the shape and fit of the surface relative to similar-sized errors in data values in areas with a denser coverage. Higher levels of non-response are often clustered in inner-city areas in the case of census data and crime data, and the underlying mechanisms are often non-ignorable because they are linked to poverty or crime levels which are attributes that the surveys may be seeking to measure. In the case of remotely sensed data, sensor failure along a scan line is not usually related to the underlying surface so such a linear structure of missing values need not necessarily violate the missing-at-random assumption. If unemployment data are not recorded for a group of adjacent areas because of local strike action this does not necessarily imply a non-ignorable missing-data problem. These examples also illustrate the need to consider why values are missing before deciding on the approach to take.

The second issue is that, whilst some forms of spatial data (such as Census attributes) might display considerable levels of spatial heterogeneity over quite short distances, there is an underlying continuity even in these data that can be exploited to estimate missing data. More generally the presence of spatial correlation in attribute values means that neighbouring attribute values provide an information source for missing-data prediction. This may also allow some of the difficulties created by the first issue to be overcome by drawing on local subsets of the data close to the area that contains a spatial grouping of missing data.

Finally, estimating or predicting missing values, or identifying values at locations where measurements have not been taken, often assumes particular importance in spatial analysis because the analyst wishes to provide a map of the spatial variability of some characteristic. Although some warning would need to be attached to such areas of the map (such as an additional map showing an estimate of prediction error or sampling density) this might still be preferable to leaving areas blank.

As part of an exploratory investigation of the data, cases with missing values should be mapped to check for spatial bias and also examined to see they are not different with respect to those attributes for which data are available. This will provide some evidence as to whether the missing-data mechanism is ignorable or not. Unwin et al. (1996) have developed the software system MANET that displays properties of datapoints with missing values. In MANET, missing-data information is provided in chart form for each variable to show the proportion of missing values. MANET also adds an extra 'missing data' bar on a histogram, includes missing data as an extra category in a mosaic plot (Friendly, 1995) and in a scatterplot represents cases with missing-data values on either X or Y as a projection on to the Y or X axes respectively of the scatterplot.

(a)   Approaches to analysis when data are missing

Approaches to the analysis of data sets with missing values fall broadly into three types. We illustrate with reference to the case where the data are needed for regression modelling. In the first approach the analyst uses only those data records that are complete. In regression analysis this means discarding any case where the response value is unknown and/or one or more explanatory variables have missing values. With small amounts of missing data this may be a reasonable strategy. However, the presence of missing values in a data matrix can have effects out of proportion to the number of missing values. For example if missing values are scattered randomly through a data matrix and if every case with at least one missing value has to be discarded from an analysis, a relatively small proportion of missing values can lead to the discarding of a large amount of valid data. This leads to inflated variance estimates, a loss of power in hypothesis testing and a loss of precision in deriving confidence intervals. If the analyst intends to fit several different regression models to the data and the pattern of missing data means that different subsets are used for different models, this may create problems of comparability.

The second and third approaches produce *single* quantities that provide an estimate or prediction of the missing value and, in the case where the missing value is a drawing from an underlying probability model that has been specified, an estimate of its prediction error. These approaches can be used whether the analyst's interest is in the missing values themselves or in further statistical analysis.

The second approach is based on *imputation* in which the missing data are predicted, the data matrix is filled in and then analysis proceeds by standard methods. Four of the methods of imputation described by Little and Rubin (1987, pp. 43–7, 60–7) are relevant here. *Mean imputation* substitutes the mean of the responding units for the missing values. *Hot deck imputation* involves a variety of different practices based on constructing an empirical distribution of values usually based on the set of observed values. The substitute for each missing value is drawn from this empirical distribution. In the case of *regression imputation*, a regression model for the variable with missing values is estimated based on data cases with a full set of observed values (for the response and all the explanatory variables). Each missing value is then predicted using the model by substituting values for the observed variables into the prediction equation values (see, e.g., Buck, 1960). *Stochastic regression imputation* adds random noise to each imputed value to reflect the uncertainty associated with each prediction. To reflect the regression-based nature of the approach the noise might be drawn from a normal distribution with a mean of zero and a variance estimated from the regression residuals.

This approach does not draw on any underlying model of the data and these methods are only suited to the situation where the missing-data problem is 'sufficiently minor' (Dempster and Rubin, 1983). Inference in the regression model is based on the number of degrees of freedom (the amount of independent information available in the data set) which is a function of sample size (n). However, missing-data values have been imputed using other data in the database so cannot be considered as data in the sense of providing independent observations. Inferences about variables that have a complete data record can be based on n, but inferences relating to variables where incomplete records have been used in the estimation (such as the regression parameters) should be based on $(n - m)$ where m is the number of cases with missing values.

There are two extended forms of imputation: multiple and iterative imputation. The *single* imputation methods, described above, aim to arrive at a single valid inference of the missing value which then allows standard complete-data methods to be used. *Multiple* imputation constructs M 'possible' data matrices. These M data matrices, that may be obtained by bootstrapping for example, can be used to fit the regression model M times from which the variability associated with parameter estimates can be computed (Little and Rubin, 1987, pp. 255–9). A claimed advantage of multiple imputation is that it can reflect sampling variability if only one model is specified for the missing data, and it can reflect the variability associated with choice of model if more than one model is specified. This is sometimes cited as an advantage of multiple imputation in the case of non-ignorable missing-data mechanisms which can arise with Census data, because it means the analyst can assess the sensitivity of findings to different assumptions about the mechanism. Under any one model, variability can be decomposed into average within-imputation variance and between-imputation variance (Little and Rubin, 1987, pp. 256–7).

Iterative imputation involves replacing missing values with predicted values, estimating parameters using the filled data matrix, obtaining a new set of predicted missing values assuming the parameter estimates are correct, re-estimating parameters and so on until convergence. This extended form of imputation provides the background to the third, *model- or distribution-based*, approach to missing-data prediction. This method specifies a model for the full data (observed and missing data) and uses the EM algorithm to fit the model and 'estimate' missing values. Little and Rubin (1987, pp. 129–30) discuss the background and development of the EM algorithm which generalizes and formalizes the iterative imputation approach to handling missing-data problems (see also Orchard and Woodbury's (1972) 'missing information principle'). The M step is a maximum likelihood estimation of the parameters given the completed – observed plus imputed – data set. The E step finds the conditional expectation of the 'missing data' given the observed data and the current estimates of the parameters and then substitutes these for the 'missing data'. It is this step that distinguishes the EM approach from iterative imputation. The 'missing-data' values are not actual 'estimates' of the individual missing values but rather functions of the missing values which are the missing sufficient statistics.

Little and Rubin (1987, pp. 152–7) describe different versions of this procedure for regression modelling. Values may be missing from the response variable (Y) and/or from one or more of the explanatory variables. In the case where it is only the response that suffers from missing data, the E step involves finding an expected value given the current estimates of the regression parameters and the corresponding levels of the explanatory variables. So if $y(i)$ is missing and $\mathbf{y}_{obs}$ is the vector of observed values on the response variable, at the tth iteration of the algorithm, the E step gives:

$$E\left[y(i) \mid \mathbf{X}, \mathbf{y}_{obs}, \hat{\boldsymbol{\beta}}^{(t)}, \hat{\boldsymbol{\sigma}}^{2(t)}\right] = \hat{\boldsymbol{\beta}}^{(t)} \mathbf{X}(i)$$

where $E[|]$ denotes conditional expectation, $\mathbf{X}$ is the matrix of data values on the explanatory variables (including the constant term) and $\mathbf{X}(i)$ is the (column) vector of values on the explanatory variables for case i. The row vector $\hat{\boldsymbol{\beta}}^{(t)}$ denotes the current estimate of the intercept and slope parameters and $\hat{\boldsymbol{\sigma}}^{2(t)}$ the current estimate of the error variance. Each are based on the updated data matrix using the values of the 'missing values' from the previous iteration.

(b)   Approaches to analysis when spatial data are missing
The problem addressed here is as follows. Data on k variables have been collected for n areas. Some cells of the $n \times k$ data matrix are empty. In order to undertake further analysis missing values need to be estimated to complete the data matrix. Alternatively the analyst wants to fit a model making the best use of the available data.

Each of the methods of imputation reviewed in the preceding section have spatial analogues in the sense that there is an equivalent methodology which utilizes the extra information provided by the spatial distribution of values. However there may be particular circumstances to take account of with spatial data. Values may be missing for some spatial units but the sum across all the areas is known or even for subgroups of areas. County-level data values are missing or suppressed but state-level totals are known. In these cases missing value imputation should preserve 'volume' by, for example, scaling the imputed values. Since the spatial distribution of values will be utilized, what are the implications if missing-data values occur in geographical clusters?

(i) Spatial mean imputation with equal weights assigned to each included data value. These methods substitute the missing value with the arithmetic mean of the data values within some search neighbourhood or spatial window defined around the area with the missing value. A robust version of this would be to use the median rather than the mean but in either case the analyst needs to consider the size of the window. The areas that share a common border with the area with the missing value could be used (Bernstein et al., 1984).

The choice of window size is clearly a matter of importance – whether to just use the adjacent neighbours or to include higher-order neighbours. There may be no strong reason to prefer one sized window over another, other than the fact that larger windows have a greater smoothing effect. There may be advantages in taking several and examining how missing-data imputations are affected.

(ii) Spatial mean imputation with unequal weights assigned to each included data value. The previous methods are vulnerable to various clustering effects in the distribution of irregular shaped areas. To avoid some of these effects a weighted mean can be employed where the weight $a(i, j)$ might be the proportion of the total border of area i occupied by area j (Kennedy and Tobler, 1983; Tobler and Kennedy, 1985). Thus if $y(i)$ is missing and $N(i)$ denotes the set of neighbours of i:

$$\hat{y}(i) = \left[\sum_{j \in N(i)} a(i, j) y(j)\right] / \sum_{j \in N(i)} a(i, j) \qquad (4.28)$$

There are problems with an estimating equation like (4.28). Some large nearby areas may only share a narrow border with the area with the missing

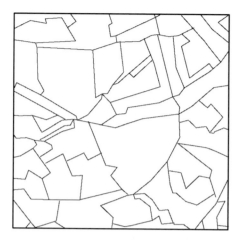

Figure 4.6 ED boundaries from an area of Sheffield to illustrate the range of different types of boundary adjacencies

value or even be 'hidden', whilst some smaller areas may occupy a large proportion of the boundary (see figure 4.6). Missing-data prediction using these methods appears to perform best when the prediction equation concentrates on the near neighbours and when spatial autocorrelation is strong (Upton, 1985). Spatial heterogeneity in the form of a spatially varying autocorrelation structure can also undermine the performance of these methods unless allowance is made by using weights that vary across the map to reflect that heterogeneity. Upton (1991) draws attention to the need to recognize that treating all rates and ratios as equally robust, even though precision may be a function of the population size, can be a source of missing-data imputation error. He recommends incorporating a measure of area size into the estimating equation in order to reflect different levels of precision.

Missing data on an areal partition raises the problem of what to do if the missing values are themselves clustered so that values are not available at some (or all) near neighbours of the partition. In Kennedy and Tobler (1983) averaging methods are adapted to this situation through a process of iteration (see also Tobler and Lau, 1978). A weighted least squares function of the difference between area values (observed and missing) is minimized where the weights reflect the lengths of shared common boundaries.

An equation such as (4.28) has the property that missing-data values cannot exceed the maximum value nor be less than the minimum value of the set of observed values so missing-data predictions cannot follow gradients. A solution to this is to include an estimate of trend (linear, quadratic or higher) in $\hat{y}(i)$. This trend could be extracted first for the whole data set (see chapter 9). An estimator such as (4.28) is then applied to the residuals from the trend and then the trend re-introduced at the end. Ripley (1981, p. 37) describes the method of distance weighted least squares of Pelto et al. (1968) and McLain (1974) which fit linear or higher-order trends to individual points. The methodology is the univariate version of what in the geographical literature has been referred to as geographically weighted regression (Fotheringham et al., 2000).

Before leaving these methods note that if each area in the partition is represented by its area or population-weighted centroid then missing-data imputation could proceed using the methods to be described below in section 4.4.2.

(iii) Spatial hot deck imputation. This approach constructs the empirical distribution from a spatial window of values. Instead of computing the mean as in spatial mean imputation a value is drawn from the empirical distribution. If imputation is required for households in a Census tract, the empirical distribution is constructed from the set of all observed values in the same Census tract and then particular values imputed by sampling this distribution. Values can be drawn for each missing value by sampling from the set of observed values without going through the process of fitting a distribution to the observed values. This approach is likely to be attractive if there is local spatial heterogeneity so that it is safer to use data values from a local area within the Census tract close to the missing value rather than the entire tract. An extreme form of this type of approach is simply to substitute the nearest neighbour observed value for the missing value as described above. Such a method does not qualify as hot deck imputation since there is no sampling (Little and Rubin, 1987, p. 60 call this *'substitution'*) but illustrates how different methods merge into one another. Sampling from the observed data values again constrains the imputed value not to be more than the local maximum value nor less than the local minimum. This is not a constraint if samples are from a distribution fitted to the data. Hot deck methods in common with all methods that impute on the basis of taking a single observed value may produce discontinuities in the spatial distribution of values. Such discontinuities are an artefact of the method not necessarily any real property of the spatial distribution of values.

(iv) Spatial regression imputation. This approach extends regression imputation as described by Buck (1960) to the case where neighbouring values of the variable (X) without missing values appear in the model specification. For example a model of the form:

$$Y(i) = \beta_0 + \beta_1 X(i) + \beta_2 \mathbf{W}^* \mathbf{X}(i) + e(i)$$

is fit to the data cases with complete records and then used for missing-data prediction on Y. The variable $\mathbf{W}^* \mathbf{X}(i)$ is a spatial average of $X(i)$ as defined in section 2.4 and the approach can be generalized to include more predictors and more spatial average terms. This type of spatial regression model will be discussed in chapter 9. It does not raise special parameter estimation problems, unlike some forms of spatial regression model. Once the data values on the variable $\mathbf{W}^*\mathbf{X}$ have been constructed the model can be fit by least squares using standard statistical software.

(v) A maximum likelihood approach. Martin (1984, 1989), Haining et al. (1984) and Griffith et al. (1989) developed a maximum likelihood approach to the problem of missing spatial data. The approach involves the iterative estimation of model parameters and *prediction* of missing values and is similar to the EM algorithm. The approach has similarities with the method of *simple* kriging, in that the weights are not required to sum to 1, and also *universal* kriging, in the sense that the mean is not assumed known and the prediction equations have similarities. Kriging is discussed in section 4.4.2. The full approach is described and then an approximation suggested which is similar to proposals found in geostatistics.

Suppose a region has n areas and observations are missing on a variable Y on k of these. Y is assumed multivariate normal with mean $\mathbf{A\theta}$ and variance–covariance matrix $\sigma^2\mathbf{V}$. The elements of the matrix $\mathbf{A}$ are the co-ordinates and powers of those co-ordinates for each datapoint for a pre-specified order of trend surface. So there are n rows and as many columns as necessary for the order of trend surface. If the trend surface is linear, $\mathbf{A}$ has three columns, if quadratic, six, and so on (see section 9.1.1 for details). The vector $\mathbf{\theta}$ denotes the parameters of the trend surface to be estimated. The scale parameter σ^2 is to be estimated and $\mathbf{V}$ depends on a set of unknown 'spatial' parameters that also need to be estimated. Permissable models for $\mathbf{V}$ will be discussed in chapter 9.

The column vector of observed values $\mathbf{y}$ may be partitioned, after suitable permutation so that $\mathbf{y}^T = [\mathbf{y}_o^T \mid \mathbf{y}_m^T]$ where $\mathbf{y}_o$ denotes the $(n-k)$ dimensional column vector of observed values and $\mathbf{y}_m$ denotes the k dimensional column vector of missing values.

Let the matrix $\mathbf{V}$ be partitioned after permutation so that:

$$\mathbf{V} = \begin{bmatrix} \mathbf{V}_{oo} & \mathbf{V}_{om} \\ \mathbf{V}_{mo} & \mathbf{V}_{mm} \end{bmatrix} \qquad \mathbf{V}^{-1} = \begin{bmatrix} \mathbf{V}^{oo} & \mathbf{V}^{om} \\ \mathbf{V}^{mo} & \mathbf{V}^{mm} \end{bmatrix}$$

The function to be minimized which is minus twice the log likelihood function is (Martin, 1984):

$$\ln L(\mathbf{\theta}, \mathbf{\sigma}, \mathbf{V}, \mathbf{y}_m \mid \mathbf{y}_o) = (n-k)\ln(2\pi) + (n-k)\ln(\sigma^2) + \ln(|\mathbf{V}_{oo}|)$$
$$+ \sigma^{-2}(\mathbf{y} - \mathbf{A\theta})^T \mathbf{V}^{-1}(\mathbf{y} - \mathbf{A\theta}) \qquad (4.29)$$

Note that an alternative expression for $|\mathbf{V}_{oo}|$ is $|\mathbf{V}^{mm}|/|\mathbf{V}^{-1}|$. The predictor for the set of missing values is:

$$\hat{\mathbf{y}}_m = \mathbf{A}_m \hat{\boldsymbol{\theta}} + \hat{\mathbf{V}}_{mo}(\hat{\mathbf{V}}_{oo})^{-1}(\mathbf{y}_o - \mathbf{A}_o \hat{\boldsymbol{\theta}}) \qquad (4.30)$$

where $\mathbf{A}_m$ and $\mathbf{A}_o$ are submatrices of $\mathbf{A}$ referring to the missing and observed segments of the data respectively. An alternative expression for $\mathbf{V}_{mo}(\mathbf{V}_{oo})^{-1}$ is $-(\mathbf{V}^{mm})^{-1}\mathbf{V}^{mo}$.

The maximum likelihood estimators for the unknown parameters are:

$$\hat{\boldsymbol{\theta}} = (\mathbf{A}^T \hat{\mathbf{V}}^{-1} \mathbf{A})^{-1}(\mathbf{A}^T \hat{\mathbf{V}}^{-1} \mathbf{y}) \qquad (4.31)$$

$$\hat{\sigma}^2 = (\mathbf{y} - \mathbf{A}\hat{\boldsymbol{\theta}})^T \hat{\mathbf{V}}^{-1}(\mathbf{y} - \mathbf{A}\hat{\boldsymbol{\theta}})/(n - k) \qquad (4.32)$$

where $\mathbf{y}^T = [\mathbf{y}_o^T \mid \hat{\mathbf{y}}_m^T]$. The parameters of $\mathbf{V}$ are obtained by minimizing:

$$|\mathbf{V}_{oo}|^{1/(n-k)}(\mathbf{y} - \mathbf{A}\hat{\boldsymbol{\theta}})^T \mathbf{V}^{-1}(\mathbf{y} - \mathbf{A}\hat{\boldsymbol{\theta}}) \qquad (4.33)$$

At the first iteration the missing values $\mathbf{y}_m$ might be computed using one of the methods of imputation discussed earlier. Setting $\mathbf{V} = \mathbf{I}$ the parameter vector $\boldsymbol{\theta}$ is estimated and then the parameters of $\mathbf{V}$ and σ^2. At the second iteration the missing values are predicted using (4.30), parameters are re-estimated and the cycle continues until convergence – which can be slow if there are many missing values.

The matrix of prediction errors for the missing values is given by:

$$\boldsymbol{\Psi}_{m,m} = \hat{\mathbf{V}}_{mm} - \hat{\mathbf{V}}_{mo}(\hat{\mathbf{V}}_{oo})^{-1}\hat{\mathbf{V}}_{om} \qquad (4.34)$$

and the (95%) prediction interval for the ith missing value $y_{m(i)}$ is:

$$\hat{y}_{m(i)} \pm 1.96[\boldsymbol{\Psi}_{m,m}(i, i)]^{1/2} \qquad (4.35)$$

where $\Psi_{m,m}(i, i)$ denotes the ith diagonal entry in the matrix $\boldsymbol{\Psi}_{m,m}$ corresponding to $\hat{y}_{m(i)}$. Note that the prediction interval for a missing value will tend to be an underestimate because it ignores the statistical error associated with the estimation of the unknown parameters.

There are complications with the methodology. If the mean is not constant the order of trend surface has to be specified. $\mathbf{V}$ needs to be modelled to ensure a valid covariance function. In practice therefore there are additional specification issues in implementing this approach and there is a further complication which is a circularity problem in estimating $\mathbf{A}\boldsymbol{\theta}$ and $\mathbf{V}$ (Cressie, 1991, pp. 165–9).

The maximum likelihood predictor (4.30) uses weights defined by $\mathbf{V}_{mo}(\mathbf{V}_{oo})^{-1}$. Because of the similarities with kriging, discussion of these weights is deferred until section 4.4.2(v). Martin (1984, 1989), Krug and

Martin (1990) and results in Haining et al. (1989) investigate the effect of the pattern of missing data in an areal system. Whether the pattern is scattered or clustered and if clustered the form of the configuration has an effect on the estimation of model parameters and the performance of missing-data prediction. The proximity of the missing data to the study area boundary can also affect parameter estimation and hence missing-data prediction.

If the mean $\mathbf{A\theta}$ is a constant ($\mathbf{\mu}$) and known, and the parameters in $\mathbf{V}$ are known then (4.30) is the best linear unbiased predictor of the missing values and the method of prediction is equivalent to *simple* kriging (Cressie, 1991, pp. 109–10). The predictor for $\mathbf{y}_m$ is a weighted sum of the observed values but weights are not constrained to sum to 1.

The full methodology is undoubtedly cumbersome. When the mean is not known there are parallels with ordinary and universal kriging except that weights do not sum to 1. Following the practice in geostatistics and suggested for example in Isaaks and Srivastava (1989, p. 532), a more practical alternative to implementing the full methodology seems to be the following. First fit a trend ($\mathbf{A\theta}$) to the data. Use the residuals from the trend surface to estimate $\mathbf{V}$ – that is obtain the spatial covariances and fit a model to the covariances. Substitute ($\mathbf{A\theta}$) and $\mathbf{V}$ into (4.30), (4.34) and (4.35) to predict the missing values and to estimate the prediction error.

Note that the maximum likelihood approach can be extended to the case where the mean $\mathbf{A\theta} = \mathbf{X\beta}$. So the mean now comprises a set of explanatory variables ($\mathbf{X}$) and their associated coefficients ($\mathbf{\beta}$). The model may be the regression model of interest to the analyst who wants to fit it using as much of the data as are available. However the method applies when missing values are only associated with the response variable. No methods appear to have been developed in the spatial literature to deal with missing values in the explanatory variables or the explanatory and the response variables. A possible direction would seem to be an adaptation of the method in Little and Rubin (1987, pp. 142–5, 153–5) treating the data on $(Y, X(1), \ldots, X(k))$ as samples from a joint multivariate distribution.

4.4.2   Spatial interpolation, spatial prediction

There are four requirements any method for estimating or predicting values on a spatially continuous surface should seek to satisfy. First, the method should exploit the spatial structure in the surface and give the most weight to data at locations close to the site where the prediction is needed. Second, observed values that are spatially close together duplicate information so that without some form of weighting for members of a cluster there is a danger that any estimate could be unduly influenced by measurements from one part of the map. Third, the method should give some indication of the likely error associated with the prediction. Fourth, the method should honour the known properties of the sample and in particular any method, when used to estimate a measured value should return that value – the closer the better. These requirements also underlie some of the (univariate) approaches to the missing-data problem. Smoothing methods (methods for reducing noise in mapped data for the purpose of identifying spatial patterns in mapped data) also share some of these requirements. Smoothing methods will be discussed in chapter 7 where the methods to be described now will again be relevant.

The term spatial interpolation is sometimes used to refer to methods that provide estimates of data values where no underlying probability model is assumed. Interpolation methods tend to be based on exploiting geometric attributes of the locations where there are observed values and in a general sense satisfy the first requirement and sometimes the second as well. Spatial or geostatistical prediction, draws on a probability model for the surface. Kriging comprises a group of methods for predicting data values. They address the weaknesses associated with geometric approaches as well as providing prediction errors. Kriging has been described as 'the logical conclusion' (Webster and Oliver, 2001, p. 37) of this area of methodological development.

In section 4.4.1 observations referred to areas, and areas can be represented by points (e.g. the area centroid). In this section observations refer to point locations but it is possible to construct areas around points using Dirichlet polygons for example (see chapter 2). There is potential therefore for methods to cross over between these two types of spatial data. What is important is that methods are applied critically with an awareness of their properties, linking choice of method to valid assumptions about the nature of the data and the underlying spatial variation in the specific data set.

(i) The Dirichlet partition. The simplest method is to assign to any location the same attribute value as its nearest observed neighbour. This is equivalent to constructing a Dirichlet partition on the sample sites. The imputed value is then given by the sample value associated with the Dirichlet polygon the site falls within. This rule gives rise to discontinuity in the estimation. The surface is implicitly made up of a set of plateaux or plates that meet along lines of discontinuity. To overcome this and problems associated with clustering, the methods reviewed in section 4.4.1 under (i), (ii) and (iii) could be used.

(ii) Cell declustering. This method divides an area about the location to be interpolated into quadrants (for example). The mean is computed for each quadrant and the four means then averaged to yield the estimate. In reported

trials cell declustering did not perform well and not as well as (4.28) (Isaaks and Srivastava, 1989, pp. 241–3).

(iii) Triangulation. The method of triangulation is also a weighted interpolator. Three nearby sites to the site to be predicted are chosen after performing a Delaunay triangulation on the set of observed sites (see Isaaks and Srivastava, 1989, pp. 251–6). The three sites, which are from the nearest neighbours in the Dirichlet partition, are chosen so they form a triangle that encloses the site to be estimated. The next step is to solve for the three unknown coefficients (a, b, c) of the equation of a plane:

$$y(j) = a E(j) + b N(j) + c$$

This is done by substitution where $y(j)$ is the observed value and $E(j)$ and $N(j)$ are the easting and northing co-ordinates of point j ($j = 1, 2, 3$). So:

$$\hat{y}(i) = \hat{a} E(i) + \hat{b} N(i) + \hat{c}$$

where $\hat{a}, \hat{b}$ and $\hat{c}$ are the estimates of the coefficients. This estimator produces a spatially smoother set of values than some of the earlier methods described, but still produces abrupt changes of gradient at the margins of the triangle (Webster and Oliver, 2001, p. 39). Sibson's (1981) 'natural neighbour interpolator' which is an extension of the triangulation method also generates discontinuities (Webster and Oliver, 2001, pp. 39–40).

(iv) Inverse distance weighting. Interpolation methods that employ distance weighting give differential weights to observations based on their proximity to the missing value. Distance weighting is introduced to capture the idea that attribute values close in distance terms tend to be similar but that the similarity weakens as distance separation increases. So it is the nearest sites that should be given most weight in any imputation. The interpolated value for $y(i)$ is:

$$\hat{y}(i) = \Sigma_{j=1,\dots,n} \lambda(i, j) y(j) \tag{4.36}$$

where for example $\lambda(i, j) = d(i, j)^{-\alpha}$, α is a positive constant and $d(i, j)$ is the distance between sites i and j. Other choices for $\lambda(i, j)$ include $\exp(-\alpha d(i, j))$ and $\exp(-\alpha d^2(i, j))$ and $\lambda(i, j)$ is often set to zero when it exceeds some chosen distance. The coefficients are scaled so that $\Sigma_{j=1,\dots,n} \lambda(i, j) = 1$. The choice for α influences the contribution made by data from different distances and the smoothness properties of the surface (Ripley, 1981, p. 36). A large value for α means that distant sites play a small role in the imputation which tends to produce a very spiky interpolated map. A small value of α has the effect of giving equal weight over longer distances and gives rise to a smooth interpolated map. Isaaks and Srivastava (1989, pp. 258–9) illustrate the effect of varying

α on the contribution made by sites at different distances from the site to be estimated.

Distance-weighting methods like (4.36) are affected by clusters in the datapoints. A combination of methods might be appropriate. The cell declustering method could be employed in which the quadrant mean is attached to the centroid of the observed sites and then the imputation uses a distance weighting of the means as in (4.36).

Isaaks and Srivastava (1989, pp. 266–77) compare several of the above methods through a series of case studies. They remark: 'the method which is "best" depends on the yardstick we choose' (p. 272). Some methods, like (4.36) with a relatively large decay parameter (e.g. $\alpha = 2.0$) are to be favoured if the objective is to minimize the *largest* errors of estimation whilst triangulation methods have relatively good *average* levels of error. They suggest that incorporating more nearby sites improves estimates but that if the sites are clustered this can offset the benefits of taking more sites (p. 276). The best methods, they conclude, use all the nearby sites and account for the possibility of clustering. These properties will be present in the kriging predictors to be discussed below which utilize information on the spatial covariance properties of the data. By contrast the estimators described above depend only on geometric relationships between the sites and the estimates they yield are dependent on neighbourhood size and how the weights are specified. They do not provide an estimate of the possible error associated with the imputation.

(v) Kriging. It is now assumed that the data $y(1), \ldots, y(n)$ are the realization of a (weakly stationary) stochastic model with mean, $\mu(.)$, and (symmetric) variance–covariance matrix, Σ. Given a sample of size n, the BLUP of any unsampled point on the surface can be obtained by *simple kriging*. To predict the attribute value at site $o, (o)$, which is not included in the sample compute (Cressie, 1991, p. 110):

$$\tilde{y}(o) = \mu(o) + \mathbf{c}^T \Sigma^{-1}(\mathbf{y} - \boldsymbol{\mu}) \tag{4.37}$$

where $\mathbf{c}^T = (\text{cov}(Y(0), Y(1)), \ldots, \text{cov}(Y(o), Y(n)))$. Σ, as noted, is an $n \times n$ symmetric matrix with (i, j)th element equal to $\text{cov}(Y(i), Y(j))$, $\mathbf{y} = (y(1), \ldots, y(n))^T$, $\boldsymbol{\mu} = (\mu(1), \ldots, \mu(n))^T$ and $\mu(o)$ is the mean evaluated at site o. The second term in (4.37) identifies the simple kriging weights, $\mathbf{c}^T \Sigma^{-1}$, assigned to each datapoint, that yields the BLUP of the unknown attribute value.

The weights in (4.37), $\mathbf{c}^T \Sigma^{-1}$, reflect spatial covariance properties in the data, or '*statistical* distance weighting', as Isaaks and Srivastava (1989, p. 300) call it, rather than arbitrary definitions of spatial relationships. The vector $\mathbf{c}$ introduces a form of distance weighting into the prediction, because covariances tend to decrease with increasing distance so that more remote observed sites contribute less to the prediction. The presence of Σ incorporates the co-variances between each observed data value and every other. Observed values that are close together have large values in Σ, those far apart, small values. The multiplication of $\mathbf{c}$ by Σ^{-1} 'adjusts the raw inverse statistical distance weights (in $\mathbf{c}$) to account for possible redundancies between the observed values' (Isaaks and Srivastava, 1989, p. 300). The presence of Σ^{-1} therefore handles the effect of clustering in the observed values on prediction. This is achieved by downweighting the contribution from sample sites that are members of clusters. If data clustering is not a problem (as in the case where data are from a systematic spatial sample or grid of pixels) it is not surprising that simpler distance weighting or even nearest neighbour methods that approximate the elements of $\mathbf{c}$ but do not incorporate a term corresponding to Σ^{-1} have done nearly as well in some trials (Haining et al., 1989).

The prediction error for (4.37) is:

$$\sigma_{sk}^2 = \text{Cov}(Y(o), Y(o)) - \mathbf{c}^T \Sigma^{-1} \mathbf{c} \tag{4.38}$$

The first term, $\text{Cov}(Y(o), Y(o)) = \text{Var}(Y(o))$ measures the variability of the attribute Y. The second term is the weighted sum of the covariances between the samples and the value to be predicted ($\mathbf{c}$) where the weights are given by the simple kriging weights ($\mathbf{c}^T \Sigma^{-1}$). Prediction intervals can be constructed from (4.38). In the case of 95% intervals the interval is given by $y(o) \pm 1.96\, \sigma_{sk}$. The reader should compare (4.37) and (4.38) with (4.30) and (4.34) recalling though that in section 4.4.1(b)(v) the mean and variance–covariance matrix are to be estimated and the notation $\Sigma = \sigma^2 \mathbf{V}$ was used.

In practice the mean is not usually known and for mapping and interpolation the weights are usually required to sum to one. This last condition ensures uniform unbiasedness (Cressie, 1991, p. 120). Assume the mean is constant and the covariance matrix Σ is known at least up to a scalar σ^2. Optimal prediction in the sense of minimizing the prediction error is achieved through *ordinary kriging*. Ordinary kriging provides the prediction in one step and does not require any explicit identification of the mean. If the mean is unknown but follows some order of trend surface (see section 9.1.1), optimal prediction is achieved through *universal kriging*. This too is a one-step prediction which does not require explicit identification of the mean although it does require specification of the order of the trend surface. The following description assumes weak stationarity. Expressions for (4.39) and (4.41) assuming only intrinsic rather than weak stationarity (see section 9.1.2(a)) and expressed in terms of the semi-variogram are given in Cressie (1991, pp. 122 and 153–4).

Ordinary and universal kriging predictors under weak stationarity are given by (Cressie, 1991, pp. 123 and 154):

$$\tilde{y}(o) = [\mathbf{c}^T \Sigma^{-1} + (\mathbf{Am})^T \Sigma^{-1}]\mathbf{y} \tag{4.39}$$

where:

$$\mathbf{m} = (\mathbf{A}^T \mathbf{\Sigma}^{-1} \mathbf{A})^{-1} (\mathbf{a} - \mathbf{A}^T \mathbf{\Sigma}^{-1} \mathbf{c}) \tag{4.40}$$

$\mathbf{A}$ is an $n \times p$ matrix where p denotes the number of trend surface parameters which is determined by the order (q) of the trend surface (see below section 9.1.1). The p dimensional column vector $\mathbf{a}$ identifies the corresponding spatial co-ordinates for site o. For the constant mean case (ordinary kriging) $p = 1$, $q = 1$; $\mathbf{A}$ is a column vector of 1s and $\mathbf{a}$ is equal to 1. For a first- or higher-order trend surface (universal kriging) the ith row of $\mathbf{A}$ refers to sample point i. For a second-order trend surface for example, $q = 2$, $p = 6$, and the ith row of $\mathbf{A}$ is:

$$(1, s_1(i), s_2(i), s_1(i)^2, s_1(i)^2, s_1(i)s_2(i))$$

where $(s_1(i), s_2(i))$ denotes the co-ordinate position of the ith point. The p dimensional column vector $\mathbf{m}$ contains the p Lagrange multipliers that ensure the weights sum to 1.

The kriging prediction error is (Cressie, 1991, pp. 123 and 155):

$$\sigma^2 = \text{Cov}(Y(o), Y(o)) - \mathbf{c}^T \mathbf{\Sigma}^{-1} \mathbf{c} + \mathbf{m}(\mathbf{a} - \mathbf{A}^T \mathbf{\Sigma}^{-1} \mathbf{c}) \tag{4.41}$$

Now 95% prediction intervals can be constructed using: $\bar{y}(o) \pm 1.96 \, \sigma$. The first two terms of (4.41) also appear in (4.38). The third term is new and represents the variance arising from the estimate of the mean, however it is usually very small (Webster and Oliver, 2001, p. 179). To further consider (4.41) we take the case of ordinary kriging $(p = 1)$, then (4.41) becomes:

$$\text{Cov}(Y(o), Y(o)) - \mathbf{c}^T \mathbf{\Sigma}^{-1} \mathbf{c} + \mathbf{m}[1 - (\mathbf{1}^T \mathbf{\Sigma}^{-1} \mathbf{c})] \tag{4.42}$$

where:

$$\mathbf{m} = (\mathbf{1}^T \mathbf{\Sigma}^{-1} \mathbf{1})^{-1} [1 - \mathbf{1}^T \mathbf{\Sigma}^{-1} \mathbf{c}]$$

First, prediction error will increase the further site o is from sample data. If there are few nearby sample observations, prediction errors will be larger than if there are several. The closer the elements in the vector $\mathbf{c}$ are to 0 the smaller the second term in (4.42) and the larger the prediction error. This effect is also present in the third term in (4.42) where $\mathbf{c}$ again appears. However consider now the term in square brackets, $[1 - (\mathbf{1}^T \mathbf{\Sigma}^{-1} \mathbf{c})]$, and also the term $(\mathbf{1}^T \mathbf{\Sigma}^{-1} \mathbf{1})^{-1}$. The term in square brackets is the difference between the sum of the simple kriging weights and 1. If the simple kriging weights are close to 1 then this term (which appears twice) will be close to 0 and so the third term in (4.42) will be close to 0. If sample points are clustered this will yield higher prediction errors than if the sample points are spread out. This is because with spatially correlated phenomena, two sample points that are very close together are not much better than one in predicting values at a nearby site – they carry similar information for the purposes of predicting the value at any unsampled location. In the context of (4.42) this effect is measured through the sum of the covariances between all the sample pairs: $(\mathbf{1}^T\Sigma^{-1}\mathbf{1})^{-1}$. The bigger the spatial covariance in the sample the bigger $(\mathbf{1}^T\Sigma^{-1}\mathbf{1})^{-1}$. However its effect in (4.42) depends on the size of the discrepancy between the sum of the simple kriging weights and 1. These points are illustrated for the case of ordinary kriging ($q = 1$), using a different perspective, by Isaaks and Srivastava (1989, pp. 497–9). Webster and Oliver (2001, chapter 8) provide further illustrations. Irregularly spaced data generate patches with larger prediction errors (i.e. in areas with few sample points), than regularly spaced data for a given density of sample points and given spatial covariance structure.

Ordinary kriging appears to be a more widely employed method of spatial prediction in geostatistics than universal kriging. Spatial prediction is based on local information and within such a window the assumption of a constant mean may be appropriate. Isaaks and Srivastava (1989, p. 532) suggest an alternative to universal kriging which they suggest is more practical for geostatistical interpolation where Σ is usually not known. First fit a trend ($\mathbf{A\theta}$) to the data. Next remove the trend and use the residuals from the trend to estimate Σ. Apply the method of ordinary kriging to the residuals adding the trend back in at the end in making the final set of predictions.

There are other forms of kriging to apply in situations where the data are non-Gaussian such as categorical or count data and for which non-linear geostatistics is needed. See Journel (1983) for indicator kriging and Matheron (1976) for disjunctive kriging both of which are reviewed in Cressie (1991, pp. 278–84).

An irregular distribution of seven sites is shown in figure 4.7 together with the Dirichlet polygon for site o. It is the value at site o that is to be predicted. Tables 4.1(a)(a) to (d) give the data values and intersite distances, the upper triangular elements (including the diagonal) of the matrices Σ and Σ^{-1} and the elements of the vector $\mathbf{c}$. These covariances are based on an assumed isotropic covariance model of the form:

$$C(h) = 5.0 \exp(-0.2|h|)$$

In practice the full data set (of which the area shown is a subarea) would be used to estimate the spatial covariances (or semi-variogram) at various distances and then a model fitted to the plot.

Table 4.2(a) shows the results of a selection of different methods described in sections 4.4.1 and 4.4.2 for predicting the value at the site marked o. In the

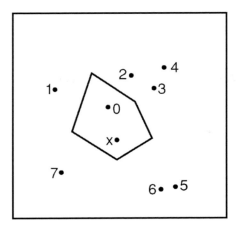

Figure 4.7 Pattern of sites for the worked examples in tables 4.1 and 4.2

Table 4.1(a) *Distances* ($|h|$) *between sites in figure 4.7; data values recorded at the seven sample sites*

	0	1	2	3	4	5	6	7
0	0.0	2.0	1.5	2.0	2.5	4.0	3.7	3.0
1		0.0	2.9	3.7	4.2	5.7	5.5	3.2
2			0.0	1.0	1.3	4.5	4.5	4.6
3				0.0	0.8	3.8	3.9	4.8
4					0.0	4.6	4.7	5.5
5						0.0	0.5	4.3
6							0.0	3.7
Values		55	45	41	49	75	78	80

Table 4.1(b) *Variance–covariance matrix between the sample* (1–7): Σ

5.00	2.799	2.385	2.158	1.599	1.664	2.636
2.799	5.00	4.093	3.855	2.032	2.032	1.992
2.385	4.093	5.00	4.260	2.338	2.292	1.914
2.158	3.855	4.260	5.00	1.992	1.953	1.664
1.599	2.032	2.338	1.992	5.00	4.524	2.115
1.664	2.032	2.292	1.953	4.524	5.00	2.385
2.636	1.992	1.914	1.664	2.115	2.385	5.00

case of the kriging predictions note that the severest downweighting in Σ^{-1} to compensate for the effect of clustering is associated with sites 5 and 6 (–0.998), 3 and 4 (–0.524) and 2 and 3 (–0.365). Sites 4 and 5 both receive negative kriging weights that reflect the fact that 4 is 'hidden' by 2 and 3 and 5 is 'hidden'

Table 4.1(c) *Inverse of variance–covariance matrix* $\Sigma : \Sigma^{-1}$

0.347	−0.148	−0.005	0.010	−0.010	0.013	−0.127
−0.148	0.725	−0.365	−0.180	0.011	−0.013	−0.009
−0.005	−0.365	1.002	−0.523	−0.084	−0.020	−0.015
0.010	−0.180	−0.523	0.777	−0.002	0.005	0.006
−0.010	0.011	−0.084	−0.002	1.129	−0.998	0.033
0.013	−0.013	−0.020	0.005	−0.998	1.175	−0.134
−0.127	−0.009	−0.015	0.006	0.033	−0.134	0.324

Table 4.1(d) *Covariances* (**c**) *between sample points and prediction site o* ($\mathbf{c}^T$)

	1	2	3	4	5	6	7
covariances	3.351	3.704	3.352	3.033	2.247	2.386	2.744

Table 4.1(e) *Distances and covariances* (**c**) *to sample points and prediction site x*

	1	2	3	4	5	6	7
distance to x	3.0	2.5	2.5	3.4	2.9	2.5	2.4
covariances	2.744	3.032	3.032	2.533	2.799	3.032	3.093

Table 4.2(a) *Weights associated with different interpolation methods for site o*

Site	a	b	c	d	e	f	g
1	0.143	0.2	0.24	0.250	0.17	0.19	0.16
2	0.143	0.2	0.22	0.083	0.23	0.32	0.18
3	0.143	0.2	0.16	0.083	0.17	0.19	0.16
4	0.143	0	0	0.083	0.14	0.12	0.15
5	0.143	0	0	0.125	0.09	0.05	0.11
6	0.143	0.2	0.16	0.125	0.09	0.05	0.11
7	0.143	0.2	0.22	0.250	0.11	0.08	0.13
Interpolation	60.43	59.8	59.74	64.125	56.1	52.57	58.04

Notes: **a:** Arithmetic mean. **b:** Dirichlet neighbours. **c:** Dirichlet neighbours weighted by length of shared common border. **d:** Cell declustering (N–S/E–W axes used for quadrant borders). **e:** Inverse distance weighting (4.36: $\lambda(i, j) = |h|^{-1}$). **f:** Inverse distance weighting (4.36: $\lambda(i, j) = |h|^{-2}$). **g:** Negative exponential weighting (4.36: $\lambda(i, j) = \exp(-0.2 |h|)$).

Table 4.2(b) *Triangulation method applied to site o*
(i) Co-ordinates of sites

	0	1	2	3	6	7
E–W	35	16	44	53	55	18
N–S	43	49	55	50	12	17

(ii) Coefficient estimates and interpolated values

Triangulation sites	a (E–W)	b (N–S)	c	Interpolated value
(1,2,6)	−0.182	−0.814	97.816	56.416
(1,2,7)	−0.187	−0.792	96.850	56.249
Average				56.332

4.2(c) *Weights associated with simple kriging and ordinary kriging (sites o and x)*

	Simple kriging		Ordinary kriging	
Site	site (o)	site (x)	site (o)	site (x)
1	0.286	0.129	0.278	0.125
2	0.387	0.190	0.385	0.189
3	0.119	0.244	0.120	0.244
4	−0.004	−0.107	−0.013	−0.112
5	−0.041	−0.024	−0.048	−0.028
6	0.136	0.302	0.133	0.301
7	0.151	0.282	0.144	0.278
Sum	1.04	1.02	1.00	1.00

4.2(d) *Predictions and prediction errors for simple and ordinary kriging (sites o and x)*

	Simple kriging		Ordinary kriging	
	(o)	(x)	(o)	(x)
Predicted value	57.523	64.875	55.25	63.675
Prediction error	1.571	1.872	1.574	1.873
Lagrange multiplier (m)			−0.10	−0.05

Note: The predicted value for simple kriging is obtained from (4.37) for ordinary kriging (4.39). The prediction error for simple kriging is obtained from (4.38) for ordinary kriging (4.42).

by 6. Although both sites 1 and 3 are at equal distances from the site to be predicted the weighting for site 1 is the larger since there are no other sites close to 1. The ordinary kriging error is only slightly larger than the simple kriging prediction error because the simple kriging weights are only slightly greater than 1.0. To illustrate how prediction error is affected by the distribution of sample sites in relation to the site where the prediction is needed, a second site (x) has been evaluated. The distances (from site x to the sample sites) and the corresponding elements of the vector **c** are shown in table 4.1(e). Note the increase in the prediction error at x relative to o (table 4.2(d)). The closest neighbour of x is 2.4 units, compared to 1.5 for o, which has three neighbours less than or equal to 2.0 units away.

Apart from inverse distance weighting ($\alpha = 2$) where the weights decay very rapidly with increasing distance all the methods provide predictions that are larger than the prediction provided by ordinary kriging which should be considered the 'gold standard'. Note this remark also applies to the distance weighting method using the exponential function ($\exp(-0.2|h|)$). The cell-declustering method (based on a north–south/east–west partition of the area) gives the largest value which differs most from the 'gold standard'. Two triangulations are examined and the average of the two computed because there seems nothing to choose between them. This table of results should not be taken as indicative of which methods come closest to approximating the gold standard since with other data sets other rank orderings might arise. The purpose is to show the range of predictions arising from different methods.

Figure 4.8 summarizes the approaches to missing-data estimation and spatial interpolation and prediction.

4.4.3   Boundaries, weights matrices and data completeness

In conclusion we briefly draw attention to some other aspects of model completeness that ought to be considered for spatial data analysis. The application of some spatial statistical techniques, because they draw on data from spatial neighbourhoods around each case, may require data on variables where the data values refer to spatial units beyond the boundary of the study area. If these boundary data are not available this represents a form of data incompleteness. Figure 4.9 classifies sites. An *interior site* lies inside the study area so its value is observed as are the values for all its neighbours. An *observed boundary site* lies inside the study area so its value is observed but the values for at least one of its neighbours lies outside the study area and so has not been observed. An *unobserved boundary site* may be a neighbour of an observed boundary site but

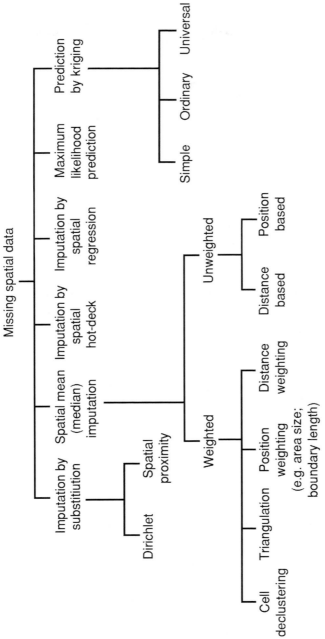

Figure 4.8 Approaches to missing-data estimation, interpolation and prediction

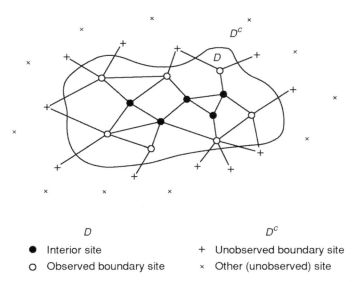

D D^c

● Interior site   + Unobserved boundary site
○ Observed boundary site   × Other (unobserved) site

Figure 4.9 Classification of sites according to their relationship to the boundary and the definition of their neighbours

because it lies outside the study area its value has not been observed. The classification of sites depends on the graph or connectivity structure that has been assumed to represent spatial relationships between the sites (see 2.4).

Martin (1987) defines two aspects of the boundary value problem: the effect of different boundary formulations on the variance–covariance matrix (**V**) of the process; the effect of boundary forms on the properties of estimators of model parameters. The effect of the boundary on the conduct of data analysis depends on how the boundary is handled and this depends on how the underlying process is conceptualized. One approach is to treat the process observed within the study region as a subset of a process that is operating over a wider area. The model covariance, for example, between an observed and unobserved boundary site is the same function of distance as between any pair of observed values and may be estimated accordingly using the data that are available. A second approach is to define one model for the interior sites and another model for the boundary. This would seem appropriate where there are different processes operating on either side of the boundary. This could arise if the boundary is natural in process terms such as in the case of an island where the coastal boundary affects the spread of some disease. It could also arise if the boundary is a political boundary with different policies or regimes operating on either side of the boundary. Let $\mathbf{y} = (y(1), \ldots, y(n))$ denote the observed values (interior and boundary sites) whilst $(y(n + 1), \ldots, y(n + h))$ denotes the unobserved boundary values (D^c). Let $\mathbf{y}^* = (y(1), \ldots, y(n + h))$ and $\mathbf{V_{y^*}}$ denote the variance–covariance matrix for $\mathbf{y}^*$. The variance covariance matrix for $(y(1), \ldots, y(n))$, denoted $\mathbf{V_y}$ is obtained by deleting the rows and columns associated with $(y(n + 1), \ldots, y(n + h))$ in $\mathbf{V_{y^*}}$. From the theory of partitioned matrices if:

$$
\mathbf{V}_{y^*}^{-1} = \begin{bmatrix} \mathbf{V}^{oo} & \mathbf{V}^{ob} \\ \mathbf{V}^{bo} & \mathbf{V}^{bb} \end{bmatrix}
$$

where o refers to the observed sites (interior and boundary) and b refers to the unobserved boundary sites in D^c then:

$$
\mathbf{V}_y^{-1} = \mathbf{V}^{oo} - \mathbf{V}^{bo}(\mathbf{V}^{bb})^{-1}\mathbf{V}^{ob} = \mathbf{V}^{oo} - \Gamma
$$

(see Martin, 1987). The matrix $\Gamma = 0$ if, for example, the process outside the region is 'disconnected' from the process inside the study region so that $\mathbf{V}^{bo} = \mathbf{V}^{ob} = 0$. Martin gives examples. Different boundary assumptions can be modelled through the specification of Γ. The contrasting cases mirror the 'stochastic' and 'fixed' boundary value assumptions used in time series analysis with the added complication that now the boundary encircles the study region. Now boundary assumptions will influence a larger proportion of data values and boundary effects only dissipate towards the interior of the study region providing the study region has a large number of interior sites that are distant from the boundary.

The specification of connectivity relationships between sites or areas (the elements of the matrix $\mathbf{W}$) is a modelling assumption but it can be informed by relevant data on interactions between sites or areas. The objective of analysis may be to explain spatial variation in per capita income levels between regions within a country. But such regional variation is unlikely to be independent and a minimal specification will need to allow for spatial dependencies between the regions. Ideally factor flow data are needed either to build explicitly into the specification of the model or to inform the specification of the spatial relationships, described by the matrix $\mathbf{W}$. The absence of interaction data in these circumstances may be seen either as a form of model incompleteness or data incompleteness. It might be seen as a form of data incompleteness because, whilst it may be possible to specify a plausible connectivity structure from other information about the geography of the regions, the data necessary to undertake this carefully cannot be considered complete if real interaction data are not available.

4.5   Concluding remarks

This chapter has considered the implication of spatial data quality for the conduct of spatial data analysis. Many of the techniques of spatial data

Table 4.3 *The dimensions of spatial data quality in relation to stages of spatial data analysis*

	Accuracy	Resolution	Consistency	Completeness
Data collection and preparation of final database	Concerns about the presence of gross errors	Creating a common spatial framework for data collected on different frameworks	Incompatible data values (e.g. disease cases reported in areas without population)	Presence of missing data; need to interpolate or predict
Form and conduct of statistical analysis	Choice of error model. Need for robust and resistant statistical methods of analysis	Differences in variable precision across spatial units. Sensitivity of results to different methods of areal interpolation		Modelling in the presence of missing data.
Interpretation of results		Ecological bias. Forming invalid individual level hypotheses from aggregate analyses		Concerns about spatial variation in undercounting. Model misspecification due to the effects of missing variables

analysis have evolved in response to data quality issues that regularly arise in handling spatial data. The purpose of this chapter has been to show how various techniques can be ordered and classified in relation to this need. In conclusion we return to the earlier remark about how different facets of spatial data quality may impact on particular stages of spatial data analysis. Table 4.3 provides a cross-classification by data quality dimension and stage of analysis with some indicative examples.

The exploratory analysis of spatial data

5

Exploratory spatial data analysis: conceptual models

Chapters 6 and 7 consider methods used to explore spatial data. This chapter covers some preliminary ground by defining in section 5.1 exploratory spatial data analysis (ESDA) with reference to exploratory data analysis (EDA). It then considers in section 5.2 some conceptual models for describing spatial variation. Conceptual models are identified because they help to characterize spatial variation in a non-formal way and so help to clarify what the analyst might look out for (in terms of spatial variation) in undertaking ESDA. Formal, statistical, models are not needed for exploratory spatial data analysis, although some of the exploratory forms of analysis discussed in chapter 7 that involve hypothesis testing need to specify a model for spatial randomness. These are discussed as they are needed. The discussion of statistical models for more general forms of spatial variation is left until chapter 9.

5.1   EDA and ESDA

Good (1983) defines exploratory data analysis (EDA) to be a collection of techniques for summarizing data properties (descriptive statistics) but also for detecting patterns in data, identifying unusual or interesting features in data, detecting data errors, distinguishing accidental from important features in a data set, formulating hypotheses from data. EDA techniques may also be used to examine model results, provide evidence on whether model assumptions are met and whether there are influential data effects in model fits.

EDA techniques make no assumptions about the population from which the data are drawn and hypothesis testing is often excluded. The set of techniques that are employed are visual (including charts, graphs and figures) and/or numerical in the sense of being quantitative summaries of the data (Tukey, 1977; Hoaglin et al., 1983, 1985). Statistical quantities such as the median and quartiles of a distribution are computed and then displayed in a boxplot or added to a dotplot. A scatterplot is constructed and the data are summarized by some numerical smoothing operation such as a loess curve (see section 7.1.1).

Many exploratory techniques stay 'close' to the original data in the sense they use only simple transformations of the raw data and do not employ inference theory. The statistics that are used are typically 'resistant' which means they are not greatly affected by the presence of a small number of extreme values. The median and inter-quartile range are resistant estimators of the centre and spread of a distribution of numbers respectively, whereas the mean and standard deviation are not. The median is a resistant method of identifying the centre (or location) of a set of data values because it defines the middle value (the 50th percentile), when the data values are ordered, as the centre. Even if a few values on either tail of a distribution are much larger or smaller than the rest, this will not have a major affect on the calculation of the median. By contrast the mean, as a measure of the centre of a distribution, is affected by extreme values since each data value in a set of n values contributes $(1/n)$th to the value of the mean. The inter-quartile range, the difference between the upper and lower quartiles, is a resistant measure of the spread of a distribution of values, whereas the standard deviation is based on the mean (which is not resistant) and each value's squared difference from the mean contributes $(1/n)$th to the value of the square of the standard deviation.

ESDA comprises techniques for exploring *spatial* data – summarizing spatial properties of the data, detecting spatial patterns in data, formulating hypotheses which refer to the geography of the data, identifying cases or subsets of cases that are unusual given their location on the map (Cressie, 1984; Haining, 1990). As with EDA, ESDA techniques are visual and numerically resistant. Now the map, which identifies for the analyst where cases are and their spatial relationship to one another, assumes a particularly important role in the analysis of the data or examining model results. It will be important to be able to answer questions such as: 'where do those extreme cases on the histogram fall on the map?'; 'where do attribute values from this part of the map fall on the scatterplot?'; 'which cases fall in this subregion of the map *and* meet these specified attribute criteria?'; 'what are the spatial patterns and spatial associations in this data set?'; or 'what are the spatial patterns and spatial associations in this *geographically defined subset* of the data?'. In the case of regression modelling we will want to see a map of the positive and negative residuals and ask: 'is there any evidence of spatial pattern in the arrangement of the residuals?' The view here is that the set of ESDA tools includes those used in EDA but also includes additional methods that address the special queries that arise as a consequence of the spatial referencing of the data.

5.2   Conceptual models of spatial variation

(a)   The regional model

In geography the regional model has a long history as a conceptual model for spatial variation (Grigg, 1967). The term 'region' has been given many meanings and many different types of regions have been specified in the literature reflecting the diversity of need when seeking to theorize about phenomena that have a geography (see Bunge, 1962; Burrough and Frank, 1995; Raper, 2001). Here the emphasis is on the definition of regions as spatial units used for the purpose of spatial data analysis. Three types of regions are of particular interest.

Formal (or *uniform*) regions are constructed by the sharp partitioning of space into homogeneous or quasi-homogeneous areas. Formal regions are a classification of space into areas based on attribute similarity and contiguity in space. As such, formal regions should be derived by operations that satisfy the usual rules for classification (Grigg, 1967). Borders between regions are based on changes in attribute levels. Such a partitioning or segmentation into formal regions is easier to justify when the number of variables used in the 'regionalization' (the process of partitioning space) is small and becomes progressively more difficult as the number increases unless there is very strong covariation amongst the variables. It is also difficult to sustain the formal regional model when attribute variation is continuous across space. Automated methods have been developed to assist the analyst to specify formal regions as in the case of land cover maps but in social science applications it is still common to see them constructed by a mixture of formal methods and local knowledge (Haining et al., 1994; Sampson et al., 1997). Spatial statistical approaches to formal region building are reviewed in section 6.2.2.

Functional (or *nodal*) regions are demarcated using interaction data. Whereas formal regions are defined by the uniformity of attribute values, functional regions are bound together by the pattern of social or economic interaction that occurs within them and which sets them apart from neighbouring functional regions. In the UK, labour market regions are functional regions defined in terms of the patterns of commuting by workers (travel-to-work areas) and search areas adopted by firms when recruiting staff (Coombes and Openshaw, 2001). Some regional definitions combine formal and functional criteria. Community and neighbourhood areas may be defined in terms of attribute similarity, such as housing type, but also through reference to social networks and the common use of local facilities such as shops or General Practitioner surgeries (e.g. Gordon and Womersley, 1997). Landmark linear features (such as roads and railway lines) may be important in specifying the boundaries of socially defined functional regions particularly if they act as barriers to interaction across them.

Administrative regions are a consequence of political and other forms of decision making. Administrative regions usually have precise boundaries and are constructed by governments and public and private sector organizations in order to manage space. They provide a framework for collecting data, delivering services, distributing government funds, and in the case of private sector firms that operate nationally they provide a framework for product marketing and implementing price-setting policies. More than in the case of the other two regional types, administrative regions justifiably have sharp boundaries in those cases where policies are implemented that differentiate between places on the basis of which administrative region they lie within.

Notwithstanding the evident limitations of the regional model it is a widely used framework or building block for the analysis of spatial variation in many areas of social, environmental, physical and biological sciences (Sokal et al., 1993). However, even if it is accepted that partitioning space into discrete units constitutes a building block for spatial analysis, the representation of spatial variation it provides (a sharply demarcated set of areas) is limited. The conceptual models that we turn to now add further structure to what is meant by spatial variation.

(b)   Spatial 'rough' and 'smooth'

For the purpose of exploratory data analysis, Tukey (1977) assumes two components to any data value. Tukey's model is:

$$\text{data} = \text{smooth} + \text{rough} \tag{5.1}$$

Any observed data value can be partitioned into two mutually exclusive components. The smooth (also called fit) is the regular or 'predictable' part of the data value whilst the rough is the irregular, 'unpredictable' component. The regular component for data value $z(i)$ may be identified by applying data operations to the entire data set or local subsets close to and including $z(i)$ and then:

$$\text{rough} = \text{data} - \text{smooth} \tag{5.2}$$

This conceptual partition can be translated to the case of spatial data (Haining et al., 1998). The smooth element of a spatial distribution is that which is spatially regular so that, for example, knowing some part of this element of a spatial distribution it is possible to extrapolate it at least locally. What is left over after the smooth has been extracted is the spatially unpredictable (rough) component which cannot be used for extrapolation. The rough component may be predictable (or explainable) in terms of underlying processes but is defined as rough in respect of the map pattern because in purely spatial terms it has no recognizable structure and has no predictable attributes.

In spatial terms then, the Tukey model is:

$$\text{spatial data} = \text{spatial smooth} + \text{spatial rough} \qquad (5.3)$$

These elements are not mapped to any formal model nor is there any one-to-one association with particular statistical techniques. Equation (5.3) provides a conceptual decomposition of spatial variation into (spatially) regular and (spatially) irregular components. This model is useful in the context of exploratory spatial data analysis where the analyst might be interested in detecting spatial structures in the data. Such 'smooth' structure might include the following: (i) blocks or clumps of similar values including 'topographical' features such as 'ramps', valleys and ridges with edges and zones of rapid transition between them (Mungiole, 1999); (ii) an overall (whole map) propensity for similar data values to be found close together, dissimilar values apart (spatial autocorrelation); (iii) contrasts between different segments of a map (e.g. north/south or east/west contrasts, regions); (iv) trends or gradients across the map. Spatial 'rough' might include localized hot or cold spots, spatial outliers and localized areas of discontinuity (Haining et al., 1998).

The components of spatial data values that are deemed rough by the original Tukey definition may be *spatially* smooth if they are all found together in one area of a map. Regression residuals might be considered rough but if all positive residuals are in one part of the map and all negative residuals in the other then in spatial terms the residuals (after categorization) are smooth.

(c)   Scales of spatial variation

A more specific representation explicitly recognizes different *scales* of spatial variation. (This representation will provide a bridge into the formal statistical models to be described in chapter 9.) Spatial variation for data from a continuous surface of values might be represented in terms of:

$$\begin{aligned} \text{spatial data} = {}& \text{large (macro) scale variation} \\ &+ \text{medium/small (meso) scale variation} + \text{error} \end{aligned} \qquad (5.4)$$

The term 'error' includes measurement error that may be independent noise or it may have spatial structure (chapter 4). The same model can be applied to data for areas but now 'error' may also include intra-area variability which may be a function of area or population size (section 4.2.1). The 'error' component also includes that scale of variation which is present at the level of the individual spatial unit. In terms of the previous conceptual model (5.3), the first two components in (5.4) are smooth and the last is rough.

These different scales of variation are defined for the study area and are conditional on its size. Large- or macro-scale variation refers to trends or gradients present across the study area. For example, in studying heart disease mortality in England a broadly linear trend has been described showing an increasing rate from the south-east to the north-west of the country. In the case of respiratory disease across a city, any trends in risk might be broadly quadratic. Rates might peak at the city centre and decline with increasing distance from it. This conjecture is based on assumptions about levels of air pollution which tend to be highest near the city centre and in areas of poorer quality housing.

Medium- and small-(meso) scale variation refer to more localized spatial structures in the data and are conceptualized as being superimposed on any large-scale pattern that may be present. In the case of the risk of urban respiratory disease such scales of variation may be associated with medium- and small-scale socio-economic structures in the city that link with lifestyle and mobility patterns. Finally there might be highly localized disease risk hot spots close to a congested road intersection. There might be risk cold spots – small pockets of neighbourhood gentrification near the city centre where housing is better and residents have lifestyles that mean they spend less time in the area than other groups or they have different life histories.

For any particular variable certain components in (5.4) may dominate its spatial variation. The map of a variable which is dominated by small-scale variation plus error is said to display *spatial heterogeneity*. Spatial heterogeneity may originate in the underlying processes responsible for the observed variation, giving rise to spatial variation in means, variances and covariances (Getis and Ord, 1996). This might be referred to as *process-induced* spatial heterogeneity. This source of heterogeneity may be compounded in the case of regional data by measuring attributes through spatial units of different size (see section 4.2.1). This might be referred to as *measurement-induced* spatial heterogeneity because it is a product of how attributes are observed and measured (Bao and Henry, 1996). If small-scale variation dominates or the spatial units through which data are collected are small and of variable sizes and the attribute of interest is relatively rare the effect in both cases is likely to be of a mosaic-like surface when values are mapped.

The extent to which different scales of variation dominate will have implications for the analysis of spatial data. Where process-induced heterogeneity is strong, data analysis based on 'local' statistics may be preferable to data analysis based on global or 'whole map' statistics. Local statistics analyse the data by examining spatial subsets using a moving window. 'Whole map' statistics use all the data to compute spatially invariant statistical measures. In the case of data referring to surfaces there may be local anomalies and lines of discontinuity associated with geological processes which may introduce local features into the data and call for special care in estimating the variogram and in kriging (Isaaks and Srivastava, 1989, pp. 107–39; Cressie, 1991, pp. 42–6).

In the next chapter links are suggested between different ESDA techniques and different components of spatial variation. That is, certain ESDA tools are appropriate, either individually or in combination, for indicating the presence of certain types of spatial structure in a variable. Amongst the most important structures are: local structures including clusters or clumps of similar values, ridges and valleys; a 'whole map' propensity for similar values to be found together (spatial autocorrelation); spatial contrasts (say between values in different parts of a map); spatial gradients such as linear or higher order trends in data values. A map of variation in a single variable might be composed of one or more of these elements.

6

Exploratory spatial data analysis: visualization methods

This chapter focuses on visualization tools for exploratory spatial data analysis whilst chapter 7 deals with numerical tools and fitting. Visualizing data has come to mean both the *construction* of graphs and fits but also a *process* of interacting with the parameters of the graph. Visualization has acquired a distinctive status in the conduct of exploratory data analysis for at least two reasons. First, advances in computing have made it possible to visually examine data quickly and easily and in more informative ways. Second, graphical tools are often more intuitive for the non-specialist. This means people who are not statisticians but who have knowledge of the subject area and can bring subject matter context to the analysis may be better able to participate in the process of getting data insights.

Section 6.1 discusses visualization, identifying the data analytic tasks it is used for, classifying data visualization approaches and describing some of the techniques that are available for EDA. Section 6.2 considers the visualization of spatial data including discussion of some data preparation issues and the special problems associated with the visualization of spatial data. Section 6.3 discusses visualization techniques that have been developed for spatial data and these are classified by the types of spatial pattern the analyst might be interested in looking for as part of ESDA (see chapter 5). The chapter concludes with a short illustrative example.

6.1 Data visualization and exploratory data analysis

Visualization has a natural role to play in exploratory data analysis. Visualization supports the process of detecting data properties, which is a central objective of EDA and ESDA. Many forms of graphic display are provided as single, end-product views of the data. They are intended to convey publicly to the many viewers (who are not familiar with the data) features of the data that have been identified. Typically such *presentation graphics* provide static views of the data and represent a permanent, selective, record of what is known. These graphics are not designed to support data analysis.

Data visualization or *scientific visualization* however is concerned with the provision of many graphical views of a data set as part of an on-going process of understanding and gaining insight into the data – that is identifying properties of the data (Earnshaw and Wiseman, 1992). The user of data visualization tools is probably familar with the data – perhaps because he or she collected them. In coming to understand the data the user wants to be able to 'interact' easily with the data in the sense of generating multiple, dynamic but usually temporary views of the data many of which may be used once and then discarded. Visualization of data often means retaining each individual data-point in a graph whilst also applying some (resistant) smoother in order to help detect patterns in what may be a complex array of data. Data visualization comprises a range of approaches and individual tools to support data analysis rather than with the provision of graphics for a final report (Unwin, 1997; Wise et al., 1999).

Data visualization in the statistics and computer science literature is principally concerned with graphic tools. Spatial data visualization also needs to employ cartographic tools – different forms of map display – referred to as *cartographic visualization*. The concerns of cartographic visualization include but extend beyond those of ESDA (see for example Monmonier and MacEachren, 1992; Hearnshaw and Unwin, 1994; Orford et al., 1998). Spatial data visualization also extends to multi-media and visual reality (virtual landscape) representations but these are not discussed here (see, Câmara and Raper, 1999). In the immediately following section the focus is on issues underlying all aspects of data visualization for ESDA and draws on the work of Cleveland (1993, 1994) and Buja et al. (1996) in particular. There is considerable diversity of effort in this area and it is useful to summarize the work that has tried to bring order to this diversity.

6.1.1   Data visualization: approaches and tasks

The taxonomy proposed by Buja et al. (1996) for classifying work in data visualization divides the *approaches* to data visualization (in contrast to the individual tools themselves) into two areas: *rendering* and *manipulation*. Rendering refers to the decision as to what to show in a plot and in particular in deciding what type of plot to construct. In the case of univariate data this includes for example techniques for displaying distributions (histograms, boxplots, Q–Q and rankit plots) and time series (plots). In the case of multivariate data Buja et al. (1996, p. 79) identify scatterplots (where cases are depicted as points), traces (where cases are depicted as functions of a parameter as in the case of parallel co-ordinate plots) and glyphs (where cases are depicted as complex symbols). Scatterplots are the basic visual tool for bivariate data. Glyphs are usually positioned in a layout in ways which assists their interpretation.

Manipulation refers to how to operate on individual plots and how to organize multiple plots in order to explore data. Plot manipulations can be organized in terms of what data exploration tasks they are meant to support. Buja et al. (1996, p. 80) identify three tasks associated with data exploration: finding gestalt, posing queries and making comparisons. 'Finding gestalt' is the task of identifying patterns, shapes and other properties in the data set. The querying of individual cases or subsets of the data set is undertaken in order to better understand and explore in more detail the gestalt features that have been identified. The third task is to make comparisons either between variables or projections of the set of variables or between subsets of the data. Particular manipulations are suited to particular tasks. Manipulations that involve deciding which variables to include, what projections to adopt, what type of magnification and detail to adopt are suited to finding gestalt. A parallel can be drawn here with the process of using and in particular focusing a camera. Manipulations that involve linking multiple views of the data and highlighting data subsets to see where the different subsets lie in each of the different views (brushing) are suitable for posing queries, whilst decisions about how to arrange plots will have important consequences for the task of making comparisons.

The effectiveness of any statistical graph with respect to a task will also depend on the quality of its design. Cleveland (1994) constructs a model that distinguishes between table look-up and pattern perception which he argues are the principal activities an observer engages in when reading a graph. These activities overlap with the classification of tasks by Buja et al. (1996). Table look-up broadly corresponds to the task of making queries on individual cases. The reader identifies the original data values that have been encoded in the graph. The task is performed slowly, taking data cases one at a time. Pattern perception is about the detection and assembly of geometric objects in order to see pattern in the encoded data. Cleveland describes this as involving 'exceedingly fast processes that appear to operate in parallel to produce objects, or gestalts' (Cleveland, 1994, p. 224) Now, in order to evaluate the effectiveness of any statistical graph we need to identify the tasks the user engages in when interpreting graphs in either of these two modes. Table look-up requires the user to undertake one or more of: scanning (identifying values by referring to the axis or legend); interpolation (such as between tick marks on an axis); matching (finding a symbol in the key). Pattern perception requires the user to undertake one or more of: detection (of the meaning of the symbols used to encode values); assembly (visual grouping of similar symbols); estimation (of relationships between assembled groups). Good graphic design will assist the user to undertake these tasks. Wise et al. (1999) comment that many of Tufte's (1983, 1990) principles of good design can be thought of as assisting the reader in undertaking one or more of Cleveland's tasks. Cleveland (1993) lists principles of graph construction that ensure clarity of vision – datapoints should stand out from other graph features; avoid clutter; avoid overlapping datapoints; avoid putting notes or other features on the graph that cannot be switched off when not needed. Graphs must be informative – not necessarily quick and easy to read.

Figure 6.1 summarizes the main features of Buja et al.'s typology together with Cleveland's model. The purpose is to sketch the interface between, on the

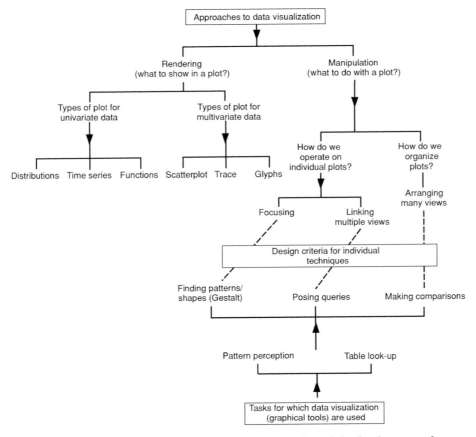

Figure 6.1 The interface between visualization tasks and the development of visualization techniques

one hand, the tasks the reader of any statistical graph wants to be able to engage in for the purpose of exploring data and, on the other, the approaches to the development of data visualization techniques. At this interface lies the question of how to design the techniques so they can undertake effectively the tasks they will be used for. Cleveland (1993) argues that the human eye and brain are good at detecting patterns and relationships from visual evidence when that visual evidence is presented and the visual displays engineered in ways that are grounded in perceptual and cognitive psychology. Bertin (1983) identifies seven variables that are available for processing information using graphs. Of these seven, shape and colour are considered to be key to the development of good visualization tools for the purposes of categorical data analysis, size, value and colour for the purposes of quantitative data analysis – that is for helping the viewer to notice patterns in data. To these variables can be added motion – made possible by advances in computing (Buja et al., 1996).

6.1.2   Data visualization: developments through computers

Technological developments in computing have made it possible to extend data visualization significantly from what was possible through drawing by hand. Graphs and maps can be produced more quickly and more consistently by computer. High-resolution graphics workstations have made it possible to develop better visualization methods for multivariate data (three dimensions and above). Traditional methods involving, for example, glyphs which can often be difficult to read and interpret have been challenged by methods made practical by the modern computer. Ideally the user issues their request for a graphics tool through some manual manipulation of an input device and there is a rapid response on a computer graphics screen. Further requests (moving a slider bar, pointing and clicking etc.) result in modifications to the original graph, new views of the same data set, and linkage to other graphs. Four types of graphical development made simpler by computer-based graphics are discussed below – those that involve interaction and multiple linkage between a displayed set of graphs, those that involve manipulation of the graphs themselves and the views they provide.

Where the analyst has two or more views of the data, the operation of highlighting cases or subsets of cases in one graph and seeing where these same cases lie in other graphs is called *brushing*. The query and the answer to the query have the advantage of both being graphical although Buja et al. (1996) note that complicated database queries involving certain clauses and many variables may be difficult to implement by brushing. In a matrix of scatterplots, brushing can be used to condition the range of values on one variable in one scatterplot in order to examine relationships amongst other variables in the other scatter-plots (see Becker et al., 1987 for examples). *Dynamic brushing* involves moving the query, for example a user defined polygon on a scatterplot, and noting the changing pattern of highlighted cases in the other graphs. The response may also include a record or display of statistics (e.g. the mean or median of some variable or the correlation between two variables) as the user-defined polygon is moved around the space of the graph.

The second type of development involves changes to the graphic display it-self. Labels can be switched on or off, cases selected (with the rest temporar-ily deleted), grid lines switched on or off, symbolism altered, overlapping cases made distinct by 'jittering' and so on to suit the needs of any particular task. The symbolic and other details of a graph are capable of *interactive modification* as part of the process of exploring the data. More profound modifications may be implemented by dynamically varying the parameters of a graph, such as chang-ing the bin widths for a histogram by moving a slider bar. Interactive modifi-cation is dynamic when changes to a graph occur rapidly in response to user instructions. Such *dynamic interactivity* is a significant property of a system be-cause of the importance of system response time to the effectiveness of scientific visualization (see Huber, quoted in Becker et al., 1987).

Modification of graphics tools can be taken further to produce animation, particularly useful in viewing multivariate data sets. 'In scatterplots . . . the choice of projection, aspect ratio, pan and zoom all have underlying continu-ous parameters that can be changed in small steps and at rapid speed resulting in animations' (Buja et al., 1996, p. 81). A plot such as a three-dimensional scat-terplot can be rotated so that the analyst can view the cube from different per-spectives (Buja et al., 1996). The grand tour is a smooth, continuous rotation of a high-dimensional, multivariate, data set (Asimov, 1985). Computers have greatly facilitated the implementation of *dynamic projection mechanisms* in which small smooth changes in the parameters of the graph performed at high speed result in user-controlled animation.

6.1.3   Data visualization: selected techniques

Table 6.1 provides a selective listing of graphical methods for visu-alizing data. Most are described in amongst other sources Tukey (1977), Cleveland (1993, 1994), Cook et al. (1995). Projection pursuit (Jones and Sibson, 1987) is concerned with providing static projections of a multivariate data set that is optimal in some sense – for example, detecting clusters in high-dimensional space (Brunsdon et al., 1998). Many of the techniques can be enhanced by the sorts of interactive and dynamic modifications described in section 6.1.2.

Table 6.1 *Graphical methods for visualizing data*

	Univariate distributions	Bivariate distributions	Multivariate distributions
Categorical	Barcharts	Trellis plots; mosaic plots; glyphs (e.g. rays and trees)	
Quantitative	Boxplots; histograms; dotplots; quantile plots; rankit plots; residual plots; level and spread plots; time series plots	Scatterplots; Q-Q plots; level and spread plots; mean difference plots	Matrix and trellis scatterplots; parallel co-ordinate plots; projection pursuit; 3-D scatterplots; coplots; glyphs (e.g. Chernoff faces); grand tour

Most of the techniques are relevant to data measured at the interval or ratio level, because such data can be represented by length or scale or position which lend themselves to graphical portrayal. Categorical data are less easily depicted by graphical methods, that is they are less easy to encode in ways that permit relatively easy visual decoding. Two interesting examples are the trellis plot developed by Cleveland (1993) and the mosaic plot developed by Friendly (1995) which generalizes the bar chart (see also the parquet diagram of Riedwyl and Schuepbach, 1994). The trellis plot displays a sequence of boxplots (or other graphical plot) in the form of rows, columns and pages – like garden trelliswork. Each panel shows the relationship of certain variables conditional on the values of other variables (Becker et al., 1996). The mosaic plot displays combinations of different categories using rectangles where the number of cases determines the size of any rectangle so that it is relatively easy to spot which combinations of cases dominate any data set.

6.2   Visualizing spatial data

This section deals specifically with the visualization of spatial data but starts by reviewing some data preparation issues that are relevant to visualizing regional data. There is further relevant discussion in chapter 7 dealing with data smoothing. This is followed by a discussion of some of the special issues and problems that arise in visualizing spatial data for ESDA.

6.2.1   Data preparation issues for aggregated data: variable values

Data for areas often come in the form of counts. Where areas differ in size, for example in terms of the population at risk in the case of a disease, then at the exploratory stage such data need adjusting so that population size effects do not distort comparisons. In converting count data to rates, the choice of denominator may sometimes be clear-cut but not always (see section 2.3.1). In comparing rates of car theft during a period of time the denominator, ideally, should be the population of cars (or car hours) at risk during that time in each area but these data are not available. Instead the area of the spatial unit might be used but this would tend to inflate city centre rates where road density is higher and there are likely to be more car theft opportunities. A better denominator might be some measure of road length plus off-street and public parking places, assembled using street maps and census data in a geographic information system.

When individual-level disease data are available for a region that identify the age and sex of each individual then age–sex specific rates can be computed by dividing the total number of cases in each age–sex cohort by the total population at risk in each age–sex cohort. Now for each area (i), the age–sex specific rates derived from the regional data can be multiplied by the area i population in each age–sex specific cohort. These components can be summed (across all the age–sex cohorts into which the population has been classified) to yield an expected number of cases for the area ($E(i)$). This expectation is based on the assumption that the region-wide rates apply uniformly throughout the region and expresses numerically the area-by-area consequences of cases occurring at random within each age–sex cohort, irrespective of area.

The (indirect) standardized ratio is the observed ($O(i)$) count divided by the expected count. It measures the extent to which each area has a number of cases more than would be predicted just on the basis of its age–sex structure ($[O(i)/E(i)] > 1.0$) or less than expected ($[O(i)/E(i)] < 1.0$) under the hypothesis of a random distribution of cases over the region. Standardized ratios for mortality and incidence are routinely constructed to display disease data by areas controlling for variation in the age and sex distributions across the areas. Similar forms of standardization could be adopted in deriving area victimization rates (the risk of residents of any particular area being a victim of crime) if the age, sex and postcode of victim is recorded. Such victimization can be differentiated into those offences that took place in the area where the victim lives and those offences that took place elsewhere.

In the case of ward-level data, the region that provides the reference for computing expected counts ($E(i)$) could be the city or region within which the wards lie. In this case the standardized ratios identify the areas that have counts that are greater than or less than expected with respect to the region within which they lie. Since the ratios are based on expectations computed using the same data, some ratios will be above 1.0 and and some below. However, the reference region does not need to be defined in this way. The reference region could be the nation (either including or excluding the study area) in which case the high and low standardized ratios are with reference to the national average. In this case it is possible that all the ratios are above or all below 1.0 if the study area has higher or lower levels of the disease than is found elsewhere in the country. How standardized ratios are interpreted depends on the choice of reference region used to compute the expected counts.

Where relevant risk factors can be identified then other forms of adjustment may be considered. Tukey (1979) lists variables such as ethnic composition, economic sector and industrial composition amongst possible adjustors. The patterns revealed by one type of adjustment may suggest other relevant adjusters. In the case of recorded burglary data, if the burglaries are differentiated say by type of dwelling, then standardization by type of house becomes possible. There may be circumstances where it would be interesting to compute ratios controlling for an area-level characteristic such as deprivation. Craglia et al. (2000) in an analysis of burglary rates in Sheffield assign enumeration districts to Townsend deprivation quartiles. So separate burglary rates were computed for each quartile using only the data taken from the enumeration districts classified as falling into that quartile. The expected count for any enumeration district was obtained by multiplying the number of households in the enumeration district with the deprivation specific rate that corresponded to the enumeration district's deprivation group. This can be used to highlight areas that have high levels of burglary given their levels of deprivation. Results must be interpreted with caution because the approach assumes that the enumeration districts are homogeneous in terms of deprivation (Morphet, 1993).

Converting counts to rates to take account of population size differences is not, however, sufficient to ensure comparability of data values for the purpose of exploring the data (Smans and Esteve, 1992). There are two reasons. The first reason is that rates based on small populations are less robust to data errors than rates computed from large populations. The addition or subtraction of a case of a disease or a burglary (as a result of reporting error for example) will have a bigger effect on the computed rate when the population denominator is small than when it is large.

The second reason is that rates computed for areas with small denominators have larger variances than rates computed for areas with large denominators and there is a mean–variance dependence in rates. Take the case where the denominator is population size ($n(i)$). Rates are observed counts in area i ($O(i)$) divided by population size and it follows from the binomial model for $O(i)$

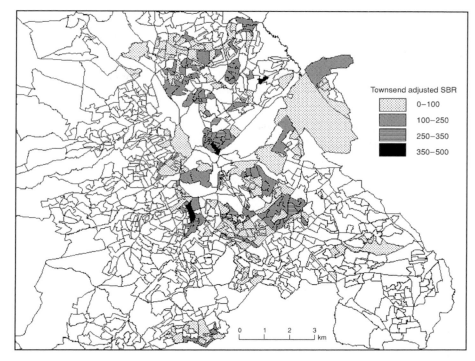

Figure 6.2 Townsend adjusted standardized offence ratio for burglary (SBR) by
enumeration district: Sheffield (Craglia et al., 2000)

that the mean, **E**[ ], and the variance, **Var**[ ], are:

$$\mathbf{E}[O(i)/n(i)] = (1/n(i))\mathbf{E}[O(i)] = p(i); \tag{6.1a}$$

$$\mathbf{Var}[O(i)/n(i)] = ((1/n(i))^2\,\mathbf{Var}[O(i)] = p(i)(1 - p(i))/n(i) \tag{6.1b}$$

where $p(i)$ is the probability that any individual in area i has the characteristic
that is being counted. So, as $n(i)$ increases, $\mathrm{Var}[O(i)/n(i)]$ decreases.

The mean and the variance in (6.1) are not independent. For $p(i)$ close to
0 or 1 and given $n(i)$ the variance is proportional to the mean. This suggests
the need for a variance-stabilizing transformation (particularly for small $n(i)$)
in order to remove this mean–variance dependence. Cressie and Read (1989)
suggest that for spatially dependent data on $O(i)$, the Freeman–Tukey square-
root transformation:

$$(O(i))/n(i))^{1/2} + ((O(i) + 1)/n(i))^{1/2} \tag{6.2}$$

is preferable to the usual square root transformation:

$$[(O(i) + 1)/n(i)]^{1/2}$$

It also follows from (6.1) that the standard error of the estimate of the rate $p(i)$ is:

$$[p(i)(1 - p(i))/n(i)]^{1/2}$$

which is inversely related to the population denominator. It follows that any real spatial variation in rates may be confounded by variations in $n(i)$ (the sample size) or alternatively spatial variation in rates could be an artefact of spatial structure in $n(i)$ (see Gelman and Price, 1999 who give examples from disease mapping in the USA). However the Freeman–Tukey transformation does not equalize variances (important in regression) for which further weighting would be needed (see Cressie, 1991, p. 398; section 6.3.1).

Standardized ratios, $SR(i)$, equal to $O(i)/E(i)$ where $E(i)$ is an expected count in area i and is assumed to be fixed are used for many types of mapping, particularly for ranking areas in relation to $E(i)$. Alternative ratios have been proposed, some of which have better properties in certain situations. For example, chi-square measures based on squared differences have been suggested (Visvalingam, 1983):

$$[O(i) - E(i)]^2/E(i) \qquad (6.3)$$

and:

$$[|O(i) - E(i)| - 0.5]^2/E(i) \qquad (6.4)$$

Other Poisson-based statistics used in epidemiology include (Wray et al., 1999):

$$(O(i) - E(i))^2 - O(i) \quad \text{(for } O(i) \geq E(i)) \qquad (6.5)$$

$$[O(i)(O(i) - 1)/E(i)^2] \qquad (6.6)$$

$$[O(i)(O(i) - 1)/E(i)] - E(i) \qquad (6.7)$$

These statistics reduce the risk of extreme values in the distribution of values that are an artefact of small population size. Areas with small populations (and hence small $O(i)$ and $E(i)$) have to be very unusual before they attain extreme values, but differences in areas with larger populations are exaggerated relative to standardized rates. Table 6.2 shows the values of these statistics for two

Table 6.2 *Statistics for comparing observed and expected counts*

	$O(i)$	$E(i)$	$O(i)/E(i)$	(6.3)	(6.4)	(6.5)	(6.6)	(6.7)
Region 1	20	15	1.33	1.66	1.35	5.0	1.688	10.33
Region 2	200	150	1.33	16.66	16.33	2300	1.768	115.33

regions that differ in size. The UK's (former) Department of the Environment, Transport and the Regions' (DETR's) index of local need at the enumeration district and ward levels was computed using a signed chi-square statistic.

Another approach is to attach confidence intervals to each standardized ratio. Take the case of disease mapping where $O(i)/E(i)$ is the maximum likelihood estimator of the unknown true area-specific relative risk ($r(i)$) of the disease, under the assumption of an independent Poisson model for the observed counts with parameter ($E(i)r(i)$) (see section 9.1.4). It follows from the properties of the Poisson distribution that:

$$\mathbf{Var}(O(i)/E(i)) = (1/E(i))^2\, \mathbf{Var}(O(i)) = (1/E(i))^2\, (E(i)r(i)) = O(i)/E(i)^2$$

So, the standard error of the standardized ratio is $O(i)^{1/2}/E(i)$. Using a normal approximation for the sampling distribution of SR(i), approximate 95% confidence intervals can be computed:

$$\text{SR}(i) \pm 1.96\, [O(i)^{1/2}/E(i)]$$

An exact method for computing the confidence interval of a standardized ratio is given in Ulm (1990) together with other methods. However there are problems here too. The standard error tends to be large for areas with small populations and small for areas with large populations because of the differing effect of population size on $O(i)^{1/2}$ as compared with $E(i)$. So extreme rates tend to be associated with small populations, but rates that are significantly different from 1.0 tend to be associated with areas with large populations (Mollie, 1996). If variation in population size by area is not a problem, each case can be classified in terms of the relationship between the span of the confidence interval and where 1.0 lies in relation to that interval.

6.2.2   Data preparation issues for aggregated data: the spatial framework

Spatial data analysis often starts from data recorded for small spatial building blocks – for example enumeration districts in the case of the UK Census and in future user-defined 'output areas'. In some cases the spatial unit is well defined and meaningful – for example constituency boundaries when analysing electoral outcomes. Often however the spatial framework is an artificial filtering of the event data. Underlying social and economic processes relate more closely to 'neighbourhoods' or 'communities' rather than Census units.

The view the analyst has of any regional data set is dependent on the areal partition. This refers to both the scale of the partition and its particular configuration with respect to the underlying reality it represents. Do boundaries cut across homogeneous areas or bundle together within spatial units areas that are very different in terms of variables that are important in the study? Further aggregation of such small spatial units will, of course, exacerbate the situation but may be necessary. There are several reasons for aggregating spatial units (Wise et al., 1997). The analyst may want areas with large enough populations so that rates are robust and permit meaningful comparisons to be made between areas. There may be a need to reduce the effects of suspected locational or attribute inaccuracies or to make some forms of spatial data analysis feasible by reducing the number of areas. 'Generalizing' a map, by removing noisy detail including merging units that have similar attribute values facilitates map visualization (Wang et al., 1997; Jaakkola, 1998). If aggregation is required then at least at the exploratory stage of analysis it might be desirable to construct other similar or equally valid aggregations according to specified criteria in order to assess the sensitivity of findings to the chosen framework.

Regionalization is a special form of classification where basic spatial units are grouped together on the basis of a set of defined criteria. Where regionalization differs from classification is that the grouping of the units introduces spatial contiguity or adjacency constraints. One of the problems with undertaking classification, and hence regionalization of basic spatial units, is that even for modest numbers of spatial units the number of possible partitions soon becomes enormous. It is theoretically possible to enumerate all the partitions to find, according to some objective function, the best. However where the number of spatial units is over 1000 no answer could be obtained in a reasonable length of time (Cliff et al., 1975).

(a)   Non-spatial approaches to region building

Ordinary (non-spatial) classification routines, or grouping strategies, can be used to construct regionalizations. Observations are typically grouped in terms of their similarity on one or more attributes in order to minimize within-group variance (intra-group homogeneity) and to maximize between group variance (inter-group heterogeneity). Classification routines adopt either an hierarchical or heuristic approach (Everett, 1974). Hierarchical classifiers either proceed by merging the n observations one at a time according to some similarity measure or proceed by splitting groups starting from the situation in which all the n observations are assigned to a single large group. Each move (to split or merge) is the best of the set of possible moves but this does not guarantee that the final classification is optimal. Non-hierarchical or heuristic methods attempt to find the 'optimal' partition and of these the K-means method (MacQueen, 1967) is the best known. The method starts with an initial partition of the n observations into K groups and then swaps the

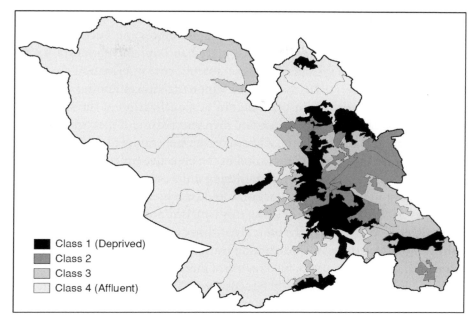

Figure 6.3 Townsend deprivation regions: Sheffield 1991 Census data (Haining et al., 1994, p. 433)

individual observations experimentally, employing an objective function to determine whether one grouping is better than another.

These non-spatial classifiers will induce a regionalization 'by default'. But when the classified units are mapped, if there are K groups or classes there will be many more than K regions with some 'regions' perhaps consisting of only one or two of the initial basic spatial units. This means the user cannot control the number of regions being generated. Haining et al. (1994) used a non-spatial classification routine based on information theory to construct Townsend deprivation regions for Sheffield. The default regionalization provided by the method was then modified by carrying out additional mergers based on the similarity of adjacent regions in terms of quantitative evidence as well as local knowledge. The approach combined quantitative data and qualitative knowledge about the areas. It was described as similar in spirit to the methods of factorial ecology and social-area analysis used to define 'natural' socio-economic areas in cities. It was similar to the methods used by some health authorities to construct community neighbourhoods for the reporting of health statistics (see Greater Glasgow Health Board, 1981; Haining, 1990).

(b)   Spatial approaches to region building

Another class of regionalization algorithms is based on the same one used for grouping except that now the splitting or merging of groups or the swapping of units between groups is not allowed to violate spatial contiguity constraints associated with the spatial units (Webster and Burrough, 1972; Perruchet, 1983). These algorithms are suited to implementation in a geographical information system since these systems store information on spatial adjacency. Openshaw (1978) developed a regionalization or zoning algorithm (AZP) that started with an initial 'random' regionalization of the study area with the required number of regions and then selected one of the regions at random. Spatial units bordering the selected region were sampled and moved into the region if there was no deterioration in the objective function. The process of sampling by region and then by bordering units continued until no further improvement in the objective function could be achieved. Openshaw and Rao (1995) review the experience of using this heuristic approach to region building and describe variants to the basic algorithm designed to speed up the algorithm and reduce the risk of premature convergence on local optima (see also Openshaw and Wymer, 1995, for a review of methods).

Presumably any method could be sensitive to the choice of initial regionalization so this should be assessed. Rather than selecting a 'random' regionalization, the user might want to start from an initial regionalization that recognized the presence of any homogeneous areas on the surface. Wise et al. (1997) suggest possibilities and there may be a role for local statistics in specifying starting points (see section 7.3). Womble (1951) proposed a method of regionalizing for data from a continuous surface based on a function (the systemic function) that measures the weighted average change with distance of a set of variables. The regionalization proceeds therefore by identifying the areas of the map where sharp gradients are to be found. Identifying such gradients may not create regions (there is no guarantee that such changes of slope form a partition) but could provide a starting point or constraint within a regionalization algorithm. Thus rather than looking for homogeneous 'flat spots' the focus is on identifying gradients that might form the borders of regions. Oden et al. (1993) have extended Womble's approach to categorical data by computing the proportion of category mismatches between adjacent areas. They show how their 'categorical wombling statistic' can be used for regionalization. Jacquez et al. (2000, pp. 225–33) review Wombling and other boundary detection methods.

Methods like the one developed by Openshaw and colleagues only allow units to be in the same region if they are spatially contiguous. This is also true of another approach to regionalization described by Berry (1966) and Spence (1968) in which the geographical co-ordinates of the spatial units are included as an additional pair of variates in the objective function. Oliver and Webster (1989) suggest that in the case of soil mapping the identification of unique regions in this sense does not reflect the nature of soil variation. Their method is a non-hierarchical version of an approach described in Webster (1977). The attribute dissimilarity $(d(i,j))$ of two spatial units (i and j) is measured by the weighted difference of their attribute values. The weighting can be adjusted to reflect the importance to be attached to each of the attributes used in the dissimilarity measure. At the next step this dissimilarity score is multiplied by a function (f) of their geographical separation to produce a new dissimilarity index:

$$d^*(i, j) = d(i, j)f(|\mathbf{s}(i) - \mathbf{s}(j)|) \tag{6.8}$$

where $|\mathbf{s}(i) - \mathbf{s}(j)| = u(i, j)$ denotes the distance separation between the two units or sites whose locations are given by $\mathbf{s}(i)$ and $\mathbf{s}(j)$. Oliver and Webster suggest that the function f could be defined by the geostatistical model that best describes the spatial autocorrelation in the data. For example in the case where that model is the exponential:

$$d^*(i, j) = d(i, j)\{[c/(c_0 + c)][1.5(u(i, j)/a) - 0.5(u(i, j)/a)^3]\}$$
$$+ d(i, j)\{c_0/(c_0 + c)\} \quad 0 < u(i, j) \le a$$
$$d^*(i, j) = d(i, j) \quad u(i, j) > a$$

where c and c_0 are the sill and nugget variance respectively of the geostatistical function and a is the range of the function (see section 2.4). The modified dissimilarity index $d^*(i, j)$ is deflated relative to non-spatial dissimilarity index $d(i, j)$ up to the range of the function. Note that if the nugget variance is large relative to the sill variance then $d^*(i, j)$ is close to $d(i, j)$. The modification is greatest when there is strong spatial autocorrelation in the data. Beyond the range of the function $d^*(i, j) = d(i, j)$. As the range increases the map smoothing increases and the number of unique regions decreases (see figure 6.4). For details on the choice of geostatistical model see Oliver and Webster (1989).

(c)   Design criteria for region building

Regionalization becomes more problematic if there are several design criteria. For example it might be required that the regionalization creates homogeneous regions in terms of a set of attribute values (for example deprivation), and where the regions are approximately of equal size in terms of population and are relatively compact. Wise et al. (1997) suggest that these three criteria might be appropriate in the context of designing regions for analysing health service provision – the first two criteria being appropriate for data analysis, the third in the context of actual service delivery. If only data analysis is envisaged the compactness criterion is unlikely to be of much importance.

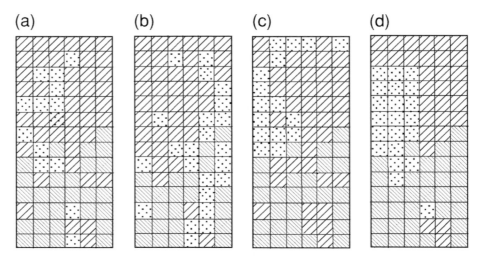

Figure 6.4 A method of regionalization based on a distance-weighted measure of
dissimilarity and showing the effects of increasing the range of the weighting
function (Oliver and Webster, 1989)

Horn (1995) develops a regionalization algorithm similar to that of Rossiter
and Johnston (1981) for electoral districting. The aim, to avoid the charge of
gerrymandering, is to construct compact and connected districts subject to a
constraint that limits the deviation of district populations from the average.

The regionalization approach developed by Wise et al. (1997) is based on
forming a single objective function that incorporates the three design functions (see also Cliff et al., 1975 and Martin, 1998). Their objective function is
a weighted sum of three functions. First, a homogeneity function which minimizes the within-group variance of one or more attributes. Second, an equality function which minimizes the difference between the value of an attribute
such as population size across the regions. Third, a spatial compactness function that minimizes the sum of squares of within-group variance for the co-
ordinates of area centroids. The user can vary the importance of each through
the specification of weights. A criticism of this approach is that it combines
properties measured in different units to a common system through standardization and the user must specify weights without necessarily understanding
the implications for the regionalization of doing so. The method is a K-means-
based classification which only allows swaps at the boundary of any region.
Swaps are allowed, even if one or two of the individual criteria become worse,
providing the overall objective function improves and providing the functions
that do become worse do not deteriorate by more than a user-defined threshold. This aspect of the algorithm is introduced to try to reduce the risk of premature convergence on a poor-quality local optima. Trial runs have shown that

the algorithm can create regions that are aberrant in some sense. One or two individual regions may have very large populations so that some additional splitting may be needed after the algorithm has produced its best solution (see Wise et al., 1997; Haining et al., 1998).

An alternative approach to handling multiple design criteria is to specify one of the criteria as the objective function to be optimized (for example homogeneity) subject to equality or inequality constraints specified for the other criteria. Alvanides et al. (2000) describe a zone design algorithm (ZDES) based on optimizing an objective function with constraints. A penalty function is added to the objective function to represent the 'cost' of violating the constraints. Weights have to be specified for the penalty function and different approaches have been suggested (see Alvanides et al., 2000, for a discussion). It is not clear which if either of these approaches yields 'better' solutions and some comparative trials of these and other methods are called for. The inclusion of a zone design criterion in the objective function or in the set of constraints depends at least in part on the nature of the problem. If a zone design criterion is a key element of a proposed regionalization (such as equality of population size) then it would seem better placed in the objective function rather than simply included as a constraint.

Regionalization methods for exploratory spatial data analysis should be relatively quick to implement, particularly if the analyst wishes to view the data through many different partitions at the same scale or through a range of different scales. Macmillan (2001) has implemented a regionalization algorithm (SARA) that uses a fast switching point method to implement contiguity checks and achieves considerable gains in processing time for large data sets over the usual connectivity checking procedures.

Speed of implementation was one of the criteria behind the regionalization procedure of Wise et al. (1997). New regionalizations have to be constructed by re-running the programme. This eliminates interactivity and the analyst would prefer to be able to dynamically vary the spatial framework to see the effects of scale and partition in the same way that histogram intervals can be varied to see how interval size affects the shape of the histogram. Whilst not achieving this degree of interactivity, Cislaghi et al. (1995) have developed a type of regionalization procedure they call the method of iterative random partitions (IRP) for the exploratory spatial analysis of disease data at different spatial scales. The method forms K aggregations of the basic spatial units starting from K randomly chosen seed points. Each basic spatial unit is assigned to its closest seed point by a measure of distance and assigned a standardized score computed from all the basic spatial units lying in the same region. The process is replicated j times using different randomly selected seed points. This creates a distribution of values for each basic spatial unit and the mean or median used to represent the basic spatial unit at the scale specified by the choice of K. Different values of K can be chosen to examine patterns at different scales. This is not a regionalization procedure in the sense of earlier methods but rather a method of developing a spatial filter which uses all possible groupings of the basic spatial units. Choosing random seeds given K does not ensure regions of similar size. There will be some large and some small regions so if the intention is to create randomly generated regions of comparable size to control for scale, stratification should probably be added to the criteria for selecting seed points.

As a final point relating to the spatial framework note that several of the specialist graphics tools for ESDA depend on the specification of the weights matrix ($\mathbf{W}$). The ability to explore, through graphical or numerical tools, regional data making different assumptions about the underlying graph has been noted as a potentially important aspect of ESDA (Haining et al., 1998).

6.2.3   Special issues in the visualization of spatial data

The graphical tools discussed in section 6.1 still have an important role to play in the visualization of spatial data for ESDA, but in addition new graphical tools are needed to explore for *spatial* patterns. Some of these will be described in section 6.3. In addition maps become an essential element in visualization. One of the consequences of this has been summarized by MacEachren and Monmonier (1992, p. 197): 'For cartography, the single biggest challenge is to redirect our attention from an emphasis on maps that present answers to maps that foster a search for questions.' This is the same type of distinction that was drawn between presentation graphics and scientific visualization. The questions that arise are: how can maps be made to support visualization objectives and what sorts of tools should be provided with them? (MacEachren, 1995; Dykes, 1997). MacEachren and Monmonier (1992) draw attention to the importance of multiple map views, trying out a 'variety of map measurements, categories, symbolization schemes, scales, scopes, generalizations, azimuths, elevations, times or time periods and juxtapositions' (p. 198). Interaction with graphs and graph parameters are as important to ESDA as to EDA. In addition ESDA requires the ability to be able to quickly draw appropriate maps, the parameters of which can be interactively modified to provide different views, and there needs to be dynamic links between cartographic and statistical displays (Haslett et al., 1990, 1991; Dorling, 1992; Dykes, 1997, 1998; MacDougall, 1992; MacEachren and Kraak, 1997).

Linking maps and statistical graphics tools through the operation of brushing has been incorporated into ESDA software (see Anselin, 1998 and Wise et al., 1999 for selective reviews). Monmonier (1989) referred to this as *geographical brushing*. Graph-to-map linkage enables the user to query where particular cases on a graph (e.g. the outliers on a boxplot) are located on the map of the study area, map-to-graph linkage enables the user to query if a selected area (e.g. an inner-city area) is distinctive in terms of attributes. Dynamic graph-to-map linkage allows the user to move a polygon over a graph and see the cases highlighted in the map window. Dynamic map-to-graph linkage allows the user to move a specified polygon or line trace over the map surface and highlight cases on the graph, or report statistics computed just for those cases that fall within the boundary of the moving window (Craig et al., 1989). The map is more than just a repository of information on where cases are located. Majure et al. (1996) compute a spatial cumulative distribution function (SCDF) computed from a forest defoliation index. They overlay population density data in the map window in order to explore links between differences in defoliation picked up by the SCDF computed on different subareas of the map and a possible source of tree stress. In this way the map holds both locational information and also data of relevance to the analysis that is underway in other graph windows. High city crime rate areas identified using a boxplot can be highlighted on a map of city census tracts that contain socio-economic data. Areas near to one another having similar levels of deprivation but markedly different crime rates would be of interest to detect and investigate more closely.

There are circumstances where map-to-map brushing may be helpful. A population cartogram provides a map where areas are proportional to the number of people in each area and may provide a better spatial framework through which to view, for example, the spatial aspects of social structure (Dorling, 1994, 1995). To achieve this, the cartogram transforms the physical area the user may be familiar with. Contiguous and non-contiguous forms of cartogram have been developed (Schulman et al., 1988; Olson, 1976). However the viewer may lose their orientation and sense of where they are on a cartogram so that in Cleveland's terminology the cartogram, on its own, fails the table look-up test. However, linkage to a 'familiar' map (where the linkage is switched on and off as requested) will restore that orientation.

Section 6.1.2 noted the importance of dynamic interactive manipulation of graphics parameters to scientific visualization. The display responds immediately to user-specified changes that can be implemented through a slider bar. Dykes (1997) and Andrienko and Andrienko (1999) illustrate the role of dynamic interactive manipulation of the display parameters of maps to support the exploration of spatial data. Flexibility can be achieved for example in terms of choice of symbols, class intervals for choropleth maps and use of colour. One of the benefits is being able to make 'dynamic visual comparisons' from which spatial patterns may be more easily detected. Andrienko and Andrienko (1999) illustrate the use of dynamic shifts in the intervals for different colour shadings on a choropleth map to reveal patterns in the distribution of mean household size in Bonn. They extend this to the analysis of relationships between two variables.

There are additonal parameters to consider when manipulating graphic views of spatial data. These include: variation in the direction of any plot to explore for anisotropy (e.g., deciding to compare structures in the north–south direction with those in the east–west); variation in lag distances and distance bands as well as the tolerances associated with these parameters. Majure and Cressie (1997, p. 157) comment in the context of geostatistical ESDA: 'It is the dynamic control of the angle and distance classes that will provide a truly useful exploratory data analysis environment.'

Cleveland's model (see section 6.1.1) also categorizes many of the perception tasks involved in reading a map. His model seems to provide a framework for evaluating the effectiveness of particular map designs for the purpose of visualization, although Olson (1987, p. 89) has argued that 'maps are more complicated forms of graphics'. Indeed, maps do seem to raise some special problems for pattern perception. Consider the following. Most statistical graphs re-sort data so that cases with similar values are brought together on the graph. The graph prioritizes the similarity of variable value in the way it presents data to the eye. If the data values have to be sorted by some other (categorical) variable then it may still be possible to lay out the various plots in ways that facilitate comparison (as in a trellis plot, for example). Maps however prioritize the spatial arrangement property of data values. Pattern perception on maps seems to involve a more complex process of assembly that has to deal simultaneously with 'value similarity' and 'fixed layout' in a visual form (the map) that has prioritized the second of these two. This of course is one of the reasons why maps and statistical graphs complement one another in ESDA. It is also one of the reasons why statistical graphs have been developed such as spatial scatterplots, 'spatial' trellis plots and cloud plots for example (Cressie, 1984; Haining, 1990, pp. 197–228). These statistical graphs embed information about spatial relationships into the graph. In this way they bridge the gap between, on the one hand, non-spatial graphics tools that emphasize value similarity and, on the other hand, maps that emphasize the spatial layout of data values in ways that can be difficult to decode because of the rigidity of the layout. One implication of this is that in order to visualize spatial data and construct relevant graphical tools, additional data layers are needed that store data on distances between sites on a continuous surface or on the spatial weights that define the adjacency relationships between spatial objects (Haining, 1994).

There may be other problems when reading a map for the purposes of ESDA. For interval and ratio data, choropleth maps based on classes discard information; unclassed maps based on shading or colour intensity can be difficult to decode, particularly if more than eight class intervals are adopted (Tufte, 1983). Choice of colour shading and interval choice for choropleth maps can have a profound impact on visual appearance. Areas on a map are not usually identical in size either in terms of areal extent or population. Larger areas will tend to visually dominate a map and small areas become invisible. If the map displays rates and the large areas are rural areas with sparse populations these rates may be amongst the least robust or precise (section 6.2.1). To cope with this, shading may be limited to say the inhabited areas, a cartogram used or symbols or shapes superimposed on the maps that are proportional to population size (Dunn, 1987). Smaller more densely populated areas in towns and cities may carry more robust, more precise evidence on rates but be difficult to see on the map. Dunn (1989) suggests a joint representation of attribute rate and population size as a two-variable colour map with colours or shading becoming stronger only as both attribute value *and* population size increase. This problem extends to a graphic rendering of data such as a boxplot or scatterplot. Not all the rates may be equally reliable and particular groups of datapoints on a graph that look interesting or striking in some way may need to be checked that they are not associated with the less-reliable rates.

Although missing data are not unique to spatial data, the absence of certain data values (or concerns about persistant undercounting) could be particularly problematic where interest focuses on the arrangement property of data values or on making comparisons between places. Each missing value refers to a particular area and a collection of missing values might obscure an entire subarea – not as in some statistical contexts a missing value from an experimental set that may contain replication. There is a particular need as part of interactive graphics to track missing data in spatial data sets (Unwin et al., 1996).

Good map design will still facilitate the table look-up and pattern perception tasks associated with visualizing spatial data for ESDA. Traditional map design rules are discussed in Cuff and Mattson (1982) and Dent (1985). Buttenfield and Mark (1994) provide an overview. However more importance may now attach to particular forms of abstraction and generalization in order to prevent clutter (Brassel and Weibel, 1988; Monmonier, 1991). This is in keeping with the conduct of scientific visualization where, for example, 'jittering' is introduced to reveal overlapping datapoints on a graph (Cleveland, 1993, pp. 21–2). Some of the detail (compass points, scale bars, etc.) that is appropriate for *presentation cartography* can be dispensed with because the analyst will be familiar with this aspect of the data. The user is not designing the single 'best' map and several maps might be displayed, switching on and off, as needed, different elements of any display.

Openshaw and Abrahart (1996) argue that in many applications the volumes and complexity of geo-referenced data create problems for the implementation of visualization methods and that geo-computational tools based on artificial neural networks and genetic algorithms offer more insights than visualization (and exploratory statistical techniques) for ESDA. Gahegan (1999) identifies barriers to the development of visualization methods for data from continuous surfaces. One of Gahegan's barriers is the speed with which plots can be implemented (rendering speed) within any system that has to handle large complex data sets since interactivity is deemed critical to the success of visualization. A second is the complexity of the mapping from geographical data sets to the visual domain and the enormous number of views of spatial data that are possible. These and the other barriers identified by Gahegan are reflections of the problems identified from other areas of visualization (Wegman, 1995). The problems are particularly acute for spatial data however for three reasons. First, there are the extra demands associated with mapping large data sets. Second ESDA draws on extra spatial relational data layers (distances and/or spatial weights) that to compute involve operations of the order n^2 since they typically involve all pairs of cases. Third ESDA involves the detection of both spatial and non-spatial data properties. Anselin (1998) comments that simple extrapolation of methods that have been developed for small to medium sized data sets is not feasible. 'New approaches are needed that use efficient algorithms, implement sparsity and possibly focus on subsets of the data based on careful spatial sampling' (p. 90). Knowledge-based systems may assist (Andrienko and Andrienko, 1999) or focusing on the design of systems that are developed for particular types of problems. Gahegan (1999) discusses these issues and provides illustrations of these and other problems.

6.3   Data visualization and exploratory spatial data analysis

In this section some graphical tools for exploring spatial data are identified together with references. The organizing framework for the list of tools for univariate spatial data (tables 6.3(a) and 6.3(b)) consists of the different elements of the conceptual models of chapter 5 – where the terms are explained. The purpose is to link graphical tools for univariate data that are based on simple data operations to types of spatial pattern detection tasks. The classification should not be interpreted too rigidly but is intended to bring some order to what is a range of different techniques (and variants of those techniques). The other elements of the classification (table 6.3) are similar to those used in table 6.1. We conclude this section with a short illustrative example.

6.3.1   Spatial data visualization: selected techniques for univariate data

The techniques are divided into those that are mainly associated with data for point or area objects (table 6.3(a)) and those mainly associated with data from continuous surfaces (table 6.3(b)). Mapping, including three dimensional scatterplots in which the attribute value is plotted on the vertical axis against the two spatial co-ordinates defining the location of each case can provide useful insights into the presence of whole map ('global') as well as local map properties. Individual outliers and anomalous cases (described in the terminology of chapter 5 as 'rough' elements of the data) are often identified as departures from the smooth or regular features associated with the techniques listed here. For example, a boxplot provides a natural criterion for identifying an extreme value (see section 4.1.2) whilst a scatterplot helps to reveal data pairs that differ from patterns in the rest of the data.

(a)   Methods for data associated with point or area objects

Table 6.3(a)   identifies some graphical plots that may help to reveal different pattern properties in data. Boxplots or CDFs for example computed on separate spatial subsets and then overlaid on one another or otherwise compared may indicate differences between areas. The individual spatial units comprising each subarea may not be equivalent, for example they may

Table 6.3(a)   *Graphical methods for area and point object data*

Spatial contrasts	Trends	Local clusters and clumps	Spatial autocorrelation
Boxplots; cumulative distribution functions	Spatially lagged boxplots; comaps	Comaps; spatial scatterplots	Spatial scatterplots

Table 6.3(b)   *Graphical methods for data from a continuous surface*

Spatial contrasts	Trends	Local clusters and clumps	Spatial autocorrelation
Boxplots; cumulative distribution functions	Data postings; contour maps; symbol maps; indicator maps; level-spread traces; comaps	Comaps; spatial scatterplots; pocket plots	Spatial (or $h-$) scatterplots; square-root differences clouds

vary in size, and weighting might be considered (Mungiole et al., 1999). In the case of the boxplot, let $z(1),\ldots z(n(r))$ denote the $n(r)$ values for region r ($r = 1,\ldots, R$). The purpose is to make comparisons over R subregions ($R \geq 2$). Let $w(1),\ldots, w(n(r))$ denote the corresponding weights for each of the $n(r)$ individual spatial units in r. To find the weighted median, for example, sort the observations so that $z([1]) \leq z([2]) \leq \cdots \leq z([n(r)])$. The weights are then re-ordered to correspond with the ordering on the $\{z(i)\}$. The cumulative sums of the re-ordered weights is:

$$S(k) = \Sigma_{j=1,\ldots,k} w([j]) \quad k = 1,\ldots, n(r) \tag{6.1}$$

The index, t, of the weighted median is the smallest value of k such that $S(k)$ is at least half the total sum of the weights:

$$t = \min\{k \mid S(k) \geq S(n(r))/2\}$$

If $S(t)$ is strictly greater than $S(n(r))/2$ then the weighted median is $z([t])$. If $S(t)$ is equal to $S(n(r))/2$ then the weighted median is the average of $z([t])$ and $z([t+1])$. The weighted upper and lower quartiles can be obtained using the same principle and the boxplot constructed.

Weighted medians are used in various contexts including smoothing spatial data (see chapter 7). The weights might be based on area (larger weights for larger areas), distance from a given point (larger weights assigned to nearby locations) or population size (larger weights for areas with larger populations) for example. Larger weights may be given to data values with smaller standard errors, which, in the case of rates, are areas with larger populations (see Cressie and Read, 1989) The CDF suggested by Majure et al. (1996) includes weights:

$$F_{n(r)}(z;r) = \Sigma_{j=1,\ldots,n(r)} w(j) I(z(j) \leq z)/\Sigma_{j=1,\ldots,n(r)} w(j) \tag{6.2}$$

where $I(z(j) \leq z)$ is the indicator function scoring 1 if $z(j) \leq z$, 0 otherwise. The CDF is the plot of $F_{n(r)}(z;r)$ on the vertical axis against z on the horizontal axis.

By 'spatially lagged boxplots' is meant a series of ordered boxplots. The ordering is with respect to a chosen site or area. Each boxplot uses that subset of the data which is a given distance (or 'lag order', see section 2.4) from the chosen area. Again each boxplot could be based on weighted statistics as described above. Figure 6.5 shows unweighted boxplots of standardized mortality ratios (SMRs) for all cancers for community medicine areas in Glasgow (Haining, 1990, p. 224). Distance bands refer to distances from the city centre. Individual area SMRs are shown within each of the boxplots. Box widths can be scaled to reflect the number of data cases used in the construction of the boxplot since

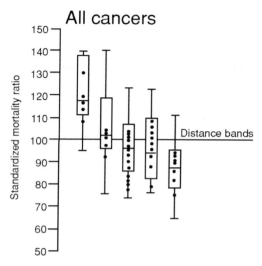

Figure 6.5 Unweighted boxplots showing standardized mortality ratios by community medicine areas grouped into distance bands from the centre of Glasgow: all cancers

boxplots close to the pre-selected area will tend to be based on fewer numbers of areas than boxplots for the areas further away.

Comaps are an adaptation of Cleveland's (1993) coplots to spatial data (Brunsdon, 2000). Data on the variable of interest (the conditioning variable in the case of a coplot) are partitioned into six partially overlapping subsets across the range from the minimum to the maximum value. Each subset has the same number of cases. The six panels of the comap then display the locations of the cases in the corresponding subset. In a coplot the six panels would be scatterplots on two other variables. Brunsdon suggests smoothing the plots using a kernel density estimator to help draw out any patterns that might exist in the spatial distribution. The technique is similar to brushing using a moving window that slides across the distribution of data values on the data value of interest, except that comapping ensures equal numbers of cases in each view.

Spatial scatterplots have been suggested in two forms. In one form, n pairs of values $\{\Sigma_{j\in N(i)}w^*(i, j)z(j), z(i)\}_{i=1,...,n}$ are plotted where $w^*(i, j)$ denotes the (i, j)th entry of the row standardized adjacency or weights matrix and $N(i)$ denotes the set of neighbours of case i ($i \notin N(i)$). The vertical axis is used for the spatially averaged neighbouring values and the horizontal for the value for the area at the centre of the spatial average. Anselin has suggested calling this the Moran scatterplot (Anselin, 1996; Anselin and Bao, 1997). If the values $\{z(i)\}$ are deviations from the mean the Moran coefficient of spatial autocorrelation is the ordinary least squares estimator of the slope of the regression of $\{\Sigma_{j \in N(i)} w^*(i, j) z(j), z(i)\}_{i=1,...,n}$ on $\{z(i)\}_{i=1,...,n}$; hence the suggested name. In the other form of spatial scatterplot the axes are swapped. This is the type of plot used to explore the relationship between $z(i)$ as a function of the average of its spatial neighbours. This corresponds to the scatterplot for a particular type of simultaneous spatial autoregression (see section 9.1.2(b)) in which values for the response variable (Z) are plotted against values for the average of the adjacent neighbours of the same variable (W^*Z). Different orders of

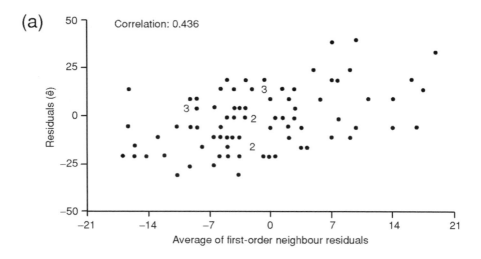

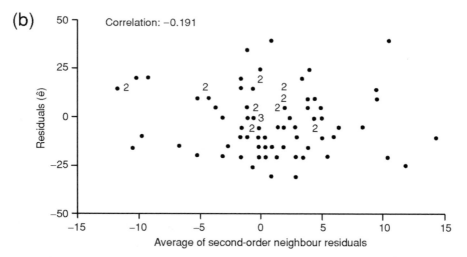

Figure 6.6 Scatterplot of residuals against the average of the neighbours:
(a) first-order neighbours; (b) second-order neighbours

scatterplot can be constructed to see how spatial autocorrelation effects decay with increasing lag distance by defining different orders of neighbours through the matrix $\mathbf{W}$ (see Haining, 1990, pp. 273–4). Figure 6.6 shows such a plot for regression residuals for first-order and second-order nearest neighbours.

Positive spatial autocorrelation for the chosen definition of adjacency is indicated by spatial scatterplots that are increasing upwards to the right, whilst negative spatial autocorrelation is indicated by plots that are downward decreasing to the right. No trend in the scatterplot is indicative of no global or whole-map spatial autocorrelation. The scatterplot can be divided into quadrats when the $\{z(i)\}$ are standardized to a mean of 0 and standard deviation of 1. Cases in each quadrat are highlighted with a particular colour or using a particular symbol and then mapped to reveal clusters of high values, low values, relatively high values surrounded by relatively low values and vice versa. Anselin et al. (1993) also suggest constructing what they call spatial lag pies (for a positive valued variable) and spatial lag bar charts which can then be mapped. These too are based on visualizing the relative sizes of $z(i)$ and $\Sigma_{j\in N(i)}w^*(i,j)z(j)$. Other smoothings of spatial scatterplots to help draw out pattern properties are described in the next chapter.

(b)   Methods for data from a continuous surface

Table 6.3(b) provides a list of graphical plots. Again boxplots and CDFs may be useful for identifying spatial contrasts. As for regional data, weighting may again be a good idea, this time if there is bias in the sampling design with say intense sampling in some parts of the study region and sparse coverage in other areas. In that case weights may be attached in order to downweight the contribution from cases that form clusters in one area of the map (e.g. cell declustering) as discussed for example in section 4.4.

Isaaks and Srivastava (1989, pp. 40–66) list a number of simple graphical tools with examples for detecting trends, clusterings (of high or low values) and spatial continuity (or autocorrelation). Data posting means locating each datapoint along with its data value on a map of the region. High and low values can be highlighted and this might reveal trends or clusters. Contour mapping is good for highlighting trends as is symbol mapping where data values on a data posting are grouped into ordered categories and then symbols assigned to the categories that make it easy to recognize the order relations. Indicator maps are symbol maps with just two categories. Passing a moving window along a trace line across a map in a direction of interest and plotting median and interquartile range statistics against location as the window moves can also help to reveal trend and show how level and spread characteristics change locally. Trends may be present in either, both or neither of the two statistics (Isaaks and

Srivastava, 1989, p. 49). These techniques will also be useful for exploring area and point object data.

Spatial scatterplots in geostatistics are typically a function of distance rather than lag separation and plot data pairs separated by distance h plus or minus a tolerance band defined by δ, that is $\{z(i), z(j) \mid d_{|i-j|} = h \pm \delta\}$. This is instead of plotting a data value paired with the average of its adjacent neighbours as in the case of regional data. If the data are on a lattice grid then plots can be constructed for different directions (north; east; north-east; south-east) and there is no need for a tolerance band. The spatial scatterplot (or h-scatterplot as it is called by Isaaks and Srivastava, 1989, p. 52) is rather limited if the aim is to describe spatial continuity in the data at a range of distances. However Majure and Cressie (1997, p. 135) note that the scatterplot is of interest. This is because the squared distance from each point on the scatterplot to the $45°$ line is half the squared differences used in the classical variogram estimator. This means that outlier points on the scatterplot may influence the estimate of the variogram in later stages of analysis.

Cressie (1991, pp. 41–6) describes the pocket plot and square-root differences cloud. The square-root differences cloud is a plot of $|z(i) - z(j)|^{1/2}$ against $d(i, j)$, the distance between sites i and j, for all data pairs. The plot can be summarized using a series of boxplots computed on the basis of the data pairs separated by given distances or distance bands. An example of such a plot is shown in figure 6.7. This indicates the spatial continuity or spatial autocorrelation at successive distances and gives a fuller picture of the scale at which spatial autocorrelation is present in the data than is possible from a spatial (or h-) scatterplot. These plots can be read like semi-variogram plots. Rising boxplots (medians increasing) that then level off are suggestive of spatial autocorrelation at short distances that then decays. The length or span of the boxplot is indicative of the spread of differences in the data at that distance band and hence the extent to which there may be outliers and anomalies in the data that might influence the estimate of the variogram.

The construction of a pocket plot is based on dividing the area into R subsets – Cressie's example (pp. 42–6) is based on rows of a data set on a grid. Assume in what follows that all distances are defined as average distances between the subsets, not between the individual datapoints. Take all data pairs i and j that are in different subsets and separated by a distance d_1 and compute the quantity $|z(i) - z(j)|^{1/2}$ for each pair and then compute the average of this quantity across all the data pairs. Call this $A(d_1)$. This calculation is repeated for all distances, $d_1, d_2, \ldots, d_k$. This produces a sequence of averages of the square roots of the absolute differences $A(d_1), A(d_2), \ldots, A(d_k)$ – one for each of the distances. Now take all data values $\{z(i)\}$ in the first subset of the data and compute

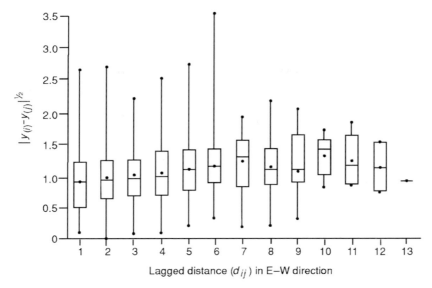

Figure 6.7 Square root difference plot summarized using boxplots at different distance bands (after Cressie, 1984)

the square root of the absolute difference based on all datapoints $\{z(j)\}$ in the second subset. Compute the average of this quantity $(A(1, 2))$. Suppose subsets 1 and 2 are a distance d_1 apart, compute:

$$P(1, 2) = A(1, 2) - A(d_1)$$

Repeat this for all subsets that can be paired with subset 1, noting of course that distance separation will vary and the appropriate $A(d_1)$ or $A(d_2), \ldots, A(d_k)$ will be needed for the calculation:

$$P(1, j) = A(1, j) - A(d_{|1-j|})$$

where $d_{|1-j|}$ denotes the distance between subsets 1 and j. The pocket plot plots the set of values for subset $1\{P(1, R)\}_R$ on the vertical axis against the position 1 (distance 1) on the horizontal axis. This vertical scatter can be summarized in a boxplot, as suggested by Cressie (1991, pp. 44–5). These calculations are repeated for all the other subsets, 2, 3, . . . , R:

$$P(i, j) = A(i, j) - A(d_{|i-j|})$$

and the resulting boxplots are in sequence from subset 1 to R. Subsets that are atypical will tend to have boxplots that are centred above 0 on the pocket plot and these subsets will also appear as extreme values in the boxplots of the other subsets.

6.3.2   Spatial data visualization: selected techniques for bi- and multi-variate data

In non-spatial data, visualization is used to explore for patterns and relationships between variables in data sets with two or more variables. The same holds true in exploring spatial data, although there are additional aspects to the exploration, so some new types of plot are needed. Spatial dependency may exist *between* variables and so more general plots are needed that compare variables Z_1 and Z_m at different distance (and angle) separations (Majure and Cressie, 1997). Each component has to be standardized to mean 0 and unit variance to prevent spurious dependencies being detected that are simply a consequence of the relative sizes of these two properties of the two distributions. Majure and Cressie generalize the variogram, the absolute differences cloud plot and the lagged scatterplot, to the situation of analysing two variables for cross-spatial dependency.

Scatterplots of $\{z_1(i)\}$ against $\{\Sigma_j w^+(i,j) z_m(j)\}$ can be used to examine the relationship between two variables. As before the weights matrix is row standardized but now $w(i, i)$ is non-zero. This can be used to explore for a relationship between two variables where Z_1 is a response variable, the level of which in area i is thought to be influenced by the levels of the other variable not just in area i but also in neighbouring areas. An example might arise in exploring the spatial pattern in rates of an acute respiratory condition (Z_1). The rate in area i ($i = 1, \ldots, n$) is plotted as a function of average air quality (Z_2) in i and the neighbouring areas of i in order to reflect the movement patterns of individuals in a typical area (i) and hence their pattern of exposure (Tjostheim, 1978). Another strategy would be to plot $\{z_1(i)\}$ against $\{z_2(i)\}$ but also to look at $\{z_1(i)\}$ against $\{\Sigma_j w^+(i,j) z_2(j)\}$ as described above and $\{z_1(i)\}$ against $\{\Sigma_{j \in N(i)} w^*(i,j) z_2(j)\}$ since these plots are all giving slightly different perspectives on the spatial relationships between air quality and the rate of the respiratory condition in the population.

If glyphs are used to display multivariate spatial data then the map of the area provides a suitable layout for displaying and interpreting the data. Haining (1990, p. 226) mapped Chernoff faces that had been drawn to describe attributes of air quality for a group of US cities. Andrienko and Andrienko (2001) link parallel co-ordinate plots to a map of Portugal to explore the distribution of areas with particular population profiles. Carr et al. (2000) describe the use of linked micro-map plots and conditional choropleth maps for visualizing multivariate spatial data. Micro-map plots involve a process of progressive disclosure of information. This consists of maps identifying, for example, states above the median, in terms of disease rates, and then for the states with the most extreme rates, revealing county-level details through dotplots and boxplots. This is repeated for the state with the median rate and then for those with the lowest rates. The layout of the maps and graphics on the page facilitate comparison across states and the identification of county-level attributes. Conditional choropleth maps allow visualization of a dependent variable, (e.g. disease mortality) against two explanatory variables, the levels of which can be varied through slider bar operations. Xgobi is a dynamic multivariate visualization software system that can be used to explore multivariate spatial data sets (Symanzik et al., 1996)

Wegman et al. (2001) construct an *image* grand tour where a grand tour is implemented on multispectral data. 'The desired goal is to combine the multivariate vector into a single value which can be rendered in grey-scale as an image' (p. 1). A continuously changing grey-scale map of the multispectral scene is generated and the observer watches to see if interesting configurations appear. They apply the technique to the detection of mines in a minefield.

6.3.3   Uptake of breast cancer screening in Sheffield

Some of the ideas of this section are illustrated using data from the Sheffield call–recall breast cancer screening programme in the early 1990s following the completion of one full cycle. The data were originally recorded at the enumeration district level (1159 EDs in Sheffield) but the areas have been grouped into approximately 300 areas to aid reproduction here and to give more robust rates. ED grouping used the regionalization method described in section 6.2.2 and due to Wise et al. (1997). The new areas are of comparable population size (equality criterion) and similar Townsend deprivation score (homogeneity criterion). A low priority was attached to compactness in the regionalization. Uptake rates were obtained for the new units and results here are based on these rates which have not otherwise been modified because the areas are of broadly similar population size (see section 6.2.1 for further discussion). The software package SAGE that seamlessly links a spatial analysis module to the ArcInfo GIS was used for the visualization work (see appendix I).

Are there any areas with particularly low uptake rates ('cold spots') relative to the overall distribution of uptake rates? This is of interest from a service delivery perspective. The following are three perspectives on geographical aspects of this question. Figure 6.8 shows a boxplot of the rates with those that comprise small outliers highlighted on the graph and on the map. Figure 6.9 uses spatially lagged boxplots (lag 1, 2, . . . , 9) to see if there is any evidence of a fall off in rates with 'distance' from the screening unit. Areas at lag 1 are those areas that are adjacent to the area with the screening unit; lag 2 are the areas reached in two steps (neighbours of the neighbours excluding other first-order

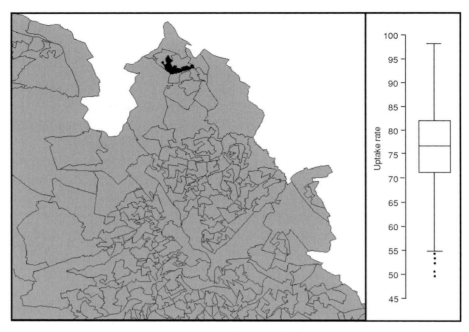

Figure 6.8 Simultaneously highlighting outliers (low uptake rates) on a boxplot and one of the areas to which they refer (Haining et al., 2000)

neighbours and the area with the screening unit itself) and so on. All NHS screening in Sheffield was undertaken at this time from a single site near the Hallamshire Hospital just south of the city centre. Boxplot widths are proportional to the number of areas, and, although there is a gentle fall in the median over the first three lags, there is no evidence to suggest a link between distance and aggregate attendance rates. The lag 3 boxplot has been highlighted, and on the map as well. This is done because the link between lag order and distance can become tenuous – and there is some evidence that this is the case at lag 3, although not a serious discrepancy.

Figure 6.10 highlights on the map an area of North Sheffield where there might be concerns about levels of uptake. It corresponds to an area that comprised one of the more deprived areas of public housing in Sheffield in terms of material well-being, mortality rates and exposure to high crime levels. The boxplot of uptake rates for the highlighted area is compared against the boxplot of rates for the whole of Sheffield. There is no marked differences in the two boxplots suggesting that the groups of women residents in this area do not appear to differ from other groups across the city in their participation in the programme. Figure 6.11 shows a scatterplot with uptake rates on the vertical axis and a weighted average of the neighbouring rates on the horizontal. A least

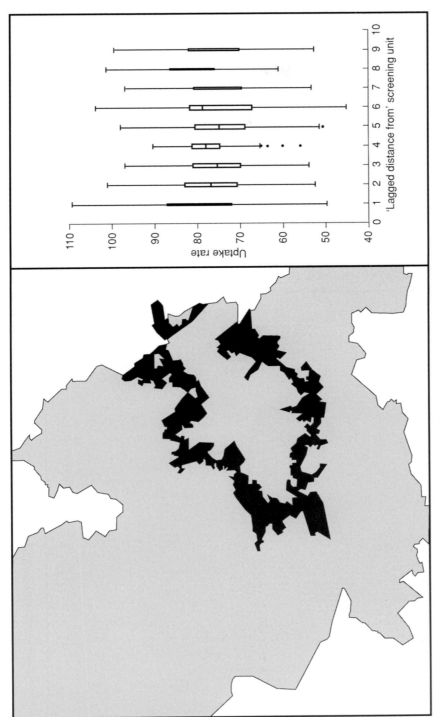

Figure 6.9 Boxplot of uptake rates by 'lag order' from the screening unit (maximum uptake rate = 100%) (Haining et al., 2000)

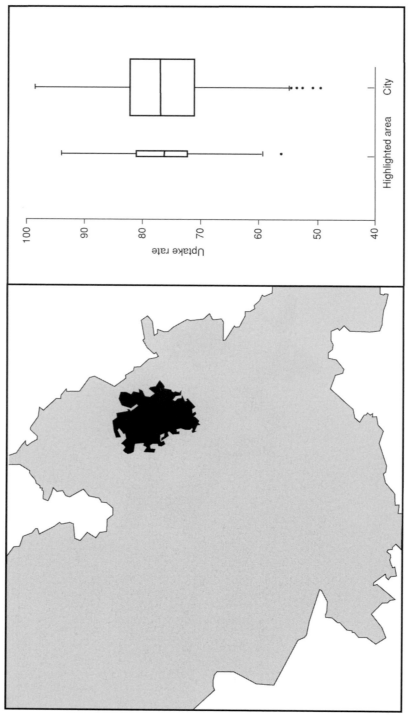

Figure 6.10 Comparing the boxplot of uptake rates for all Sheffield with that for the highlighted area (Haining et al., 2000)

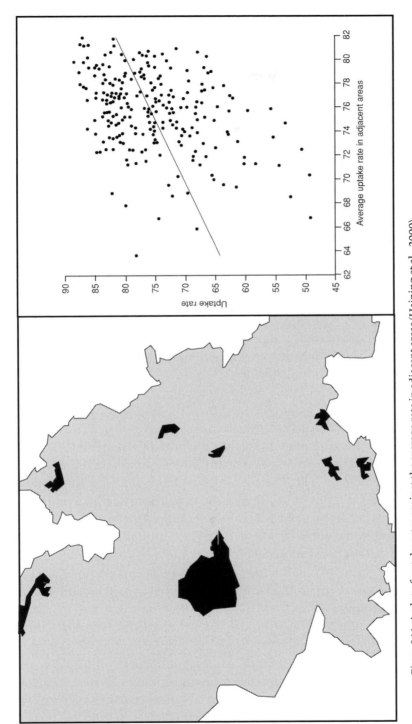

Figure 6.11 A plot of uptake rates against the average rate in adjacent areas (Haining et al., 2000)

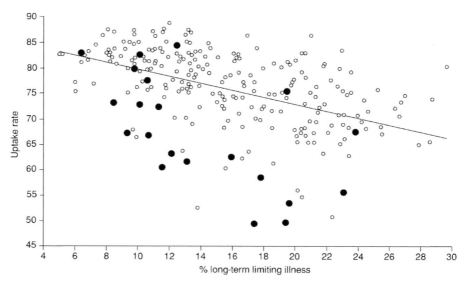

Figure 6.12 Plot of uptake rates against the % reporting limiting long-term illness in the 1991 Census (Haining et al., 2000)

squares line has been fit through the scatter. There is some evidence that areas with high or low uptake rates have neighbouring areas that also have high or low uptake rates (high/high; low/low). These can be brushed to see if there is a geography to the high/high or low/low areas. The map window shows the eight areas with the lowest uptake rates (bottom of the graph window with rates 55% or less) and these are widely scattered.

Finally, in this exploratory visualization of the data, figure 6.12 plots uptake rate against the percentage reporting limiting long-term illness (from the UK 1991 Census) by area. There is wider evidence of a negative relationship between general levels of health and the extent to which individuals adopt preventative behaviours and this is broadly supported in this case. The least squares line is downward sloping to the right. The highlighted cases are those with the smallest population denominators and many of these fall in the lowest decile of the set of uptake rates. There may be a denominator effect even in this data set where the areas have been constructed to have broadly equivalent population counts.

There are a number of applications of ESDA methods to regional data in the literature. Cressie and Read (1989) apply a number of methods to sudden infant death syndrome data by area for the counties of North Carolina, USA. Anselin and Bao (1997) examine the pattern of housing values in West Virginia and Messner et al. (1999) examine county-level homicide rates. Pereira et al. (1998) examine spatial data on the distribution of wildfires in Portugal by county.

Le Gallo and Ertur (2002) use ESDA methods to study the regional distribution of per capita income in Europe.

6.4   Concluding remarks

Visualization has a role to play in ESDA and this chapter has described how and provided examples of techniques and applications. The role of visualization tools may be further enhanced when fitting leads to a summary representation of the data. This aspect of visualization will be explored in more detail in the next chapter. However it is also important to stress the limitations of visualization within data analysis. First, without some form of summary of the data the eye can be faced by a bewildering array of images. Faced with a multitude of exploratory routes through the data it may be difficult to structure and organize ideas in a logical fashion. Monmonier (1992) proposed the adoption of graphic scripts to help structure such interrogations of spatial data. Second, pictures can mislead and be ambiguous. Different people may see different patterns and relationships in the data. This can be partly overcome, at least, by the design of better graphics and maps. However just as careful preparation of the data is essential prior to visualization, so good fitting procedures and other numerically based manipulations of the data are usually required in order to proceed further with understanding and gaining insights into the data. The next chapter continues with ESDA, drawing attention to numerical methods that complement the visual tools of this chapter.

7

Exploratory spatial data analysis: numerical methods

This chapter continues with exploratory spatial data analysis (ESDA) and discusses numerical methods. Section 7.1 reviews methods for smoothing graphs and map data. The aim of map smoothing is to remove distracting noise or extreme values present in data in order to reveal spatial features, such as trends, or ridges or zones of rapid transition in the data (Kafadar, 1999). Map smoothing is important where data values are known to be of low as well as varying precision (perhaps due to aggregation effects). In statistical terms, map smoothing involves trying to improve the precision associated with such data values whilst not introducing bias. Map smoothing may also help suggest covariates for subsequent statistical modelling of the data.

The remaining sections consider methods that involve hypothesis testing so that now the data are assumed to come from some generating model (see section 2.1.4). Section 7.2 reviews tests for detecting an overall tendency for similar values to be found near together on a map ('whole map' clustering or spatial autocorrelation). Section 7.3 discusses tests for detecting clusters ('localized' clustering). Hypothesis testing is used to identify which clusters or concentrations of an event are statistically significant. Hypothesis testing is used as an aid to sifting out potentially interesting areas of the map. The tests are exploratory in that a null model is proposed and test statistics are constructed to enable a decision to be reached as to whether to retain the null hypothesis or reject it in favour of some general non-specific alternative. No model is proposed for the alternative hypothesis. Section 7.4 describes methods for making map comparisons.

7.1   Smoothing methods

7.1.1   Resistant smoothing of graph plots

Resistant fits are applied to graphical plots to support visualization and to help identify the presence of trends and relationships. Cleveland (1993, p. 1) comments: 'there are two components to visualizing the structure of statistical data – graphing and fitting . . . Just graphing raw data, without fitting them and without graphing the fit and the residuals, often leaves important aspects of the data uncovered.' Cleveland (1994, pp. 168–80) fits smooth curves to scatterplots $\{y(i), x(i)\}$ using locally weighted regression with robustness iterations. The procedure is referred to as *loess*. The term 'local' derives from the fact that only subsets of the data are analysed at any one time.

A vertical strip is defined on the plot centred on a selected, observed, value of x. Denote this value x_a. The length of the strip on the x axis is called the window or bandwidth. The paired observations $\{(y(i), x(i))\}$ within this window are assigned neighbourhood weights $\{w_i(x)\}$ using a weight function that has its maximum at the centre of the window (the position x_a) and decays smoothly and symetrically to become 0 at the boundary of the window. A line is fitted to this subset of data by weighted least squares with weight $w_i(x_a)$ at $((y(i), x(i))$ (Weisberg, 1985, pp. 80–3). The fitted value of y at the value x_a where the window is centred is denoted $y^*(x_a)$. The window is then slid along the x axis and the procedure repeated to create a sequence of new 'smoothed' values of y: $(y^*(x_a), y^*(x_b), y^*(x_c), \ldots)$. The vertical strips overlap because the window is centred on each ordered value of x in turn. The values $\{(y^*(x_j), x_j)\}_j$ are connected by straightline segments to provide the graph summary. Procedures for fitting the loess curve at the extremes of the distribution of x values are discussed in Cleveland (1994).

A second stage of the fitting procedure involves applying robustness iterations to the initial fit to protect against outliers. Outliers, cases with residuals with a large absolute value, are identified from the initial fit and robustness weights introduced into the weighted regression procedure at the second stage. Cases with extreme residual values from the first stage of fitting are assigned small robustness weights (close to 0) thus diminishing their influence on the loess curve at the second stage. At this second stage, in specifying the weights for the weighted regression fit, the neighbourhood weights and the robustness weights are multiplied together. This has the effect that any case will have a small influence on the fit at any point on the x axis if it is close to the edge of the window and has a residual with a large absolute value. Successive robustness iterations proceed in this fashion. Cleveland (1994, p. 180) notes that four robustness iterations are usually sufficient.

The choice of weights apparently is not thought critical providing the function peaks at the central point of the window and decays smoothly. The size of the bandwidth is very important however since this determines the scale of the smoothing and a poor choice of bandwidth could distort any true patterns in the data. Cleveland (1979, 1994, p. 173) suggests procedures for selecting appropriate bandwidths.

7.1.2   Resistant description of spatial dependencies

Resistant methods have been suggested for estimating the structure of spatial dependence. Consider the estimator for the variogram. For any given distance h, the quantity:

$$2\hat{\gamma}(h) = (1/N(h))\Sigma_i \Sigma_j (z(i) - z(j))^2 \qquad (7.1)$$
$$[(i,j)|d(i,j)=h]$$

where $N(h)$ denotes the number of pairs of sites separated by distance h, is the value of the variogram at distance h (see section 2.4). Now (7.1) can be considered either as an estimator of the variance for the differences $(z(i) - z(j))$, since their mean will be 0, or as an estimator of the centre of the squared differences. A resistant estimator for the variogram is obtained for example by replacing (7.1) with the median of the squared differences (Dowd, 1984). A resistant estimator for the spatial covariance at lag h (see section 2.2) is constructed using medians rather than means (section 2.3) and taking the median of the cross products.

Section 6.3.1 described spatial scatterplots, cloud plots and pocket plots for representing spatial relationships in different types of spatial data. Methods for smoothing these descriptions were described, including the use of boxplots. The loess curve of section 7.1.1 may also assist the process of providing a smoothed representation of the structure of these plots.

7.1.3   Map smoothing

Choropleth mapping is a simple form of map smoothing for area data where the degree of smoothing depends on the number of classes into which data values are bundled. 'Resistant' choropleth maps can be constructed by selecting class intervals based on resistant measures (the median and interquartile range) of the distribution of values (Cromley, 1996). Choropleth mapping smooths by suppressing the variation in the data through assigning data values to categories. This amounts to reducing the level of measurement of the data – from interval or ratio down to the ordinal level.

The interpolation methods of section 4.4.2 can be used to smooth noisy observed data values that are assumed to derive from some true function. The methods are now applied not to estimate missing or unsampled datapoints but instead to provide an estimate of the 'true' data value associated with points or areas where values have been recorded. Providing the analyst is satisfied the interpolator is recreating as accurately as possible the true underlying function then it is meeting at least one of the performance characteristics of a good smoother specified by Kafadar (1999, p. 3170). Webster et al. (1994) provide an interesting application of kriging to obtain smoothed relative risk maps of childhood leukaemia in the West Midlands of England. It is arguable whether this should be called an exploratory method. Kriging requires specification of a model for the underlying spatial variation, so it requires the analyst to make assumptions about the data not usually associated with exploratory data analysis.

Map smoothing of area data can be thought of in terms of trying to improve the precision associated with area data values whilst not introducing serious levels of bias. This terminology is often used where each data value is thought of as a sample. A common form of smoothing centres a window of fixed size on each datapoint or area (represented by a convenient centroid) in turn. The size of the window determines the degree of smoothing. Large windows improve the precision of the statistic by borrowing large amounts of local information (in sampling terms, increasing the size of the sample) but at the cost of introducing bias because information is being borrowed from areas, further away, that may be different. Small windows reduce the risk of bias in the statistic but because little information is being borrowed the precision is not much improved. The effectiveness of 'local' or 'neighbourhood' borrowing will depend on the local homogeneity of the spatial data which will in turn depend on the size of the spatial units (or density of sampling points) in relation to the true scale of spatial variation. If adjacent areas are very different in nature then borrowing information locally may introduce bias into the smoothed map that distorts the underlying patterns through inappropriate smoothing. Constructing deprivation regions in which adjacent areas are differentiated by their level of deprivation or constructing regions based on their level of urbanization may well not be a good spatial framework on which to then smooth disease rates using spatial adjacency as the criteria for borrowing information. A better method might be to smooth by borrowing information from other areas with similar deprivation levels or similar levels of urbanization. A 'nearness' criterion can be added but the main point is to ensure that information is borrowed in an appropriate way.

Gelman et al. (2000) warn that no method of smoothing area rates can entirely eliminate the problem of unequal precision and that there is always the danger of introducing artefacts into the resulting maps. As is often done in reporting results, the best strategy is to use several smoothers and compare results.

(a)   Simple mean and median smoothers

In moving median smoothing, the value for case i, $z(i)$, is replaced by the median value from the set of values, including case i itself, contained in the window centred on i $(MM(i))$. The window size may be defined in terms of distance between data locations or adjacency criteria in the case of areas. The set of $MM(i)$ values, obtained by moving the window over the map, represents the 'smooth' component of the map from which trends or patterns at the scale of the window and above might be detected. Kafadar (1994, p. 427) notes that the process of median smoothing can be repeated (resmoothed medians) and at each iteration the map pattern inspected. The 'rough' component of the map can be extracted by computing the difference: $\{z(i) - MM(i)\}$. The moving average smoother uses the mean of the observations rather than the median and is similar to the operation of spatial differencing (Cliff and Ord, 1981, p. 192). Cressie (1991, p. 398) warns that the rough component does not contain accurate information on spatial dependency and this is because overlapping subsets of the data are used as the window is moved over the mapped data.

Whether the statistic that is used to derive the smoothed map is resistant (e.g. the median) or not (e.g. the mean) smoothers can be made resistant in the sense of Cleveland (1994) by including robustness iterations in which cases with unusually large or small residuals are downweighted in subsequent iterations. However choosing between mean- and median-based smoothers is important. In environmental science, isolated outliers are unlikely and it is desirable to smooth away spikes (although not edges). In some social applications spikes are credible (population peaks associated with cities or areas with extreme crime rates or disease rates). The smoothing method employed should preserve what is plausible in the data. Moving mean smoothers, because they are linear smoothers, tend to blurr spikes and edges. Median smoothers, because they are non-linear smoothers, tend to preserve abrupt changes (Kafadar, 1994, 1999; Mungiole, 1999). However the performance of these different smoothing methods also depends on how values are weighted, to which we now turn.

(b)   Introducing distance weighting

Distance weighting is commonly employed in map smoothing. Let $w(i, j)$ denote the weight attached to data case j in computing the smoothed value for case i. Simple disk averaging assigns $w(i,j) = 1$ if $i = j$ or if j is in the window around i (if j is within distance D of area i) and assigns $w(i,j) = 0$ if j is not in the window. However this type of neighbourhood weight function can cause values to change abruptly and smoothers based on them can have undesirable properties (Kafadar, 1996, p. 2543).

It was noted in the case of loess curves constructed on $\{y(i), x(i)\}$ that local weighting was used to downweight the contribution of cases further away in the x-domain. Smoothing univariate spatial data by distance-weighted least squares adopts the same principle but in the distance domain (Pelto et al., 1968; McLain, 1974). In the spatial application of loess, the fitted function is usually either linear, full quadratic (i.e. with crossproduct terms) or quadratic without crossproduct terms (Kafadar, 1994). Thus for a linear trend, distance-weighted least squares involves minimizing for each point i ($i = 1, \ldots, n$) with respect to the parameters $\theta_{0,0}$, $\theta_{1,0}$ and $\theta_{0,1}$:

$$\Sigma_{j=1,\ldots,n}\, w(i, j)[y(j) - [\theta_{0,0} + \theta_{1,0}\, s_1(j) + \theta_{0,1}\, s_2(j)]]^2$$

where $s_1(j)$ and $s_2(j)$ are the co-ordinates of case j. This can be implemented by weighted least squares (see Weisberg, 1985, pp. 82–3).

A distance decay function is used to downweight the contribution of data values that are further away in the distance domain. For example:

$$w(i, j) = [1.0 - (d_{i,j}/D)^\beta]^\beta \quad d_{i,j} < D \tag{7.2}$$
$$= 0 \qquad\qquad\qquad \text{otherwise}$$

where $d_{i,j}$ is the distance between cases i and j, and β is a constant. Weights are normalized to sum to one. This function is adapted from a proposal by Cleveland and Devlin (1988) in the context of loess curve construction and is an example of a kernel smoother. Values for β are commonly 2 or 3. Inverse squared distance weights are commonly used (Kafadar, 1994). Ripley (1981, p. 37) notes conditions on the choice of $w(i,j)$. The choice of weight function, providing it is sensible, is not apparently crucial (Kafadar, 1999, p. 3171).

The results of distance-weighted least squares smoothing are affected by clustering of the datapoints (Ripley, 1981, p. 35). The precise choice of weighting function is of secondary importance but the choice of D, called the bandwidth, is crucial. As noted above, the larger the value of D the greater the smoothing. There will still be a variance-bias trade-off, but by reducing the influence of areas that are further away in favour of those close by the hope is that a reduction in variance can be achieved without seriously increasing bias. Methods for estimating bandwidth have been proposed but in many applications the choice is based on either a knowledge of the nature of local mean variability on the map or a range of different bandwidths are examined. One suggestion for D is to base it on the average squared nearest neighbour distance (McLain, 1974). The aim is not necessarily to identify some 'optimal' bandwidth. Rather it is to see what insights into the data are revealed by the process of smoothing. This methodology is the univariate equivalent of geographically weighted regression (Fotheringham et al., 2000).

In the case of moving average smoothers, if $y^*(i)$ denotes the smoothed value at i then:

$$y^*(i) = \Sigma_{j=1,\ldots,n} w(i, j) y(j) / \Sigma_{j=1,\ldots,n} w(i, j)$$

In the case of a median smoother, weighted medians may be calculated for data subsets as described in section 6.3.1 using weights that reflect distance. Weights may be chosen to reflect other spatial attributes of the data if these are considered relevant. In the case of area data, for example, weights may be chosen to reflect the proportion of the window covered by each area so that greater weight is given to data values associated with large areas. Data values for large areas with large populations may be more reliable in the sense of being less susceptible to error. Some data values to be smoothed may have a large number of neighbouring values, whilst others may be located in areas with many fewer neighbouring values. One option is to adapt the bandwidth of the smoothing function to the density of nearby datapoints. So, for example, (7.2) might become:

$$w(i, j) = [1.0 - (d_{i,j}/D_i)^\beta]^\beta \quad d_{i,j} < D_i \tag{7.3}$$
$$= 0 \quad\quad\quad\quad\quad \text{otherwise}$$

where D_i is the bandwidth associated with case i. Smoothing based on fixing the number of nearest neighbours to be used will achieve the effect of adapting the bandwidth to the density of points and this is a characteristic of the head-banging smoother discussed below.

(c) Smoothing rates

The above methods can be applied to general types of data but with maps of rates there is the option to either smooth the rates or to smooth the individual components (the numerator and denominator) of the rates. Kafadar (1996) identifies when it is appropriate to smooth the individual components. This is when the means within the locally defined window are constant and when the correlation between the numerator and denominator variables satisfy certain conditions (p. 2541). Two of her cases, when it is better to smooth the components rather than the rates themselves, are of particular interest. Let $r(i) = x(i)/y(i)$ denote the rate for area i.

Case (1): $x(i)$ is binomial with parameters $y(i)$ and p and $y(i)$ is a constant. For example $y(i)$ is the population in area i and p is the probability that any randomly selected individual has a specified attribute so that $x(i)$ is the count of individuals with the specified attribute. This might be the rate of some commonly occurring offence in a well-defined at-risk population (e.g. household burglary).

Case (2): $x(i)$ is Poisson with parameter λ, $w(i)$ is Poisson with parameter v and $y(i) = x(i) + w(i)$. The parameter v is required to be much greater than λ (e.g. $v/\lambda \approx 10^3$). This arises when $r(i)$ is the incidence or mortality rate of a rare disease. $x(i)$ is the number of cases of the disease in area i whilst $w(i)$ is the number of healthy individuals in i.

Kafadar (1994) compares smoothing methods applied to directly standardized incidence ratios for areas using US county-level data on prostate cancer. In the following i denotes area, m denotes age cohort, $\pi(m)$ denotes the proportion of individuals in age group m from the standard population, $d(i, m)$ denotes the number of cases in area i and age cohort m and $n(i, m)$ denotes the number of person years at risk in area i and age cohort m. The age-adjusted rate for area i is given by:

$$r(i) = \Sigma_m \pi(m)[d(i, m)/n(i, m)] \tag{7.4}$$

The distance-weighted average (see for example 7.2) of the age-adjusted rates gives the statistic:

$$r^*(i) = \Sigma_j w(i, j) r(j)/\Sigma_j w(i, j) \tag{7.5}$$

Using distance-weighted averaging on the the numerator and denominator of (7.4) gives the statistic:

$$r^+(i) = \Sigma_m \pi(m) [d^+(i, m)/n^+(i, m)] \tag{7.6}$$

where $d^+(i, m) = \Sigma_j w(j, i)\, d(j, m)$ and $n^+(i, m) = \Sigma_j w(j, i)\, n(j, m)$. Assuming case (2) above and assuming that the bandwidth, D, is chosen so that age-specifc rates are constant within the selected window size then:

$$\text{var}(r^+(i)) < \text{var}(r^*(i)) < \text{var}(r(i)).$$

The standard errors associated with rates computed from different sized base populations under the assumption of the Poisson model are not the same. Standard errors are inversely proportional to the total population. Computing a map by smoothing across values with different standard errors is likely to be misleading and it is necessary to try to eliminate the effects of population size through a variance equalizing transformation of the original data values (see section 6.2.1 and Cressie, 1991, p. 395). Kafadar (1996) considers a double weighting for distance and population size:

$$w(i, j) = [n(j)]^\alpha [1.0 - (d_{i,j}/D)^\beta]^\beta \quad d_{i,j} < D \quad (7.7)$$
$$= 0 \qquad\qquad\qquad\qquad \text{otherwise}$$

where $n(j) = \Sigma_m n(j, m)$ is the total population in area j. She notes that with these weights the target area may not receive the largest weighting. Kafadar's results indicate that (7.5) using simple disk averaging and (7.6) using simple disk averaging, (7.3; $\beta = 3.0$) and (7.7; $\alpha = 1.0$; $\beta = 3.0$) all indicate the presence of similar patterns in the data but differ in the intensity with which patches are displayed. Other choices for α and β may be made (Kafadar, 1999). On the evidence of her analysis it seems that if the aim is to ensure patterns stand out then (7.6) with weights defined by (7.7) performs well.

(d)   Non-linear smoothing: headbanging

Linear smoothers like the moving mean tend to blur what might be meaningful sudden changes in a surface and smooth out real small-scale features. Where abrupt features are expected then non-linear smoothers are to be preferred (Kafadar, 1999). Median smoothing has been described above but we now turn to some other forms of non-linear smoothing.

Median-based head banging, is a non-linear smoother proposed by Tukey and Tukey (1981) and implemented by Hansen (1991) for spatial data. It can be used on data from a continuous surface or from a region partitioned into sub-areas. This smoother adapts the concept of a one-dimensional moving median smoother to the spatial situation where left and right are not well defined. This smoother has been advocated for geological data because it tends not to over-smooth zones of sharp transition, although it does tend to remove spikes. It also performed well in Kafadar's (1994) evaluations on non-gridded data, but the tendency to smooth out spikes is not necessarily a good property in the context of disease maps. Gelman et al. (2000) have warned that it tends to produce high rates on the boundaries of study areas.

The basic head banging algorithm is described, for example, in Kafadar (1999, p. 3173). To smooth any data value $z(i)$ at location i, the head banging smoother starts by identifying J triples of datapoints near to i from amongst its N nearest neighbours. Each triple includes $z(i)$ plus a *pair* of z values taken from the set of nearest neighbours. Members of a triple are chosen so that the two locations from the set of nearest neighbours form with the location for i an angle within $\pm 45°$ of $180°$. If there are more than J triples that satisfy the criterion choose those closest to $180°$. Now, the 'high screen' is the centre of the highest values in the neighbourhood of i whilst the 'low screen' is the centre of the lowest values. In particular, the 'high screen' is the median across all the J

triples of the larger of the two *pair* values. The 'low screen' is the median across all J triples of the smaller of the two *pair* values. If the current value, $z(i)$, lies between these two medians it is left unchanged. If $z(i)$ is lower than the 'low screen' it is changed to the value of the 'low screen'. If $z(i)$ is higher than the 'high screen' it is changed to the value of the 'high screen'. Smoothing is carried out for all points which are then all updated at the end of a cycle so that the order of the smoothing does not affect the result. The cycle is then repeated either for a fixed number of times or until there are no further changes. Special rules are needed to cope with boundary effects.

Mungiole et al. (1999) have developed a weighted form of head banging using the *weighted* median as described in section 6.3.1 for computing the low and high screens. Weighted head banging, using weights inversely proportional to the standard errors of rates, on the evidence in Mungiole et al., preserves spatial structure better than unweighted head banging, particularly when there are high levels of noise. They also discuss other modifications such as: do not change the original value $z(i)$ if it is 'reliable'. If for example the high screen change could be implemented $z(i)$ might be deemed to be 'reliable' if:

$$\text{rel}(i) = (J \times w_i) \geq \Sigma_{j=1,\dots,J} \left(w_{h(j)} \right) \tag{7.8}$$

where w_i is the weight associated to the data value at i and $w_{h(j)}$ is the weight associated to datapoint $h(j)$, the high value of the jth triple. Recall that when weights are inversely proportional to the standard error of the rate this criterion is equivalent to comparing precisions.

The local smoothing methods of this section are designed to help reveal the presence of legitimate structures in the data. Choropleth maps are made of the smoothed values. It will be important to check whether any apparent spatial structure, particularly the distribution of isolated extreme values, are not simply a reflection of the underlying spatial framework creating statistical artefacts. The geographical distribution of areas with relatively small populations, for example, often correlates with the distribution of extreme rates. If there is a geography to the distribution of tracts with small populations this is what may be evident in the geographical distribution of smoothed values. Gelman and Price (1999) give examples where 'statistical artefacts lead to misleading maps' (p. 3233) citing the identification of extreme disease rates in areas of the USA where counties have small populations. Even structures and local trends such as urban/rural structures, may be due to this effect if there are systematic differences in area population sizes for example. After mapping, the effect may be visually re-inforced if the areas with small populations are physically large. Gelman and Price suggest that one way to avoid drawing inappropriate conclusions is to construct two maps – one that displays the parameter of interest and one that displays the uncertainty associated with the value. An alternative is to construct a bivariate map that combines these two types of information.

If tests are needed or some other evidence to support a visual impression of the presence of some spatial structure in the data then further analysis will be needed on the smoothed data. Methods are described in later parts of the book, including methods for the detection of local clusters which are discussed in section 7.2. The identification of 'real' spatial trends is often of particular interest.

(e)   Non-linear smoothing: median polishing

Median polishing is used on lattice data to detect trends and to produce residuals which contain accurate information on spatial dependence which may be important for other analyses (Cressie, 1991, p. 398). Median polishing operates on entire rows and columns of the spatial data set rather than on lo-calized subsets and is designed to provide an explicit decomposition of the data into two components (fit + residual).

The principle behind median polishing is to treat the spatial data as if it were a two-way table (Tukey, 1977). The underlying model is an additive decomposition of each data value, of the form:

data value = fit + residual;

fit = common effect + row effect + column effect   (7.8)

If $z(s_1(i), s_2(i))$ is a data value then the four components are:

$$z(s_1(i), s_2(i)) = z(.,.) + [z(s_1(i),.) - z(.,.)] + [z(., s_2(i)) - z(.,.)]$$
$$+ [z(s_1(i), s_2(i)) - z(s_1(i),.) - z(., s_2(i)) + z(.,.)]$$

where the 'dot' denotes taking the median over that subscript. The use of the median (rather than the mean) means the decomposition is resistant to out-liers. The fitting algorithm involves operating first on rows by obtaining the row medians and subtracting from every member of a row the corresponding row median. The same procedure is repeated, using the table produced at the first iteration, on the columns. Then the cycle of row followed by column operations is repeated until there is no change in the table. Row, column and all-data medians are accumulated at each iteration. Properties of the method are discussed in Emerson and Hoaglin (1983).

In order to check that there are no other trends in the data that cut across rows and columns a comparison plot $\{[z(s_1(i), s_2(i)) - z(s_1(i),.) - z(., s_2(i)) + z(.,.)],$ $[z(s_1(i),.) \times z(., s_2(i))/z(.,.)]\}$ should be obtained. This is the plot of each residual against its corresponding row effect times column effect divided by the common effect. This amounts to a check for higher-order non-additive trends in the data. If the slope of the scatter is close to 0 then the purely additive model is adequate. If the slope of the scatter is not close to 0 this indicates the presence of higher-order trends that involve row and column interaction terms (see section 9.1.2). In the case of spatial data, this additional check is important because the orientation of the grid used to collect the data will not have been selected so as to ensure trends only follow the axes of the grid. There is further discussion of median polishing with an example using remotely sensed data in Haining (1990, pp. 215–20).

Median polishing is not obviously transferable to the case of non-lattice data, but since the method is not sensitive to missing values Cressie and Read (1989) suggest overlaying a rectangular lattice and transferring the data to it. In their study of rates for sudden infant death syndrome recorded for the 100 counties of North Carolina, county rates are associated with those lattice nodes that are closest to the county seat. Grid spacing was chosen to ensure no more than one county per lattice node. Since there is mean–variance dependence in the rates a variance stabilizing transformation was applied (see section 6.2.1). Since rates are not of equal precision weighted medians were used for the polish using weights $n(i)^{1/2}$ where $n(i)$ is the denominator in the rate, that is, the number of live births (section 6.2.1).

(f)   Some comparative examples

Kafadar (1994) provides examples of trends in prostate cancer amongst different populations of the USA (see also Kafadar, 1996). Figure 7.1 is from Kafadar (1999). The unsmoothed map of rates is mosaic like, with no evidence of structure. The loess smoother creates broad patterns that may be too smooth to be convincing. The age-specific rate smoother lies in between these extremes, suggesting some broad patterns as well as regional features.

7.2   The exploratory identification of global map properties: overall clustering

A number of statistics test for overall clustering where the null hypothesis is spatial randomness in the distribution of data values. Tests divide into distance-based tests when individual-level data are available and area-based tests which count numbers of events in defined subareas (Upton and Fingleton, 1985; Gatrell et al., 1996). In section 7.2.1 there is a brief review of the join-count, Moran and Geary statistics as well as the Getis–Ord G-statistic for testing for spatial correlation in area data. This is one form of clustering for area data where similar sized values are found close together. This is followed by examining statistics that test for the presence of spatial correlation and

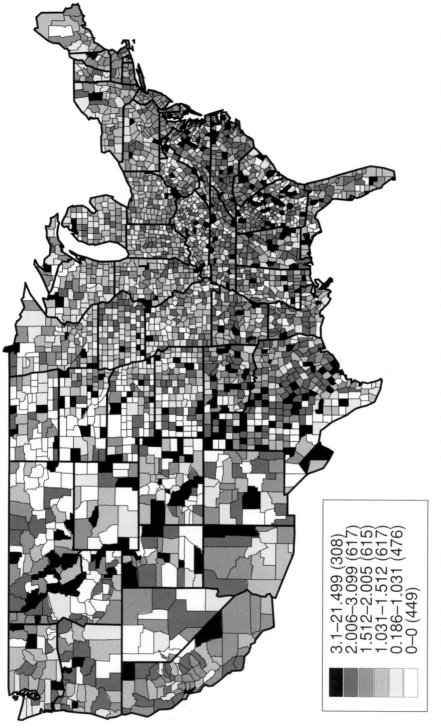

3.1–21.499	(308)
2.006–3.099	(617)
1.512–2.005	(615)
1.031–1.512	(617)
0.186–1.031	(476)
0–0	(449)

Figure 7.1(a) Map smoothing of raw raw rates for female melanoma by county in the USA 1973–1987: (a) map of raw rates; (b) age-specific rates smoothing ((7.6) and (7.7) with $\alpha = 0.5$; $\beta = 22.00$); (c) weighted head banging (100 nearest neighbours; 70 triples); (d) age- and latitude-adjusted rates, smoothed by loess (Kafadar, 1999, figure 2)

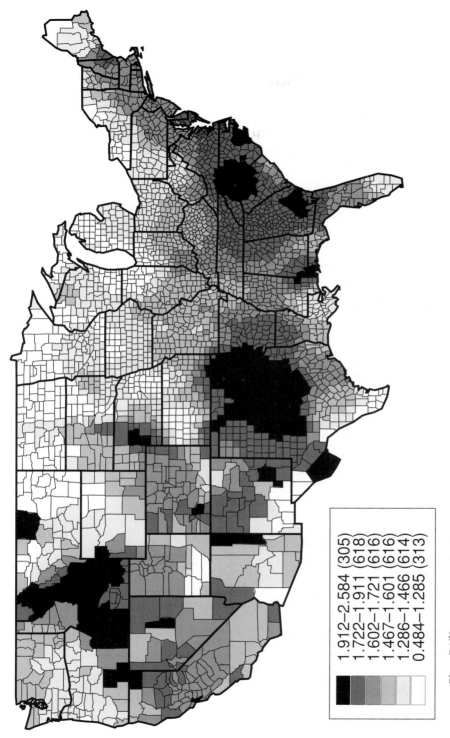

1.912–2.584 (305)
1.722–1.911 (618)
1.602–1.721 (616)
1.467–1.601 (616)
1.286–1.466 (614)
0.484–1.285 (313)

Figure 7.1(b)

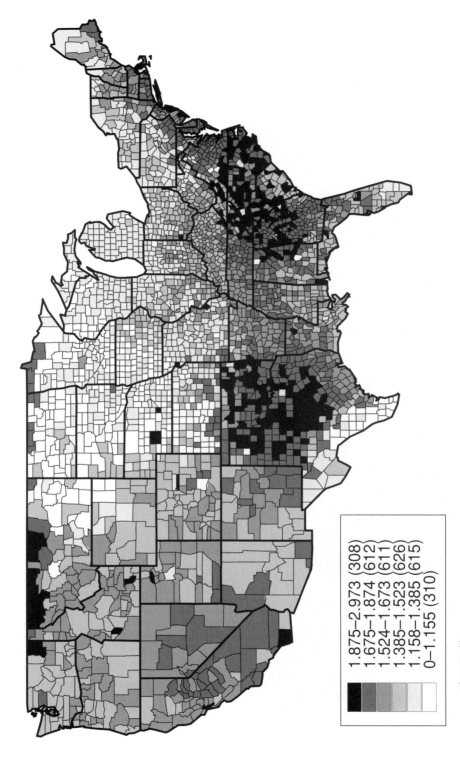

1.875–2.973 (308)
1.675–1.874 (612)
1.524–1.673 (611)
1.385–1.523 (626)
1.158–1.385 (615)
0–1.155 (310)

Figure 7.1(c)

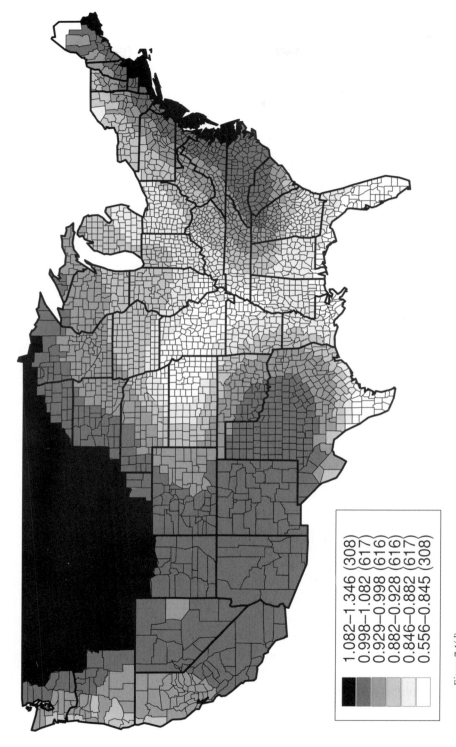

1.082–1.346 (308)
0.998–1.082 (617)
0.929–0.998 (616)
0.882–0.928 (616)
0.846–0.882 (617)
0.556–0.845 (308)

Figure 7.1(d)

spatial heterogeneity, such as Oden's test. In analysing disease rates, clustering may be present at two scales. There may be intra-area clustering of cases (spatial heterogeneity of rates). In addition areas with similar-sized rates may be found close together so areas showing intra-area clustering may themselves be near to one another (spatial correlation of rates). The term 'clustering' is scale dependent.

Section 7.2.2 discusses Diggle and Chetwynd's (1991) test for clustering of a set of objects in an inhomogeneous point pattern. The test is based on the estimation of the K-function which describes spatial dependence in the distribution of point objects at a range of distance scales (Ripley, 1981). The location of each individual in the population is fixed. Some individuals are 'marked' and the purpose of analysis is to test to see if the marked subset are spatially clustered within the population, that is controlling for the uneven distribution of the underlying population.

7.2.1 Clustering in area data

The join-count test for categorical data is based on counting the number of adjacencies of particular colours. Suppose each area is coloured black (B) or white (W). If there is a tendency for similar values to be found together (positive spatial autocorrelation) then the number of joins where adjacent colourings are the same (BB or WW) will be 'large', the number of joins where adjacent colourings are different (BW) will be 'small' relative to the counts for a random map pattern. In the case of negative spatial autocorrelation where the tendency is for *dissimilar* values to be found together BB and WW join counts will be 'small' and the BW join count will be 'large' relative to the counts for a random pattern. The test can be generalized to more than two colours and for general definitions of adjacency (in terms of weighting or order of neighbour – see section 2.4). Grimson's (1991) test for clustering is analagous to the count of BB cells.

Cliff and Ord (1981, pp. 19–20) give the moments of the join-count statistics under free and non-free sampling. With free sampling, areas are independently coloured B or W with probabilities p and $1 - p$ respectively. With non-free sampling each area has the same probability a priori of being B or W but colouring is subject to the overall constraint of a fixed number of B areas and W areas (Cliff and Ord, 1981, p. 12). These moments can be used to test the null hypothesis of randomness against a general alternative of spatial autocorrelation.

The Getis–Ord G, Moran and Geary statistics are used on data $\{z(i)\}_{i=1,\dots,n}$ measured at the ordinal level or higher. Like the join-count statistic, these statistics are special cases of a general cross-product statistic

(Hubert et al., 1981):

$$\Gamma = \Sigma_i \Sigma_j G(i, j) C(i, j) \tag{7.9}$$

where $G(i, j)$ is a measure of spatial proximity of locations i and j and $C(i, j)$ is a measure of similarity of values $z(i)$ and $z(j)$. Each of these four statistics measures the spatial dependency structure in map data.

The Getis–Ord G-statistic requires that the measured variable, Z, is positive valued and has a natural origin (Getis and Ord, 1992). Let $z(i)$ denote the observed value in area i then:

$$G = [\Sigma_i \Sigma_j w(i, j) z(i) z(j)] / [\Sigma_i \Sigma_j z(i) z(j)] \quad j \neq i \tag{7.10}$$

where $\{w(i, j)\}$ is the set of weights in the weights matrix (see section 2.4) and measures the set of assumed spatial relationships between the areas with $w(i, i) = 0$ for all i. Getis and Ord (1992, p. 195, 1993, p. 276) provide the moments to test the null hypothesis against a general alternative of clustering.

The (generalized) Moran statistic is (Cliff and Ord, 1981, p. 17):

$$I = (n/S_0) [\Sigma_i \Sigma_j w(i, j) (z(i) - \bar{z})(z(j) - \bar{z}) / \Sigma_i (z(i) - \bar{z})^2] \tag{7.11}$$

where the variable Z is measured at the ordinal level or higher. The quantity S_0 is the sum of the elements in the weights matrix so that if this matrix is row standardized to 1 then $S_0 = n$. The correspondence between (7.11) and the general form for the autocovariance function (2.4(a)) and autocorrelation function (correlogram) (2.4(a)) should be noted. Note that the numerator of (7.11) is positive providing $z(i)$ and $z(j)$ are both greater than $\bar{z}$ or less than $\bar{z}$; the numerator in (7.10) distinguishes between concentrations of high as opposed to concentrations of low values. This is a feature of the Getis–Ord statistic which distinguishes it from the Moran statistic. It is useful when analysing disease rates and the analyst wants to investigate overall tendencies towards raised or lowered incidence rates (Getis and Ord, 1992, pp. 198–9).

The (generalized) Geary statistic is (Cliff and Ord, 1981, p. 17):

$$c = [(n - 1)/2S_0][\Sigma_i \Sigma_j w(i, j) (z(i) - z(j))^2 / \Sigma_i (z(i) - \bar{z})^2] \tag{7.12}$$

The similarity between the numerator of (7.12) and the general form for the semi-variogram (2.4(a)) should be noted.

The moments of the Moran and Geary statistics, under the assumption of spatial randomness, have been derived under two situations: under the randomization assumption and under the assumption of normality (Cliff and Ord, 1981, p. 21). Under the randomization assumption the observed value of I or c is assessed relative to the set of all possible values that could be obtained by randomly permuting the observed values around the areas. It is assumed that each area is equally likely to receive a value from $\{z(i)\}$. Under the normality assumption the observed values of I and c are assessed relative to drawings of the $\{z(i)\}$ from an independent and identically distributed normal model so variances are constant over the areas.

The assumption of constant variance will be violated for some types of spatial data (see section 6.2.1), for example when testing rates where the areal system includes tracts with very different population sizes (e.g. some urban and some rural). This undermines both the normal and randomization assumptions for testing significance. If there is a regular spatial structure to the non-constant variance the map can have the appearance of being spatially autocorrelated. An uncritical use of the Moran or Geary tests can then, incorrectly, confirm this observation.

Oden (1995) has proposed an adjusted form of Moran's test to take into account the effect of varying population size. Oden's modified Moran statistic is as follows. Take the case of testing for autocorrelation in prevalence rates of a disease across n areas. In terms of the notation of (7.10), $z(i) = O(i)/N(i)$ where $O(i)$ is the number of cases of the disease in area i and $N(i)$ is the population at risk in area $i (i = 1, \ldots, n)$. Let O denote the total number of cases $(\Sigma_{i=1,\ldots,n} O(i))$ and let N denote the total population at risk $(\Sigma_{i=1,\ldots,n} N(i))$. Oden's modified Moran statistic, I_{pop}, is based on computing products of the *individual-level* data. So:

$$I_{pop} = (N/W_0)\{[\Sigma_{(2)}w(s, t)((b(s) - \bar{b})(b(t) - \bar{b})]/[\Sigma_{s=1,\ldots,N}(b(s) - \bar{b})^2]\}$$

(7.12)

where $\Sigma_{(2)} = \Sigma_{s=1,\ldots,N}\Sigma_{t=1,\ldots,N}; \ t \neq s$. The vector $\mathbf{b} = (b(1), b(2),\ldots, b(N))$ and $b(s) = 1$ if the sth individual has the disease and 0 if the individual is healthy. $\bar{b}$ is the mean of the $\{b(i)\}$ so is O/N. Now $\{w(s, t)\}$, the weights matrix, is defined to measure the degree of clustering implied by any *pair of individuals* (s and t) who are both *cases*. So, the weights matrix is N by N and W_0 is the sum of the elements in the weights matrix. Oden suggests that if two cases are in the same area then the weight might be 2, 1 if the two cases are in adjacent areas, 0 otherwise including the diagonal elements $\{w(s, s)\}$ as well as between all pairs when either both are healthy or one is healthy and the other a case. Oden (pp. 19–20) gives moments for (7.12) and a method of inference. Inference is based on the randomization assumption. Randomization refers to the 0s and 1s over all the nodes in the system of which there are N, corresponding to the number of individuals. This contrasts with randomizing the rates over the n areas in (7.10) which is inappropriate because, for example, areas with large populations are less likely to deviate as much from the mean rate as areas with small populations.

The expression (7.12) illustrates the nature of Oden's modification but is not practical for computational purposes since the double summation involves N^2 terms. Oden (p. 18) provides an alternative expression for (7.12) written in terms of the n areas so that the double summation now involves n^2 terms. The revised form is derived by setting $d(i) = N(i)/N$ and $e(i) = O(i)/O$. The new expression for I_{pop} contains the term:

$$\Sigma_{i=1,\ldots,n}\,\Sigma_{j=1,\ldots,n}\,v(i, j)(e(i) - d(i))(e(j) - d(j)) \tag{7.13}$$

where $v(i, j)$ is like a spatial weights matrix except that $v(i, i) \neq 0$. $\mathbf{V} = \{v(i, j)\}$ is the matrix of area weights.

Let:

$$v(i, j) = a(i, j)/(d(i)\,d(j))^{1/2} \quad \text{if } i \neq j$$
$$= 1/d(i) \qquad\qquad \text{if } i = j$$

where $a(i, j)$ is a declining function of the distance between i and j. Now (7.13) can be written:

$$\Sigma_i[(e(i) - d(i))^2/d(i)] + \Sigma_i \Sigma_{j;(j\neq i)}\, a(i, j)$$
$$\times [(e(i) - d(i))(e(j) - d(j))]/(d(i)d(j))^{1/2} \tag{7.14}$$

and (7.14) dominates this special version of I_{pop}. Oden (p. 18) refers to (7.14) as a 'spatialized' version of a χ^2 test. The first term is a familiar χ^2 statistic since $e(i)$ is the observed share of cases in area i and $d(i)$ is the expected share of cases in area i under the null hypothesis of a random distribution of cases. The second term is the 'spatial addition' and is a 'Moran-like' spatial covariance term. The null hypothesis of no overall clustering will be rejected if either there are sufficiently large differences within regions between the observed and expected number of cases; if there are sufficiently large differences that are geographically close together (spatial clustering); or if both occur simultaneously. Oden's test and Moran's test are not testing for the presence of the same map pattern properties.

Since $d(i)$ and $d(j)$ appear in the denominators of (7.14) then the test will tend to have more power in detecting departures from the null hypothesis when the excess cases are found in areas with small populations rather than in areas with large populations. Rogerson (1999) who examines (7.14) directly, demonstrates this. However, these areas tend to have the least robust rates.

Tango (1995) defines the statistic (7.13) but with:

$$v(i, j) = \exp(-d(i, j)/\tau) \quad \text{if } i \neq j;$$
$$= 1 \qquad\qquad\qquad \text{if } i = j$$

with $d(i,j)$ the distance between area i and j and τ a constant. Now (7.13) can be written:

$$\Sigma_i(e(i) - d(i))^2 + \Sigma_i \Sigma_{j;(j \neq i)} \exp(-d(i,j)/\tau)[(e(i) - d(i))(e(j) - d(j))]$$

(7.15)

This will tend to detect departures from the null hypothesis when excess numbers of observed cases relative to the expected number are large. This is most likely to occur when the study area has settlements of different sizes but a few large centres of population which is where the disease cases cluster.

If the null hypothesis is rejected the individual terms in expressions like (7.14) and (7.15) including the individual components in the summations can be examined to indicate which areas or groups of areas have contributed most to the rejection of the null hypothesis. This links with tests to detect clusters to be described in section 7.3.

Assunção and Reis (1999) in simulation trials show that most of the power of Oden's test is concentrated in testing rate heterogeneity (the first term in (7.14)) rather than spatial correlation (the second term in (7.14)). Another test due to Waldhor (1996), that adjusts the moments of Moran's statistic, performed poorly. Assunção and Reis (1999) propose an empirical Bayes index which equalizes variances allowing a randomization test. Their empirical Bayes index is the same as the original Moran expression replacing $z(i)$ with $z^*(i)$ where:

$$z^*(i) = \{z(i) - E[z(i)]\}/[\text{Var}(z(i))]^{1/2}$$

(7.16)

where $z(i) = O(i)/N(i)$ and $E[z(i)] = O/N$. $\text{Var}(z(i)) = a + (O/N)/N(i)$ where $a = s^2 - (O/N)/[N/n]$. n is the number of areas; $s^2 = \Sigma_i N(i)(z(i) - (O/N))^2/N$. If $\text{Var}(z(i)) < 0$ then it is recalculated setting $a = 0$. In their trials (7.16) has higher power than the original Moran test in detecting spatial correlation in rates. The expression of Assunção and Reis for the case $a = 0$ and (O/N) small is the standardization proposed by Getis and Ord (1995, p. 287) which uses the result given by (6.1) where the expected value is the regional rate (O/N) (see section 6.2.1):

$$z^*(i) = (z(i) - [O/N])/[N(i)^{-1}(O/N)(1 - (O/N))]^{1/2}$$

A problem with these tests is that in the absence of any knowledge about the areal system under study, there is a risk of arriving at decisions about spatial structure based on arbitrary assumptions about the form of the weights. Hubert et al. (1981) propose a test in which it is only necessary to specify for any site i the order relations of its weights.

7.2.2   Clustering in a marked point pattern

The aim is to determine whether some attribute is clustered in a population given its spatial distribution. Each individual either does or does not have the attribute. We use the epidemiological terminology that an individual with the attribute is called a case (c), an individual without the attribute is called a control (k). Cuzick and Edwards' (1990) provide a test that takes the inhomogeneity of the population distribution in to account. Their test counts, for each case, the number of its v nearest neighbours that are also cases. This count is compared with the count to be expected under the null hypothesis of a random allocation of cases and controls. When cases are clustered the nearest neighbours to a case will tend to be other cases so that the Cuzick and Edward's statistic will be large. A problem with this test is that nearest neighbour distances may be different across the map.

Diggle and Chetwynd (1991) propose a test based on the K function which was originally developed to analyse point processes (Ripley, 1981, pp. 144–90). $N(A)$, the number of points in an area A provides the natural basis for analysing point processes and the covariance between $N(A)$ and $N(B)$ for two areas (A and B) can be reduced to a non-negative increasing function K. If λ represents the density of the point process (the number of points per unit area) then $\lambda K(d)$ is the expected number of additional points within distance d of any randomly selected point. So for cases:

$$K_{c,c}(d) = \lambda_c^{-1} E[\text{number of additional cases}$$
$$\leq \text{distance } d \text{ of a randomly chosen case]} \qquad (7.17)$$

where λ_c is the density of cases and $E[.]$ denotes the expected value of the expression in square brackets. The K function for controls is defined in similar fashion and is denoted $K_{k,k}(d)$.

The bivariate K function for the distribution of controls in relation to cases is:

$$K_{c,k}(d) = \lambda_k^{-1} E[\text{number of controls}$$
$$\leq \text{distance } d \text{ of a randomly chosen case]} \qquad (7.18)$$

Ignoring edge effects introduced by the boundary of A, the estimator of the K function (7.17) for any distance d is:

$$\lambda_c^{-1}\left[(\lambda_c|A|)^{-1}\Sigma_1 \Sigma_{m(\neq 1)} I_d(d_{1,m})\right] \qquad (7.18)$$

where $|A|$ is the area of the study region. $I_d(d_{1, m})$ is the indicator function and scores 1 if the distance between case 1 and case m is less than or equal to d, 0 otherwise. The term in square brackets, of (7.18), is the average number of additional cases within distance d of an observed case, where the average is computed over all cases. To obtain the estimate of $K_{c,c}(d)$ this quantity is then divided by the number of cases per unit area (λ_c). The estimator of the density parameter, λ_c, is the number of cases (n_c) divided by the area of the study region $|A|$. Thus (7.18) can be written:

$$[|A|/n_c^2]\Sigma_1\Sigma_{m(\neq1)}I_d(d_{1,m}) \tag{7.19}$$

A weighting function needs to be introduced on the indicator function to correct for edge effects which if not allowed for will lead to an undercount for the indicator function (Gatrell et al., 1996, pp. 262–3). Let $w_{1,m}$ denote the weighting function which for a rectangular region may be defined as the proportion of the circumference of the circle of radius d that lies within A. Now (7.19) becomes:

$$[|A|/n_c^2]\Sigma_1\Sigma_{m(\neq1)}(I_d(d_{1,m})/w_{1,m}) \tag{7.20}$$

The Diggle and Chetwynd (1991) test for clustering of cases where the population at risk is not uniformly distributed is a test of the hypothesis of 'random labelling', that is:

$$K_{c,c}(d) = K_{k,k}(d) = K_{k,c}(d) \quad \text{for all } d$$

and they propose the statistic:

$$D_{c,k}(d) = \hat{K}_{c,c}(d) - \hat{K}_{k,k}(d) \tag{7.21}$$

If the distribution of cases mirrors the population then $D_{c,k}(d) = 0$ for all values of d. Peaks in the plot of $\{D_{c,k}(d)\}$ against d and where $D_{c,k}(d) \geq 0$ are indicative of overall clustering at that distance on the map. The significance of these peaks can be established by Monte Carlo simulation. Upper and lower envelopes for the plot of $\{D_{c,k}(d)\}$ against d are obtained by performing 99 independent simulations in which the population of cases and controls are randomly marked – ensuring that the total number of cases in each simulation equals the number of cases in the data set. The 99 simulations are used to determine the maximum and minimum values of the D function at each distance d. If the empirical value of D exceeds the maximum value for any distance d this indicates clustering of cases at distance d. If the empirical value of D is less than the minimum value for any distance d this indicates that the presence of cases inhibits other cases at distance d. This provides a 0.01 significance level for the test.

There are criminological reasons to expect that there might be a tendency for clustering of victimized offenders by household address. Victimized offenders

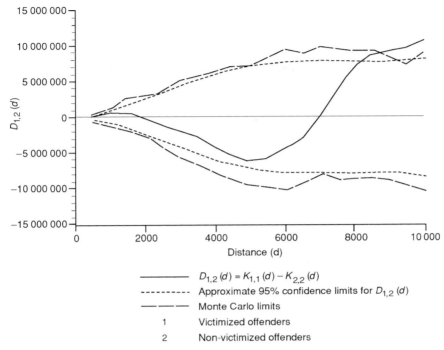

Figure 7.2 Plot of $D_{c=1, k=2}(d)$ for victimized and non-victimized offenders

are individuals who have been convicted of an offence but have themselves been the victims of a criminal offence. A test is needed to identify if the incidence of victimized offenders shows clustering or whether it simply mirrors the distribution of offenders. Figure 7.2 shows the plot of $D_{c,k}(d)$ where c now denotes victimized offenders and k denotes non-victimized offenders. The Monte Carlo limits are drawn as are the approximate 95% confidence limits (Diggle and Chetwynd, 1991). It can be seen from figure 7.2 that the only evidence of clustering is at larger distances and that there is no overall tendency to local scale clustering. This test is used by Gatrell et al. (1996) to test for whether there was overall clustering in the incidence of childhood leukaemia in west-central Lancashire (1954–92) or whether the distribution of cases simply mirrored the distribution of the child population.

Kulldorff (1998, p. 54) observes that the plot of $\{D_{c,k}(d)\}$ against d is a test of the relative clustering of two point processes and that although there are 'alternative hypotheses for which the Diggle–Chetwynd test (does) have some power. . . . It is not suitable for testing if a specific type of event shows clustering after adjusting for the inhomogeneity of another spatial point process.' Figure 7.3 from Kulldorff (1998, p. 50) shows a pattern of cases and controls

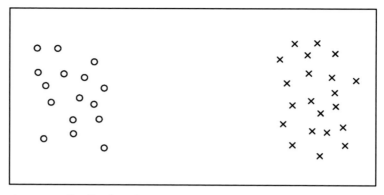

Figure 7.3 A distribution of cases and controls which would not be found to show significant clustering according to the $D_{c,k}(d)$ test (Kulldorff, 1998, figure 4.1)

for which the D-function would equal 0 for all distances and thus indicate no clustering.

Kulldorff (1998, p. 54) suggests that 'a more promising test statistic might be':

$$D^*(d) = \hat{K}_{c,c}(d) - \hat{K}_{c,k}(d) \tag{7.22}$$

which is related to a test proposed by Tango (1995).

These are just a selection of clustering techniques. Kulldorff (1998) in reviewing methods of testing for global clustering notes that 'unfortunately, there have not been many comparative power studies evaluating their various strengths and weaknesses under different alternative hypotheses' (p. 54). The power of a statistical test is the probability of rejecting the null hypothesis when the alternative hypothesis is true. Power assessments of a statistic seek to establish the types of departures from the null hypothsis the statistic is able to detect and its performance relative to other statistics. In any given situation the choice of test should be based on the type of alternative that is thought to be most likely. However, in the case of disease clustering Kulldorff (1998, p. 60) notes that the commonly encountered problem of uncertainty as to the exact locations to assign to cases (because of lack of detailed information on addresses or exactly where the disease was contracted) can severely undermine the power of any test. This problem affects other types of clustering tests, including the analysis of certain types of offence patterns for similar reasons.

7.3   The exploratory identification of local map properties

We now describe some exploratory methods for detecting local clusters, particularly important in studying disease patterns but also useful in other contexts, including the analysis of offence patterns. A cluster is an area where counts or rates are particularly large relative to some expected value based on the assumption of spatial randomness. Section 7.3.1 looks at cluster detection methods which can also be used to assess cluster alarms since they involve screening spatial data to find areas of raised incidence wherever they may be. Cluster alarms arise as a result of concerns being expressed about the presence of raised incidence in an area but usually on the basis of seeing a particular subset of the data. Assessing such a claim statistically raises, amongst others, the problems of selection bias and multiple testing. Section 7.3.2 briefly reviews focused tests which are tests for large counts or rates in the vicinity of some possible causal agent. In the epidemiological case the test might be for raised incidence for some forms of cancer near an incinerator or power station or a respiratory condition near busy roads. In criminology the test might be for excess risk of muggings close to a city's main bus terminal or burglaries of sub-urban houses along city main roads. The problem is to test for an association between what may be a higher rate and proximity to a prespecified point or line.

7.3.1   Cluster detection

(a)   Area data

Since areal units have different population sizes it is not appropriate to define clusters based on which areas have the most cases. An early test due to Choynowski (1959) computes the Poisson probability of obtaining x or more observed cases under the null hypothesis of a random distribution of cases across the study area. If $O(i)$ denotes the number of observed cases in area i, the Poisson expectation $(\lambda(i))$ for area i is the rate of cases in the total population at risk $\phi(= \Sigma_j O(j)/\Sigma_j N(j))$ multiplied by the population at risk in area i $(N(i))$. This expected value is assumed to be fixed although in practice it will be subject to sampling variation and sample size effects. The test can be readily adapted to different risk categories differentiated for example by such characteristics as sex, age, rural–urban location or even deprivation if considered appropriate and if the data are available. Separate rates are calculated for each of the C categories in the total population, $\{\phi_c\}$, and then multiplied by the corresponding population at risk in area i, $\{N_c(i)\}$. These products are then summed across all categories $c = 1, \ldots, C$ to yield the expected value of the Poisson distribution $(\lambda(i))$.

If x cases are observed in area i and if the Poisson probability:

$$\text{Prob}\{O(i) \geq x\} = 1.0 - \text{Prob}\{O(i) < x\}$$
$$= 1.0 - \Sigma_{s=0,\ldots,(x-1)} \exp(-\lambda(i)) \lambda(i)^s/s! \qquad (7.23)$$

is less than 0.05 then area i is considered to contain a cluster in area i. This version of the test applies to rare events. In the case of non-rare events (7.23) would be replaced by the binomial distribution with parameters $N(i)$ and ϕ.

Choynowski's test illustrates a number of the problems underlying cluster detection. First, the probability value of the test will be influenced by $N(i)$ so that extreme probabilities tend to be associated with areas with large populations (see section 6.2.1). Second, if there are n tests (corresponding to the n sub areas that partition the study region) then as n increases the probability of finding at least one cluster increases to one – the problem of multiple testing. The investigator analyses the problem using a particular time period and population subgroup and this further exacerbates the multiple testing problem. The number of disjoint sets from which the investigator has selected the particular data set (by area/time/population subgroup) is usually unknown, so Bonferroni corrections to determine the appropriate significance level cannot be made (see, e.g., Stone, 1988, p. 651). Third, if the reason for testing a particular area was the presence *in the collected data* of a large number of cases then the test suffers from selection bias (see also section 4.1.2). Selection bias also arises from the particular time period or population subgroup analysed. Fourth, testing is often carried out using administrative units that have no relevance in terms of the underlying process generating the cases. True clusters may be lost within an areal unit because they are small or split and hence diluted across adjacent administrative units. If clusters are small relative to the size of the areal units there may be little that can be done without collecting finer grained spatial data. However the test makes no allowance for the spatial distribution of the areas relative to each other. A group of adjacent areas may represent a significant cluster when taken together but not individually.

The spatial aspect of the selection bias problem and the fourth problem were addressed by the Geographical Analysis Machine (GAM) of Openshaw et al. (1987) by laying down multiple overlapping circles of varying size. Of course if clusters are geographically small relative to the size of the areal units for which the data are available, or if the cluster is linear, or if a cluster is split across geographical areas then these still may not be detected. GAM's 'circles' involve amalgamations of areal units and in computing the count of cases and the population at risk for any 'circle' the intra-area distribution of both ought to be used. Generally these are not known. An area is either in or out of the circle depending on whether its representative point lies inside or outside the circle.

GAM avoids the problem of spatial selection bias by testing all possible spatial subsets using the Poisson model (7.23) but as a consequence suffers from the problem of multiple testing. A Bonferroni adjustment for multiple (spatial) testing is not feasible because the divisor for the number of tests will be so large as to make the test far too conservative. Because the tests use overlapping subsets of data, the tests are not independent, so again a Bonferroni adjustment would be excessive. Although GAM provides an indication of where clusters might be located, it lacks a proper test of significance (Kulldorff et al. 1995). Its testing procedure and approach to visualization tend to pick out in the study region clusters in small, densely packed spatial units with large populations. DMAP, a test with similar strengths and weaknesses, keeps the size of the circle constant in order to allow the analyst to specify the geographic scale of clusters (Rushton and Lolonis, 1996; Rushton, 1998).

Getis and Ord (1992, 1995) develop a local version of the global G-statistic (7.10) to test for local concentrations of high or low attribute values for a set of areas. As for the global statistic, the attribute value is required to have a natural origin and be positive. The statistics, denoted $G_i(d)$ and $G_i^*(d)$, derive from the same root that supports the Diggle and Chetwynd test (see section 7.2.2), namely the K-function (Ripley, 1981) – adapting the methodology to where the data refer to counts or rates for areas. Whereas the Diggle and Chetwynd test is based on averages across the whole map, and tests for overall clustering, the Getis–Ord statistics examine each distance-defined grouping of values and tests for the presence of individual distance-dependent clusters and attempts to attach a measure of significance.

Both of the two local forms of the G-statistic evaluate the proportion of the sum of all $\{z(j)\}$ values that are within a distance d of location i. Distance can be physical distance, travel time, conceptual distance or any other measure of separation (Getis and Ord, 1995, p. 288). Whereas $G_i(d)$ excludes the value $z(i)$ in the calculation, $G_i^*(d)$ includes $z(i)$. So, whilst $G_i^*(d)$ is appropriate for detecting clusters, $G_i(d)$ is closer to (7.17) in that it evaluates the *additional* numbers of cases within distance d of i. $G_i(d)$ then has a construction similar to the earlier Diggle–Chetwynd test and the Besag–Newell test (see below).

The two Getis–Ord statistics are:

$$G_i(d) = [\Sigma_j w_d(i, j) z(j)] / [\Sigma_j z(j)] \quad j \neq i \tag{7.24}$$

and:

$$G_i^*(d) = [\Sigma_j w_d^*(i, j) z(j)] / [\Sigma_j z(j)] \tag{7.25}$$

where $w_d(i, j)$ is a symmetric zero/one matrix with ones for all entries in the matrix where area j is within distance d of i and $w_d(i, i) = 0$. In (7.25) $w_d^*(i, i) = 1$ but is otherwise identical to $w_d(i, j)$. (Results are extended to non-binary weights in their 1995 paper.) Getis and Ord (1992, with correction, 1995, p. 289) provide the expected values and variances of the two statistics. Their distribution is normal if the underlying distribution of values is normal but if this distribution is skewed, the test only approaches normality as distance increases (the sample size increases) and does so more slowly for boundary areas where there are fewer neighbours (Getis and Ord, 1995, p. 289).

The Getis–Ord local statistics avoid the problem of selection bias by considering all subsets for a given distance (d). They do not test large numbers of arbitrarily constructed circles as GAM does. This results in fewer tests so that if a Bonferroni adjustment is implemented (dividing the significance level by the number of tests performed) it does not necessarily lead to an excessively conservative testing procedure. However any such adjustment procedure will be conservative since the tests are not independent (because they use overlapping subsets of data) and become more conservative as the number of tests increases. Another concern is the effect of global autocorrelation on the significance testing (Getis and Ord, 1995, pp. 298–9, 1996, pp. 275–6). The test procedure assumes no global autocorrelation and Anselin (1995) shows that global autocorrelation has a significant impact on the distribution of local statistics such as the Getis–Ord statistics and other local statistics such as the local version of the Moran statistic. When global autocorrelation is present the expected value, variance and skewness of the distribution of Getis–Ord statistics are affected, making inference for the presence of local clusters misleading (Anselin, 1995). Bonferroni adjustments will now be still more conservative since in addition to using overlapping subsets of data, the data values themselves are locally correlated. Anselin suggests the use of Monte Carlo randomization methods that control for the level of global autocorrelation in the data to generate the sampling distribution of local statistics like the Getis–Ord and Moran statistics.

The Besag–Newell test also detects clusters of rare events over a geographical area that has been divided into small tracts, such as enumeration districts (Besag and Newell, 1991). For each tract, the number of cases of the event and the population at risk are known. The location of each tract is specified by co-ordinates, such as population weighted centroids. However unlike the previous tests, the Besag–Newell test fixes the size of the cluster rather than the population at risk so looks for significant clusters across the study region of a pre-specified size. Like GAM it uses multiple overlapping circles in order to avoid selection bias. It does not take into account the first case in any possible cluster.

The Besag–Newell test proceeds as follows. Consider any observed case and label the tract in which it occurs $A(0)$. All other tracts are ordered according to the distance to their centroids from the centroid of $A(0)$. So for this particular case the tracts are labelled $A(1), A(2), A(3), \ldots$, where the number in brackets signifies their order distance from the tract labelled $A(0)$. Count the accumulated number of *additional* cases that correspond to this ordering. The accumulated count of cases ignores the first case in $A(0)$ so $D(0)$ is the number of other cases in $A(0)$; $D(1)$ is $D(0)$, plus the number of cases (if there are any) in $A(1)$; $D(2)$ is $D(1)$ plus the number of cases (if there are any) in $A(2)$; and so on. It follows that $D(0) \leq D(1) \leq D(2) \leq \cdots$. Now $u(0)$ is the population at risk in $A(0)$ minus 1 which discounts the case that is the origin of the test. $u(1)$ is $u(0)$, plus the population at risk in $A(1)$; $u(2)$ is $u(1)$ plus the population at risk in $A(2)$; and so on.

The test is based on identifying the zones $A(0)$, $A(1)$, ..., $A(M)$, but not $A(0)$, $A(1)$, ..., $A(M-1)$ such that at least k other cases (excluding the case that was the origin of the test) occur. Intuitively a small value of M is indicative of a significant cluster of size $k + 1$ around $A(0)$. If for any particular case $M = m$ then the significance level of the test is $\text{Prob}\{M \leq m\}$ under the null hypothesis. The significance level of the test (the probability of the null hypothesis being true that there is no cluster at $A(0)$) is assessed using the Poisson distribution with parameter $\lambda = u(m)p$ where p is the total number of cases in the study area divided by the total population at risk. The significance level is given by:

$$\text{Prob}\{M \leq m\} = 1.0 - \text{Prob}\{M > m\}$$
$$= 1.0 - \Sigma_{s=0,...,(k-1)} \exp(-\lambda)\lambda^s/s! \qquad (7.26)$$

The test is repeated for all cases but if there are multiple cases in any one tract the test statistic will be the same for each case. Besag and Newell suggest that a useful diagnostic is to plot all clusters that attain the significance level (α) of 5%.

Whereas GAM constructs arbitrary 'circles' within which the population distribution is usually unknown, the Besag–Newell test only aggregates complete zones for which the population is known – similar to the Getis–Ord local statistics. It is necessary however to specify a value for k and one option is to perform the test using a range of values of k. The smallest value of k might be set with reference to Poisson probabilities so that for example start at $k = 2$ if $\lambda < 0.354$, start at $k = 3$ if $\lambda \geq 0.354$.

Even a random map pattern will have some apparent clusters of cases. Are there more clusters than would be expected in a random pattern? Besag and Newell (1991, pp. 150–1) propose two methods of testing. The first is to compare the number of significant tests against the expected number of significant tests under the null hypothesis of a random pattern. The expected count is the total number of cases multiplied by the significance level (α) chosen for the tests. They propose another statistic which is the proportion of all cases that belong to at least one significant cluster. Monte Carlo simulation can be used to provide tests of significance for these statistics. They conclude though that

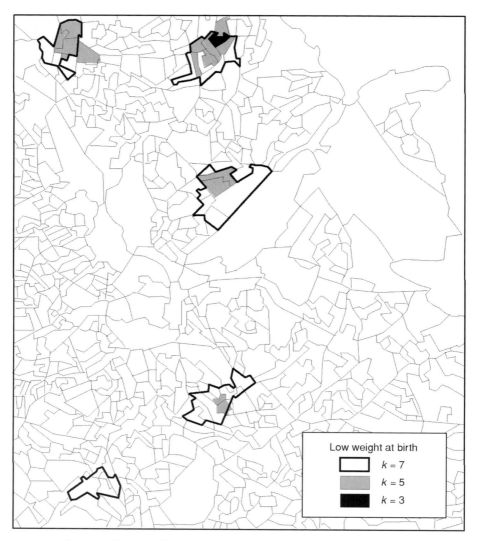

Figure 7.4 Clusters of low weight births in Sheffield using the Besag–Newell test. Cluster sizes $k = 3, 5$ and 7

if these tests are not significant this does not mean that the individual clusters that have been detected should not be investigated.

Figure 7.4 identifies clusters of low-weight births (<2.5 kg) in Sheffield using the Besag–Newell test. The study was carried out as part of an audit of children's services. The value of k is set equal to 3, 5 and 7 and in this case clusters emerge at different levels of k but each cluster detected at one level of k persists as k is increased. The clusters, of varying size, are apparently significant. Kulldorff (1998, p. 55) remarks however that like GAM and DMAP (and the

Getis–Ord statistics), the Besag–Newell technique is 'unable to do inference on statistical clusters in order to determine if the areas have an excess rate that is statistically significant or not'. Its main value is in showing where there are specific areas with high observed rates. The next test, Kulldorff's scan test, overcomes this problem, at least with respect to the most likely cluster.

The Poisson version of Kulldorff's scan test (Hjalmars, Kulldorff et al., 1996) builds on GAM and a test due to Turnbull et al. (1990). Again the number of cases in a tract and the population at risk is known and each tract is assigned a single co-ordinate denoting its location. The window of the scan statistic (a circle which increases continuously) moves over the study region. This window or zone is always centred on a tract co-ordinate and the tracts contained in any zone are those tracts whose co-ordinates lie within the radius of the zone. As with GAM, the 'circular' zones in fact will be irregular in shape when represented in terms of the underlying tracts that are included. The scan test considers all possible circles of size up to a specified percentage of the population at risk in the study region. If the maximum upper limit of 50% is chosen the scan test identifies the 'most likely' cluster in the interval (0%, 50%).

The scan test is based on a model where there is exactly one circular zone or region (R) out of the set of all circular windows **R** such that for all individuals lying within R the probability of being a case is p whereas for all individuals lying outside R (R^c) the probability is q. The study region is $T = R \cup R^c$. If tract i is inside R then $O(i)$ is Poisson distributed with parameter $pN(i)$ where $N(i)$ is again the population at risk and p can be interpreted as proportional to the relative risk inside R. If tract i is outside zone R then $O(i)$ is Poisson with parameter $qN(i)$. The null hypothesis is that $p = q$ (complete spatial randomness) and the alternative hypothesis is that $p > q$. This means people living inside R have a risk of contracting the disease p/q times higher than those people living outside R.

The scan statistic uses a likelihood ratio test: the ratio of the likelihood of the data under the alternative hypothesis to the likelihood of the data under the null hypothesis. The likelihood function is written as a product of independent Poisson distributions (see Kulldorff, 1997; Hjalmars et al., 1996, p. 710). The test identifies the maximum value of the likelihood ratio and the zone which yields the maximum is R. The distribution of the likelihood ratio is obtained by 999 Monte Carlo replications of the data set under the null hypothesis. R is a significant cluster, at the 5% level if the value of the test statistic is amongst the top 50 of the 1000 values. Note that the test only tests for the most significant cluster, although secondary clusters not part of the most significant cluster are identified. The scan test is conservative in testing for secondary clusters so that if their probability values are also less than 0.05 then it is legitimate to include

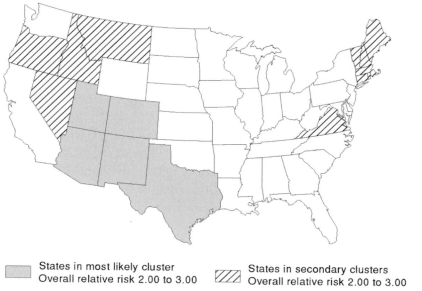

States in most likely cluster
Overall relative risk 2.00 to 3.00

States in secondary clusters
Overall relative risk 2.00 to 3.00

Figure 7.5 States in most likely cluster (significant) and in significant secondary clusters of measles cases in the USA (1962) according to the Poisson version of Kulldorff's scan test

these secondary clusters as possible clusters. Note that setting the maximum circle size to the upper limit of 50% may result in the most likely cluster occupying a very large proportion of the study area. This will not be very informative. Reducing the upper limit constrains the maximum size of R.

Figure 7.5 shows the results of Kulldorff's scan test on measles data for the contiguous states of the USA for 1962. In the early 1960s the number of measles cases was still quite high (around 400 000 cases a year) relative to today's levels which are much reduced due to high rates of vaccination. For each state the population at risk and the number of measles cases are known. Each state is represented by a co-ordinate that corresponds to the location of the state capital. A state is contained within a zone if its representative co-ordinate falls within it. A group of south-western states appear as the significant 'most likely cluster'. Three statistically significant secondary clusters are also identified with high relative risk rates in areas that include a part of the north-west and New England. As will be evident from earlier comments, the significant clusters are not necessarily the areas with the highest relative risk. Since measles is an infectious disease, this application is indicative of where the high rates of the disease were at this particular time, not long-term relative risk (Haining et al., 2002). Viel et al. (2000) apply the scan test to the distribution of soft tissue sarcomas and non-Hodgkin's lymphomas around a municipal solid waste incinerator in France which for these conditions may be indicative of long-term relative risks.

(b)   Inhomogeneous point data

The results of applying the Diggle and Chetwynd (1991) D statistic to the victimized offender data showed there was no overall tendency to clustering. There may however be local clusters. The Bernoulli version of Kulldorff's scan test can be used to detect the most likely cluster and secondary clusters of cases in an inhomogeneous point pattern – that is identify the presence of clusters of individual cases given the distribution of the individuals who comprise the population at risk.

Let $N(R)$ denote the number of individuals in zone R, $O(R)$ the number of cases in zone R. For all individuals within zone R the probability of being a case is p, whilst for all individuals outside $R(R^c)$ the probability is q. The study region, $T = R \cup R^c$. The null hypothsis is that $p = q$ and the alternative hypothesis is that $p > q$. The likelihood function is written as a product of the individual Bernoulli probabilities (Kulldorff and Nagarwalla, 1995, p. 803):

$$L(R, p, q) = p^{O(R)}(1 - p)^{N(R)-O(R)} q^{O(T)-O(R)} (1 - q)^{(N(T)-N(R))-(O(T)-O(R))}$$

$$(7.27)$$

where $N(R)$ and $O(R)$ are the population at risk and number of cases inside R respectively. $N(T)$ and $O(T)$ are the population at risk and number of cases in the study region $(R \cup R^c)$ respectively.

The likelihood ratio statistic is the maximum of $L(R, p, q)$ under the alternative hypothesis divided by its maximum value under the null hypothesis. The estimator for q under the alternative hypothesis is $[O(T) - O(R)]/[(N(T) - N(R))]$ and for p the estimator is $O(R)/N(R)$. Under the null hypothesis the estimator for p and q is $O(T)/N(T)$. Under the null hypothesis then the likelihood function (7.27), which is the denominator in the likelihood ratio statistic simplifies to:

$$L(R, p = q) = [O(T)^{O(T)}(N(T) - O(T))^{N(T)-O(T)}]/N(T)^{N(T)} \qquad (7.28)$$

The most likely cluster is the zone R which maximizes the likelihood ratio statistic (see Kulldorff and Nagarwalla, 1995, p. 803) over all possible zonings. Significance is, as for the Poisson version, determined by Monte Carlo sampling from the exact distribution of the likelihood ratio statistic. This involves randomly allocating the given number of cases to individuals and evaluating the likelihood ratio statistic. The rank order of the empirical value within the distribution of simulated values determines the significance of the cluster. A 5% significance level is usual.

Figure 7.6(a) Clusters of victimized offenders: (a) significant clusters according to
the Bernoulli version of Kulldorff's scan test

Figure 7.6(a) shows the most likely cluster ($p = 0.014$) and the presence of
two significant secondary clusters ($p = 0.042, 0.049$) for the victimized offender
data. The evidence of the scan test suggests a number of local clusters of victimized offenders which may warrant closer investigation. Figure 7.6(b) shows all
the Kulldorff clusters and the relative risk for offenders living in these areas

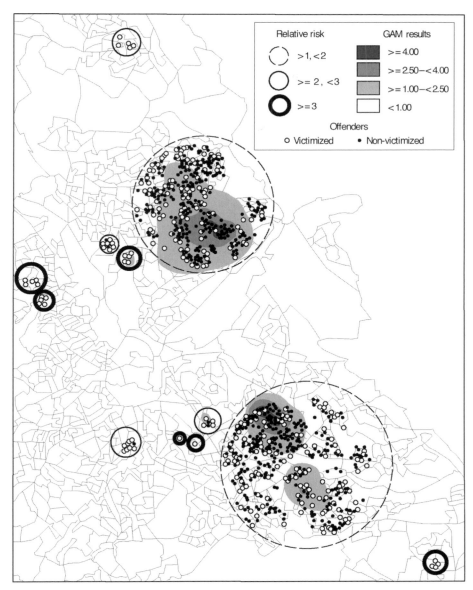

Figure 7.6(b) Areas with the highest relative risks according to Kulldorff's scan test, and the results of using Openshaw's GAM

to be victimized. Note that the highest relative risks are not associated with the significant clusters. Superimposed on figure 7.6(b) are the clusters detected by GAM. Figure 7.6(c) shows the Besag–Newell clusters and the GAM clusters.

Kulldorff (1998, pp. 59–60) comments on how data problems affect tests for randomness in inhomogeneous point data. A particular concern is location

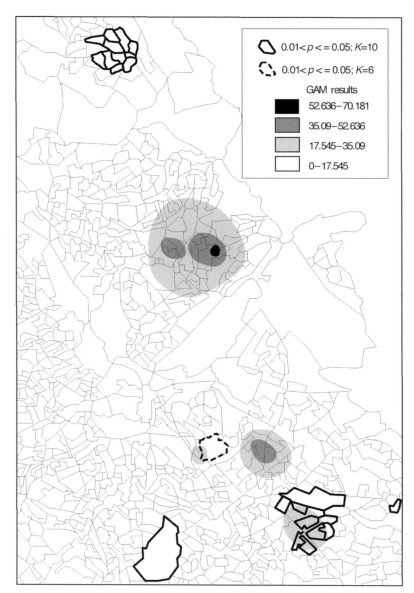

Figure 7.6(c) Areas with clusters of cases according to Openshaw's GAM and the Besag–Newell test

error, that is assigning the wrong locational reference to an event. There may be a digitizing error; the patients home address may be geo-coded but was the disease contracted at home or at work or somewhere else as the individual moved about in their daily routine? With chronic diseases, spatial analysis needs to recognize the migration history of cases. Imprecision arises when for

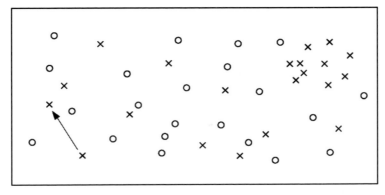

Figure 7.7 The effect of location error (case bottom left) on tests for overall clustering and tests to detect individual clusters (Kulldorff, 1998, figure 4.2)

confidentiality reasons cases are assigned to postcodes or ED centroids rather than home address. Jacquez (1994) adapts the Cuzick and Edward's test for clustering to the case where locations are inexact thus giving rise to ties. Test results for randomness will also be sensitive to other problems including classification error and missing cases – particularly if the sample size is small.

These types of data problems can affect the power of tests for overall clustering and tests for local clusters. The ability to detect a particular local cluster will be reduced if cases *in the cluster* are misclassified, wrongly located or simply missed out, although this will not affect the evidence concerning the existence of other possible clusters elsewhere on the map. The ability to detect overall clustering will be affected if *anywhere on the map* cases are, for example, incorrectly located. This is illustrated in figure 7.7 from Kulldorff (1998, p. 51). This location error will not affect detection of the cluster in the top right.

For recent developments in this area see the special edition of the *Journal of the Royal Statistical Society* (Series B), 2001, part 1.

7.3.2   Focused tests

As with cluster detection tests, focused tests can be classified into those where data are available in the form of counts and those where data are available as point data. Stone's (1988) test has become widely used and is applicable to both point and area data. We briefly review the area count version.

The test assumes that observed counts $(O(i))$ of the event are known for each of $i = 1, \ldots, N$ areas. In addition expected counts $E(i)$ are known (and assumed fixed) based on the null hypothesis of spatial randomness. The $E(i)$ can also take into account confounders such as demography and/or socio-economic characteristics of the population if necessary and if these data are available.

Each area has a location that is fixed by a representative centroid. This means that the N areas can be ordered in terms of their distance from the point (or line) source that is the object of investigation as the possible source or centre for (the suspected) raised numbers of events. Assume that the labelling of the areas $i = 1, \ldots, N$ corresponds to that spatial ordering. Stone's test assumes that the risk is strictly non-increasing with distance from the putative source and there may be circumstances where this assumption may be too restrictive at the scale of the spatial units. In the case of offences there may be a zone that is too close to a bus terminal for offenders to operate within – in this case spatial aggregation is necessary. In the case of disease, environmental factors (e.g., prevailing wind in the case of certain types of emissions) might give rise to directional bias.

Stone's test is a maximum likelihood ratio test based on the ratio of the likelihood under the alternative hypothesis to the likelihood under the null hypothesis. The null hypothesis is that the observed counts are Poisson distributed with expectation $r(i)E(i)$ where $r(i)$, the relative risk, is 1. The alternative hypothesis is that the observed counts are Poisson distributed with expectation $r(i)E(i)$ where $r(1) \geq r(2) \geq \cdots \geq r(N)$. A simple form of Stone's test is given by the statistic (Stone, 1988, p. 654):

$$T_r = \max_{(1 \leq n \leq N)} \left[\Sigma_{i=1,\ldots,n} O(i) / \Sigma_{i=1,\ldots,n} E(i) \right]. \tag{7.29}$$

and the formulae for the significance level associated with an observed value of T_r are provided. The value of the statistic T_r has a natural geometric interpretation which shows its links to (isotonic) regression. If the cumulative number of expected cases (horizontal axis) are plotted against the corresponding cumulative number of observed cases (vertical axis) then the null hypothesis is the unit slope that passes through the origin of the graph $(O(i) = E(i) = 0)$. The statistic T_r is the slope of the line that also passes through the origin of the graph and is such that all points lie on or to the lower right of the line (figure 7.8).

Stone's original formulation of the test assumes the risk function for the $\{r(i)\}$ is unknown. Bithell's linear risk score will be the most powerful test in those situations where the underlying relative risk function is known. In this case the expected value under the (simple) alternative hypothesis can be specified and a decision taken as to whether to favour it over the null hypothesis. Bithell (1995) suggests inverse distance squared or the reciprocal of distance rank will often reflect prior belief about the pattern of risk in a region. For a discussion of these and other tests see Bithell (1995). For further discussion together with the case for point data see Diggle et al. (1999).

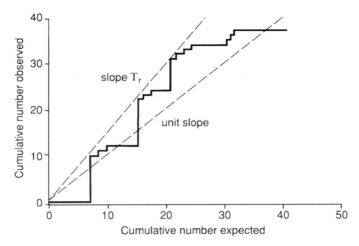

Figure 7.8 Cumulative number of observed cases against the expected number of cases and a graghical representation of Stone's test using data for Sizewell (Stone, 1988, p. 657, figure 3)

7.4   Map comparison

This section describes methods for assessing the association between two mapped distributions. We assume that data have been recorded on two variables G and H at locations in the study region. The situation may be complicated by the fact that the number of observations on G ($g(1), \ldots, g(n_g)$) and H ($h(1), \ldots, h(n_h)$) may not be the same and the two sets of data may not be recorded at the same locations or with respect to the same spatial units. Environmental data and health data frequently fall into this category since data are collected on different spatial frameworks by different agencies.

(a)   Bivariate association

In the simplest situation data are available for a common set of spatial units ($i = 1, \ldots, n$) and no data are missing. The data on G and H can be displayed using a scatterplot which will give a preliminary indication of association. The null hypothesis of no association between G and H can be tested against a general alternative that there is an association using a correlation statistic. If both G and H are spatially autocorrelated care will need to be taken in carrying out these familiar tests of association. Correlation analysis in the presence of spatial autocorrelation will be discussed in chapter 8.

We now turn to the situation where the interest is in obtaining a measure of association between two sets of polygons referring to attributes G and H. Wang et al. (1997) use resemblance tables to quantify the association between maps of earthquake, flood and drought hazards for China. Regions were

Table 7.1 *4 × 4 and 2 × 2 resemblance tables for flood and drought hazard maps of China*

		Drought					Drought	
		S	H	M	L		S+H	M+L
Flood	S	0.00	0.00	0.00	0.01	S+H	0.02	0.25
	H	0.01	0.00	0.04	0.18	M+L	0.12	0.60
	M	0.03	0.00	0.06	0.15			
	L	0.06	0.02	0.04	0.34			

Area ratio: Area ratio:
main diagonal: 0.41.   main diagonal: 0.62
opposite diagonal: 0.13   opposite diagonal: 0.37

identified for each hazard where the intensity level of the hazard was in one of four categories: severe (S), heavy (H), moderate (M) or light (L). A measure is needed of the degree of correspondence between intensity levels for different hazards.

Table 7.1 shows the 4×4 and 2×2 tables for flood and drought. Values refer to the proportion of the country under these different combinations of levels. Wang et al. (1977) sum diagonal values and compare this against the opposite diagonal to provide a simple descriptive measure of association. There appears to be an association between flood and drought because large areas of China simultaneously experience moderate and light levels of both hazards.

Is it possible to test for statistically significant association? Court (1970) (described in Norcliffe (1977, pp. 128–38)) provides an inference theory that can be used on 4×4 resemblance tables if row and column totals equal 0.25. Maruca and Jacquez (2002) consider a problem that is related to the above which is to measure the *area overlap* between two sets of polygons (I and J) each comprised of a number of small polygons ($I_1, \ldots, I_{N(I)}; J_1, \ldots, J_{N(J)}$) which each exhaustively partition a geographic region. The aim is to quantify the match between these two sets of polygons. Two of the statistics are based on computing the average maximum area of overlap of the polygon set for I with respect to J and vice versa. So, for example, for each polygon in I find the polygon in J it overlaps best with, in the sense that the intersection divided by the union of the two polygons is the largest over all polygons in J. These maxima are summed over all polygons in I and the total divided by $N(I)$. The problem is complicated by boundary effects, differences in the spatial scale of the two partitionings and where polygons differ in size. Maruca and Jacquez suggest introducing a term-by-term weighting equal to the size of the corresponding polygon from I. Monte Carlo randomization procedures are used to test for significance corresponding to a specific null hypothesis.

These tests extend earlier work by Jacquez (1995) developing statistical tests to determine whether boundaries for two or more variables coincide or overlap significantly based on *boundary overlap* statistics. The methodology can be applied to the case where different numbers of observations have been taken on G and H and the spatial locations of the observations do not correspond. A boundary detection algorithm is used to identify zones of rapid change on G and H respectively (see also Jacquez et al. (2000) and section 6.2.2(b)). Jacquez (1995, p. 2346) defines four overlap statistics that are computed on these boundaries. The first overlap statistic counts the number of locations that define boundaries on *both* G and H – the extent of coincidence. This statistic is unlikely to be appropriate if the two sets of locations on which G and H are recorded are different. The second (and third) measure the average distance from a location of rapid change on $G(H)$ to the nearest location of rapid change on $H(G)$ – the extent to which boundaries are close together. The fourth statistic is the average distance from any location on a G or H boundary to the nearest boundary for the other variable. Each statistic is sensitive to different aspects of overlap so it is necessary to examine the behaviours of the two variables on the two sides of each boundary to check for consistency of shift. The sampling distributions for these statistics are obtained through Monte Carlo simulation assuming no spatial autocorrelation. This leads to a conservative testing procedure. Other null hypotheses are discussed, including the choice between unconditional randomization (both G and H are randomized) and conditional randomization (where one is fixed, say G, and the the other, H, is randomized) and controlling for the observed spatial autocorrelation in the data (Jacquez, 1995, p. 2347). Jacquez illustrates the method on data on respiratory illness and air pollution for Southern Ontario.

These methods examine association by considering all the data and provide a single 'whole map' summary of the association between the two mapped data sets. Spatial heterogeneity can be present in the association between two variables – that is association varies from one part of the map to another. The association between house price and different house attributes could vary spatially because people living in different areas attach different price values to the same attribute. Measures of association can be computed for subsets of the data by moving windows over the region. A problem will be to know when differences are statistically meaningful and when they are simply an artefact of, for example, sample size. Fotheringham et al. (2000, pp. 107–28) adapt the method of locally weighted regression (see section 7.1.1) to fitting regression equations to spatial data. The method, geographically weighted regression (or GWR), slides a moving window over the study area and fits the same regression equation to each spatially defined subset. The fitting procedure is weighted least squares. Weights are specified in terms of the distance of each case from the middle point of the window. The procedure raises the same issues relating to choice of kernel function and bandwidth as discussed in section 7.1.3 for map smoothing. Geographically weighted regression is a method of exploring spatial heterogeneity in attribute–response relationships and output usually takes the form of maps displaying the spatial variation in parameter values.

It is necessary to be cautious in fitting and interpreting a local regression model. Since the data subsets that are taken are spatial but the model fitted is function dependent, the analyst should be satisfied that the relationships are linear when using GWR. If a single, global but non-linear relationship adequately describes the data, fitting to spatially defined data subsets could be misleading. If different geographical areas are dominated by values from different parts of the curve these would tend to generate results, suggesting that the relationship is not homogeneous across the map. Since regression fits can be sensitive to other statistical assumptions (e.g. outliers and high leverage cases), these need careful monitoring as part of interpreting results.

(b)   Spatial association

The usual correlation coefficent statistics (such as Pearson's and Spearman's) only test whether there is an association between G and H by comparing values at the same location: $\{(g(j), h(j)\}$. Map comparison however often involves more than pairwise comparison between data recorded at the same locations. The following example illustrates why. Suppose G is a measure of air pollution by area whilst H is a measure of the rate of a respiratory disease amongst the resident population by area. Even if relatively large values of G and H are not found in exactly the same spatial units, it would still be indicative of an association if relatively large values on G and H occupied locations that were close together in space. This is because spatial units are arbitrary subdivisions of the study region and people move around so will be exposed to air pollution in areas other than where they live.

There is much discussion about what is an appropriate measure of bivariate spatial association. Lee (2001) for example, develops an index (L) that combines Pearson's bivariate correlation between G and H with Moran's spatial autocorrelation measures for G and H. However the index does not seem interpretable without reference back to its three components. Historically, computing cross correlations, between $\{g(j), h(N(j)\}$, that is between G measured at location j and H measured at a neighbour of $j(N(j))$ have been used to quantify the extent to which similar values on G and H are found close together. In the case of a grid of observations then cross correlations can be computed between pairs of values and a matrix of such cross correlations computed for different orders of neighbours and different directions. In the case of an irregular spatial partition, a series of cross correlations can be computed pairing individual observations with the average value in first-order neighbours, second-order neighbours and so on.

In the case of ordinal data maps of maximum, minimum and average differences between the rank of H at location j and the ranks of G at j and the neighbours of j are one way to display spatial association between G and H. A formal test is provided by Tjostheim's (1978) index of spatial association (Λ) which measures the degree to which identically ranked values on G and H occupy positions that are close together in terms of straight line distance. If $(s_{G;1}(i), s_{G;2}(i))$ denotes the location of the ith ranked value on the variable G and $(s_{H;1}(i), s_{H;2}(i))$ is the same for H then:

$$\Lambda = \Sigma_{i=1,\ldots,n}[s_{G;1}(i), s_{H;1}(i) + s_{G;2}(i), s_{H;2}(i)]$$

When the s_1 and s_2 co-ordinates are standardized to a mean of 0 and unit variance:

$$\Lambda = 0.5[r(s_{G;1}, s_{H;1}) + r(s_{G;2}, s_{H;2})]$$

where the $r(s_{G;1}, s_{H;1})$ and $r(s_{G;2}, s_{H;2})$ are the correlations between the s_1 and s_2 co-ordinates respectively for identically ranked observations on G and H. Tjostheim derives the mean and variance of Λ under the null hypothesis of no bivariate association and no spatial association between G and H and spatial independence of both G and H. He shows how under these assumptions Λ can be tested as a standard normal deviate for large n. For small n that do not justify the normal approximation to the sampling distribution of Λ, the suggestion is to obtain it by Monte Carlo randomization methods, fixing the locations of the set of ranks for G and then assigning the ranks for H at random across the set of locations.

Doubts have been expressed as to whether the form of Tjostheim's null hypothesis is appropriate: (i) observations on H may not be spatially independent in the application; (ii) each permutation of the ranks for H generates a different value of bivariate association between G and H (Hubert et al., 1985). The investigator may wish to specify a more appropriate null hypothesis – preserving or controlling for the spatial autocorrelation in H and/or preserving or controlling for the level of bivariate association between G and H. Restricted Monte Carlo experiments will be required as discussed in section (a). To control for bivariate association the values of G and H will need to be permuted in pairs. To control for spatial autocorrelation then this property must also be preserved in the simulations. It is important to be clear on the specification of the null hypothesis because the interpretation of results depends upon it. The probability values for the test statistic under a simple null hypothesis of the sort specified by Tjostheim in his original paper will lead to a conservative test if they are (improperly) used to accept or reject a more restricted null hypothesis.

Hypothesis testing and spatial autocorrelation

8

Hypothesis testing in the presence of spatial dependence

This chapter examines a number of commonly used statistics and describes how methods of statistical inference are affected when the data are assumed to derive from a generating model (see section 2.1.4) which does not satisfy the conventional assumption of independence. Spatial data often fail to satisfy the conventional model underlying statistical inference. So in using certain tests, such as the usual tests for differences of means or correlation found in conventional statistical texts, the analyst should be concerned about the possible effect of spatial dependence on test results.

When n observations are made on a variable that is spatially dependent (and that dependence is positive so that nearby values tend to be similar) the amount of information carried by the sample is less than the amount of information that would be carried if the n observations were independent. When observations close together in geographic space are dependent in this way, taking a second measurement close to another may not provide much additional information about the process. A general consequence of this is that the sampling variance of statistics are underestimated. As the level of spatial dependence increases the underestimation increases. (This effect is reversed if spatial dependence is negative – the sampling variance of the statistic is overestimated.)

Where there is positive spatial dependence and the analyst wants to place confidence intervals on statistics then these intervals will be misleadingly small giving a spurious impression of the level of precision of the estimated quantity. Where the analyst wants to test a hypothesis the risk of committing a type I error is increased – the probability of rejecting the null hypothesis when it is true is greater than the nominal value (1%, 5%, 10%) selected for the test.

The methods that have been developed to cope with this problem are principally based on modifying the statistical test in some way in order to make allowance for spatial dependence. The modification may take the form of determining the sampling distribution of the statistic using permutations of the data that preserve the global spatial autocorrelation structure in the data. Another approach is to adjust the sample size (n) so that the adjusted value (n') measures the 'equivalent' amount of independent information in the sample. In the case of positive spatial dependence n' will be less than n. This has the effect of enlarging the estimate of the sampling variance of the statistic (in the case of positive spatial autocorrelation) under the null hypothesis.

In this chapter tests are considered with modifications that could be implemented without fitting parametric models of the sort to be described in chapter 9 relying instead on estimating the spatial autocovariance structure in the data and forming the variance–covariance matrix from these estimates. There are problems with this that are similar to those encountered elsewhere (e.g. in implementing kriging without fitting a model for the spatial covariances or semivariogram). The estimates are affected by sampling variation. The variance–covariance matrix constructed from computed estimates of the different orders of spatial autocovariance may not be permissable (see chapter 9). Further, some adjusted tests require the variance–covariance matrix to be inverted and this inversion may be unstable. On the other hand taking a model based approach can lead to 'overadjustment' if the model overstates spatial correlation at longer distances (Haining, 1991, p. 215). At this point the need to fit a model to spatial autocovariance estimates is simply noted, it will be discussed in more detail in chapter 9.

The asymptotic distribution theory for the conventional form of the test may hold in the case of autocorrelated data. Tests in conventional areas of statistics invoke the central limit theorem. Central limit theorems for spatial data have been proved under conditions that assume spatial dependence tails off to 0 at some distance (Bolthausen, 1982; Guyon, 1995).

The other approach to dealing with the problem of spatial dependence is to pre-whiten the data. Pre-whitening may involve modelling the mean of the data using some order of trend surface (see chapters 9 and 10) and then de-trending the data. The de-trended data (the original data values minus the estimated trend) are then used for subsequent analysis and conventional methods that assume independence may be appropriate. Another form of pre-whitening is variate differencing where:

$$\Delta z(i) = z(i) - \Sigma_{j \in N(i)} z(j) \tag{8.1}$$

which is analogous to removing a linear trend. Note that both forms of pre-whitening produce data that have a mean of 0 by construction so the methods are unsuitable for tests on means.

In general pre-whitening alone is rarely sufficient to cope with the problem raised by spatial dependence. However, there may be circumstances where a combination of methods is appropriate. Pre-whitening is undertaken to re-move (first-order) trend properties in the data but because this may not have re-moved (second-order) autocorrelation properties the first set of methods based on adjusting the sampling variance of the statistic need to be applied on the de-trended data. This approach has been proposed for example in implementing tests of bivariate correlation (Clifford et al. 1989; Haining, 1991).

Section 8.1 reviews the effect of spatial autocorrelation on testing the mean of a spatial data set whilst section 8.2 reviews the effects of spatial autocorre-lation on tests of *bivariate* association. The implementation of the methods de-scribed here arises as follows. In the case of testing a mean, first a test of spatial autocorrelation is carried out on the variable and the null hypothesis of no spa-tial autocorrelation is rejected (see chapter 7). In the case of a test of association, tests for spatial autocorrelation on both variables lead to the null hypothesis be-ing rejected in both cases. In these cases the analyst will need to implement the methods of this chapter rather than conventional test procedures.

8.1   Spatial autocorrelation and testing the mean of a spatial data set

Suppose n independent observations $\{z(i)\}$ are drawn from a $N(\mu, \sigma^2)$ distribution. The sample mean, $\bar{z}$, is an unbiased estimtor for μ, and the vari-ance of the sample mean is:

$$\text{Var}(\bar{z}) = \sigma^2/n \tag{8.2}$$

If σ^2 is unknown then it is estimated by:

$$s^2 = (1/(n-1)) \, \Sigma_{i=1,\dots,n} \, (z(i) - \bar{z})^2 \tag{8.3}$$

so that:

$$\text{Var}(\bar{z}) = (1/n(n-1)) \, \Sigma_{i=1,\dots,n} \, (z(i) - \bar{z})^2 \tag{8.4}$$

If the n observations are not independent then although the sample mean is still unbiased as an estimator of μ then assuming each $z(i)$ has the same variance (σ^2) the variance of the sample mean is (see, e.g., Haining, 1988, p. 575):

$$\text{Var}(\bar{z}) = \sigma^2/n + (2/n^2) \, \Sigma_i \, \Sigma_{j(i<j)} \, \text{Cov}(z(i), z(j)). \tag{8.5}$$

So, if there is positive spatial dependence and σ^2 is known then σ^2/n under-estimates the true sampling variance of the sample mean. If σ^2 is unknown and is estimated by (8.3) then if there is positive spatial dependence the expected value of s^2 is (see, e.g., Haining, 1988, p. 579):

$$E[s^2] = \sigma^2 - [(2/n(n-1))\, \Sigma_i \Sigma_{j(i<j)} \operatorname{Cov}(z(i), z(j))] \qquad (8.6)$$

so that (8.3) is a downward biased estimate of σ^2. This further compounds the underestimation of the sampling variance. Haining (1988) provides examples of the underestimation for particular situations.

Modified methods to take account of spatial dependence are based on the following argument (see, e.g., Haining, 1988). Assume the data $\mathbf{z}^T = (z(1), \dots, z(n))$, where T denotes the transpose, are drawn from a multivariate normal spatial model with mean vector given by $\mu\mathbf{1}$ and $n \times n$ variance–covariance matrix $\Sigma = \sigma^2 \mathbf{V}$. The log likelihood for the data is:

$$-(n/2)\ln 2\pi\sigma^2 - (1/2)\ln |\mathbf{V}| - (1/2\sigma^2)(\mathbf{z} - \mu\mathbf{1})^T \mathbf{V}^{-1}(\mathbf{z} - \mu\mathbf{1}) \qquad (8.7)$$

where $\mathbf{1}$ is a column vector of 1s and $|\mathbf{V}|$ denotes the determinant of $\mathbf{V}$. We assume $\mathbf{V}$ is known. The maximum likelihood estimator of μ is:

$$\tilde{\mu} = (\mathbf{1}^T \mathbf{V}^{-1}\mathbf{1})^{-1}(\mathbf{1}^T \mathbf{V}^{-1}\mathbf{z}) \qquad (8.8)$$

The estimator (8.8) is the best linear unbiased estimator (BLUE) of μ. Note that in the case of independence $\mathbf{V} = \mathbf{I}$ (the identity matrix with 1s down the diagonal and 0s elsewhere) and (8.8) reduces to the sample mean. In the case $\mathbf{V} \neq \mathbf{I}$ two modifications to the sample mean are occurring. First the denominator in (8.8) for positive spatial dependence will be less than n. Second the presence of $\mathbf{V}^{-1}$ in the numerator of (8.8) downweights the contribution of any attribute $z(i)$ where it is highly correlated with other attribute values $\{z(j)\}$ – that is where $z(i)$ is part of a cluster of observations (see also 4.4.2(v)).

The variance of $\tilde{\mu}$ is:

$$\operatorname{Var}[\tilde{\mu}] = \sigma^2(\mathbf{1}^T \mathbf{V}^{-1}\, \mathbf{1})^{-1} \qquad (8.9)$$

which reduces to σ^2/n if $\mathbf{V} = \mathbf{I}$.

Since the sample mean is an unbiased estimator of μ, one modification is to replace (8.2) with (8.9). The term $(\mathbf{1}^T \mathbf{V}^{-1}\mathbf{1})$ is proportional to Fisher's information measure (Haining, 1988, p. 586). It identifies the information about μ contained in an observation. Suppose $\mathbf{V}$ is defined by a first-order simultaneous spatial autoregression with a single parameter ρ (see p. 301) and with row sums of the weights matrix set to one. The parameter ρ lies in the interval $(-1, +1)$ but for positive spatial autocorrelation lies in the interval $(0, +1)$. Now $(\mathbf{1}^T \mathbf{V}^{-1}\, \mathbf{1}) = n(1-\rho)^2 < n$. This term might be thought of as counting the number of independent observations that possess the same amount of information on μ as the n dependent observation. Haining (1988) provides examples of this

Table 8.1 *Results of different approaches to constructing confidence intervals for the mean (μ) of a spatial population*

Estimation procedure	Mean estimate	s^2	Standard error	95% Confidence interval
(a) Sample mean; (8.2); (8.3)	0.870	0.126	0.039	(0.793, 0.947)
(b) Sample mean; (8.5); (8.3)	0.870	0.132	0.089	(0.695, 1.044)
(c) Sample mean; (8.5); (8.6)	0.870	0.132	0.089	(0.694, 1.045)
(d) (8.8); (8.9); (8.10). **V** is defined as an SAR model*	0.713	0.187	0.117	(0.482, 0.944)
(e) (8.8); (8.9); (8.10). **V** stationary **	0.827	0.127	0.097	(0.636, 1.017)

Notes: Moran coefficient for the data expressed as a standard normal deviate: 5.79.
Empirical variances and autocovariances: $C(0) = \text{Var}(Z) = 0.126$; $C(1) = 0.057$; $C(2) = 0.027$; $C(3) = 0.006$; $C(k) = 0.0$ ($k \geq 4$).
* Single parameter simultaneous autoregressive model ($\hat{\rho} = 0.221$). See (9.15).
** Substituting empirical variances and autocovariances into **V**.
For full description of procedure see Haining (1988, pp. 588–93)

information loss for different spatial models. Expression (8.9) using (8.3) to estimate σ^2 is one of the approximations examined by Cliff and Ord (1981, pp. 187–9) in a comparison of means. They suggest a quick and simple method to estimate ρ in the case of a simultaneous autoregression using Moran's I divided by its maximum value (see Cliff and Ord, 1981, p. 188).

Now (8.3) is not the maximum likelihood estimator for σ^2. This is given by:

$$\tilde{\sigma}^2 = n^{-1}(\mathbf{z} - \tilde{\mu}\mathbf{1})^T \mathbf{V}^{-1}(\mathbf{z} - \tilde{\mu}\mathbf{1}) \tag{8.10}$$

A further refinement is to replace (8.3) with (8.10) substituting the sample mean for $\tilde{\mu}$ in (8.10) where $\mathbf{V}^{-1}$ plays a role equivalent to the second term in the right-hand side of (8.6). This is another of the approximations examined by Cliff and Ord (1981, pp. 187–9). Their results, based on simulation evidence, suggest that in the case of the first-order simultaneous autoregression with positive ρ, (8.9) using (8.3) to estimate σ^2 provides a valid test of significance for $n > 25$. Results in Haining (1988) and reproduced in table 8.1 also indicate that it is the use of (8.5) that is the more important adjustment. The mean was estimated for a small (9×9) lattice of pollution data. In case (d) an SAR model was fitted to the data and this probably overcorrected for the spatial correlation in the pollution data.

The general results given by (8.3) and (8.6) are why adjustments to conventional methods are needed. The evidence suggests that it is the effect of the second term on the right-hand side of (8.5) that is the more serious, at least in the usual situation of positive spatial dependence, and that one way to deal with this is to adjust n in (8.2) thereby increasing the sampling variance of the sample mean. The size of the adjustment to n will be sensitive to the estimates of the spatial autocorrelation in the data or, if a spatial model is fitted to the data, the choice of model.

Griffith (1978) describes a procedure for adjusting for spatial autocorrelation in analysis of variance tests which requires the fitting of a simultaneous autoregression to the data. Legendre et al. (1990) develop a non-parametric approach based on randomly partitioning the the study area into contiguous regions corresponding to the size of the observed regions and then comparing the observed within group sum of squares with the Monte Carlo results obtained by randomization. Sokal et al. (1993, pp. 207–8) compare the power of these two tests. Neither method appears to provide a convincing approach but if the main aim is to avoid type I errors then the method of Legendre et al. is to be preferred.

8.2   Spatial autocorrelation and tests of bivariate association

An additional problem with spatial data is that apparently significant association may also be an artifact of the spatial properties of the data on X and Y. The equivalent problem in time series analysis has a long history (Yule, 1926; Bartlett, 1935).

8.2.1   Pearson's product moment correlation coefficient

Suppose n pairs of observations, $\{(x(i), y(i))\}_i$, are drawn from a bivariate normal distribution. Pearson's product moment correlation coefficient (r) is the statistic used to measure the association between X and Y. If the observations on the two variables are independent (there is no spatial autocorrelation in either X or Y), Fisher's z transformation can be used to carry out tests on Pearson's r. It follows that:

$$(1/2)\ln[(1+r)/(1-r)] \tag{8.11}$$

is approximately normally distributed with mean and variance given by:

$$(1/2)\ln[(1+\phi)/(1-\phi)] \tag{8.12}$$

$$[1/(n-3)] \tag{8.13}$$

respectively where ϕ in (8.12) refers to the correlation coefficient in the bivariate normal population. So:

$$z = [(n-3)^{1/2}/2]\left[\ln\left((1+r)(1-\phi)/(1-r)(1+\phi)\right)\right] \tag{8.14}$$

has approximately the standard normal distribution. So (8.14) can be used to test hypotheses. If the null hypothesis is of no association ($\phi = 0$) then (8.14) simplifies. An alternative test statistic is given by:

$$(n-2)^{1/2} |r| (1-r^2)^{-1/2} \tag{8.15}$$

which is t distributed with $(n-2)$ degrees of freedom.

These distributional results do not hold if X and Y are spatially correlated. The effect of spatial autocorrelation on tests of significance were studied by Bivand (1980), Haining (1980) and Clifford and Richardson (1985) and shown to be very severe when both X and Y have high levels of spatial autocorrelation. The problem is that when spatial autocorrelation is present the variance of the sampling distribution of r, which is a function of the number of pairs of observations n, is underestimated by the conventional formula which treats the pairs of observations as if they were independent (Richardson and Hemon, 1981).

Clifford and Richardson (1985) obtain an adjusted value for n (n') which they call the 'effective sample size'. This value, n', can be interpreted as measuring the equivalent number of independent observations so that the solution to the problem lies in choosing the conventional null distribution based on n' rather than n. An approximate expression for this quantity is:

$$n' = 1 + n^2 (\text{trace}(\mathbf{R}_x \mathbf{R}_y))^{-1} \tag{8.16}$$

where $\mathbf{R}_x$ and $\mathbf{R}_y$ are the estimated spatial correlation matrices for X and Y respectively. The second term on the right-hand side of (8.16) is an estimator for the inverse of the variance of r when the data are spatially correlated. So now the null hypothesis of no association between X and Y is rejected if:

$$(n'-2)^{1/2} |r| (1-r^2)^{-1/2} \tag{8.17}$$

exceeds the critical value of the t distribution with $(n'-2)$ degrees of freedom. Note from (8.16) that if either X or Y are not spatially correlated then no adjustment is necessary. Dutilleul (1993) has proposed an alternative estimator for n' but no comparative tests appear to have been reported on type I error rates. Besag and Clifford (1989) describe a generalized Monte Carlo approach to testing association for data recorded on a regular lattice.

The effect of making the adjustment (8.17) when spatial correlation is strong in both variables is illustrated in Clifford and Richardson (1985) and Clifford et al. (1989) based on Monte Carlo experiments. Using (8.15), type I errors for a 5% significance level ranged from 8.2% in the least correlated case to 52% when nearest neighbour spatial correlation reached 0.8. When using (8.17) type I errors for a 5% test ranged from 3.2% to 7%. Further results are given in Haining (1990, pp. 319–20).

The adjustment to n applies to stationary data with constant variance. Data with trend should be de-trended prior to using their method and allowance made for non-constant variance particularly in the case of data relating to different sized spatial units. Haining (1991) provides an example of the Clifford and Richardson approach to test for significant association between two variables. Standardized mortality ratios (SMRs) for lung cancer deaths between 1980 and 1982 for 87 community medicine areas (CMAs) in Glasgow were correlated with a measure of deprivation (percentage of households living in poor accommodation lacking basic amenities) taken from the 1981 UK Census. The data are reported in the original article. The SMRs were log transformed ($Y = \log(SMR)$) in order to stabilize variances and make the distribution of the ratios more nearly normal. The distribution of the deprivation measure is highly skewed – large numbers of CMAs have low deprivation scores. A log transformation was applied to the deprivation measure (Z): ($X = \log(Z + 1)$). The unadjusted Pearson correlation coefficient is 0.252 and the t-value is 2.40 which is significant at the 1% level for a one tailed test.

The two variables were each de-trended by fitting a low-order trend surface (see section 9.1.1). CMAs were represented by area centroids for purposes of fitting the trend surface. The method of fitting such a model will be described in chapter 10. The trend surface errors were modelled using a first-order simultaneous autoregression, for example:

$$\mathbf{X} = \mathbf{A\theta} + \mathbf{u}_x$$
$$\mathbf{u}_x = \rho\mathbf{W} + \mathbf{e}_x$$

where matrix $\mathbf{A}$ contains the CMA co-ordinates ($s_1(i), s_2(i)$) and powers of those co-ordinates (see section 9.1.1); $\mathbf{u}_x$ is a first-order spatial autoregressive model with parameter ρ, $\mathbf{W}$ is an unscaled binary connectivity matrix and $\mathbf{e}_x$ is a vector of independent normal errors with mean 0 and variance σ^2. Because of the definition of $\mathbf{W}$, ρ is constrained to lie in the interval $(-0.32, 0.16)$. Both variables can be described by a second-order trend surface which has a maximum at or near the city centre with a long axis from south-west to north-east.

For each variable, two sets of residuals can be obtained – the residuals from the trend surface ($\tilde{u}$) which are spatially correlated if ρ is non-zero and the uncorrelated residuals from the full model ($\tilde{e}$). Haining (1991, p. 217) computes the Pearson correlation on the uncorrelated residuals ($\tilde{e}_x, \tilde{e}_y$). This represents a pre-whitening of the data. The correlation is 0.115 which has a t-value of 1.06. The correlation is not significant. The Clifford and Richardson approach was implemented on the correlated residuals ($\tilde{u}_x, \tilde{u}_y$). Table 8.2 summarizes the results. The adjusted value for n (n') is 41. The estimated spatial correlations that are substituted into the correlation matices to obtain trace($\mathbf{R}_x\mathbf{R}_y$) are reported.

Table 8.2 *Pearson bivariate correlation: lung cancer SMR (Y) and housing quality (X)*

	$\theta_{0,0}$	$\theta_{1,0}$	$\theta_{0,1}$	$\theta_{2,0}$	$\theta_{0,2}$	$\theta_{1,1}$	$R^2(\%)$	ρ
				Trend surface model				
Cancer SMR (Y)	4.58	−0.55	0.79	−0.27	−1.35	1.46	40.7%	0.11
Housing amenities (X)	−1.26	4.03	4.83	−3.19	−3.13	−2.38	46.8%	0.12

Correlation structure of trend surface residuals

	R(0)	R(1)	R(2)	R(3)	R(4)	R(5)	R(6)
$\tilde{u}_y$	1.0	0.31	0.00	−0.06	−0.05	−0.06	−0.07
$\tilde{u}_x$	1.0	0.43	0.14	−0.00	−0.11	−0.20	−0.17
No. of pairs		228	471	650	738	654	481

Notes: First subscript on θ refers to the east–west (co-ordinate value increasing to the east); the second to the north–south axis (co-ordinate value increasing to the north).
ρ significant in both models.

The Pearson correlation coefficient is 0.134 and the *t*-value is 0.847 which for $n' = 41$ is not significant.

Figure 8.1(a) shows the spatial distribution of the diagonal elements of $(\mathbf{R}_x\mathbf{R}_y)$ and 8.1(b) the plot of the diagonal elements of $(\mathbf{R}_x\mathbf{R}_y)$ against the corresponding number of neighbours for each CMA – the bivariate correlation between these two quantities is 0.927. It is the highly connected sites, most of which are in the interior, that contribute most to the adjustment in the degrees of freedom.

The populations at risk for the set of CMAs range from 2800 to 22 800. As noted above (see section 6.2.1) variances will tend to be smaller for areas with large populations relative to areas with small populations. Figure 8.2 shows the plot of $\tilde{e}_x$ and $\tilde{e}_y$ against the populations of their CMAs. There appears to be a size variance problem in the case of the residuals for the lung cancer data. The evidence is the same for the trend surface residuals. Multiplying $\tilde{e}_y$ by population size ($\tilde{e}_y p$) reduces this variance size relationship. The same results from computing $\tilde{u}_y p$. The bivariate correlation between $\tilde{e}_x$ and $\tilde{e}_y p$ is 0.103 (*t*-value 0.954). The bivariate correlation between $\tilde{u}_x$, $\tilde{u}_y p$ is 0.115 (*t*-value 0.769 with $n' = 45.78$). The conclusions remain the same. There is no significant correlation between lung cancer mortality and the housing amenity measure of material deprivation once the effects of spatial correlation are allowed for and the effect of non-constant variance.

Haining (1990, pp. 321–3) also illustrates the effect of using (8.17) rather than (8.14) when carrying out tests of significance on Spearman's rank correlation coefficient applied to spatially correlated data. Improvements in type I errors are comparable to those obtained in the case of Pearson's *r*. There is a worked example in Haining (1991, pp. 224–5) using per capita residential

(a)

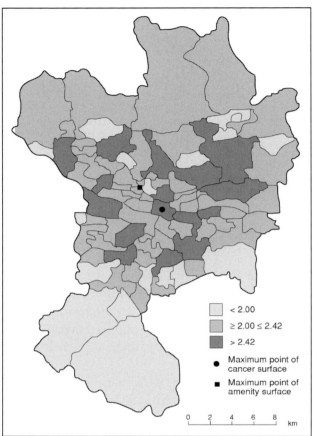

(b)

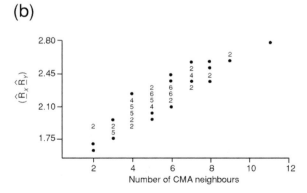

Figure 8.1 (a) Spatial distribution of the diagonal elements of $(\mathbf{R}_x\mathbf{R}_y)$; (b) plot of the diagonal elements of $(\mathbf{R}_x\mathbf{R}_y)$ against the corresponding number of neighbours for each CMA (Haining, 1991, figures 2 and 3, pp. 221, 223)

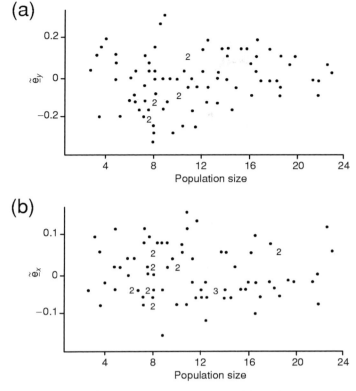

Figure 8.2 Residuals from the trend surface models with autocorrelated errors plotted against population size for (a) cancer, (b) deprivation (Haining, 1991, figure 5, p. 225)

burglary data in 1980 and 1981 for 64 police reporting districts in Santa Barbara, California. Do areas with high levels of reported burglary in 1980 have high levels in 1981? The conventional Spearman's rank correlation is 0.66 which is significant at the 1% level. Applying the Clifford and Richardson adjustment, the correlation is still significant at the 1% level. The results are reported in table 8.3.

8.2.2   Chi-square tests for contingency tables

A qualitatively similar approach to that described in section 8.2.1 has been adopted in carrying out tests for association in spatially correlated categorical data (Upton and Fingleton, 1989; Cerioli, 1997). A factor is applied to the conventional chi-square statistic that adjusts for the non-independence. The following description is based on Cerioli (1997) and refers to 2×2 tables.

Table 8.3 *Spearman's rank correlation coefficient for the 1980 and 1981 burglary data adjusted for spatial autocorrelation*

	R(0)	R(1)	R(2)	R(3)	R(4)	R(5)	R(6)
1980	1.0	0.275	0.111	0.054	0.015	−0.067	−0.273
1981	1.0	0.055	−0.000	0.059	0.010	−0.044	−0.111
No. of data pairs	64	146	286	384	390	329	249

Spearman rank correlation: 0.66. $n = 64$. $n' = 47.37$
t statistic: 6.98 ($n = 64$)
t statistic: 5.97 ($n' = 47$)

Suppose n pairs of observations $\{(x(i), y(i))\}$ are taken at n sites. Both variables are binary (0, 1). $\pi(s, t)$ denotes the probability that $x(i) = s$ and $y(i) = t$ ($s = 0, 1; t = 0, 1$). The process is assumed to be spatially homogeneous so this probability is constant across the n sites. The estimate of $\pi(s, t)$ is $p(s, t) = n(s, t)/n$ where $n(s, t)$ is the number of sites where $x(i) = s$ and $y(i) = t$.

The null hypothesis is that there is no association between X and Y. Formally, the test is of the log-linear hypothesis that:

$$\ln[\alpha] = \ln[\pi(0, 0)\,\pi(1, 1)/\pi(0, 1)\,\pi(1, 0)] = 0 \tag{8.18}$$

or

$$\ln[\pi(0, 0)\,\pi(1, 1)] = \ln[\pi(0, 1)\,\pi(1, 0)]$$

To test this hypothesis the conventional statistic expressed in terms of α is:

$$W = n \ln(\hat{\alpha})^2/\hat{\sigma}^2 \tag{8.19}$$

where $\hat{\alpha} = (n(0, 0)n(1, 1)/n(0, 1)n(1, 0))$ and $\hat{\sigma}^2$ is $n[\Sigma_s \Sigma_t (n(s, t))^{-1}]$. Under the null hypothesis of no association and assuming no spatial dependence, for large n, W is χ^2 distributed with one degree of freedom (χ_1^2).

A modified large sample test of no association between X and Y that takes account of spatial autocorrelation is obtained by Cerioli (1997) by multiplying W by a factor $(1 + \lambda)^{-1}$. This modified statistic (W^*) is also χ_1^2. Cerioli (1997, p. 622) gives two forms for $\lambda(\lambda_a$ and $\lambda_b)$ for regular lattices both of which performed well in terms of type I error rates in simulation trials. The two forms for λ assume X and Y are mutually independent and are given by:

$$\lambda_a = (2/n(1, 1)n(0, 0)) \Sigma_{h=1,\ldots,H} \gamma^x(h)\,\gamma^y(h)\,N(h) \tag{8.20}$$

and

$$\lambda_b = (2/n) \Sigma_{h=1,\ldots,H} \rho^x(h)\,\rho^y(h)\,N(h) \tag{8.21}$$

$N(h)$ denotes the number of pairs of data values separated by distance h and:

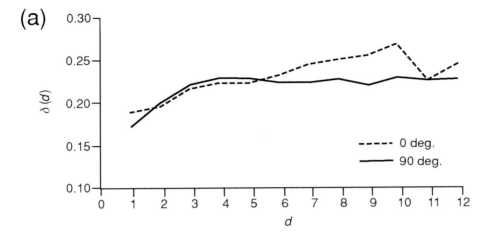

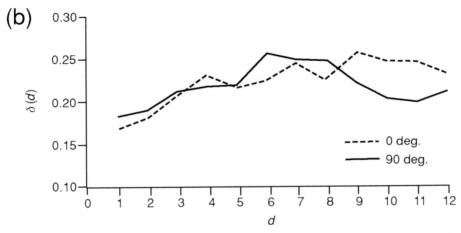

Figure 8.3 Directional sample semi-variograms for (a) hickories, (b) maples (Cerioli, 1997, figure 2, p. 625)

$$\gamma^*(h) = (1/N(h)) \sum_{\{s,t \mid d(s,t)=h\}} (x(s) - E(x))(x(t) - E(x)) \tag{8.22}$$

$$\rho^*(h) = \gamma^*(h)/(E(x)(1 - E(x))) \tag{8.23}$$

and similarly for the variable Y. In (8.22) the notation on the summation, $\{s, t \mid d(s, t) = h\}$, denotes all pairs (s, t) separated by distance h. In (8.22) and (8.23) $E(x)$ is the expected value of X and is given by $(n(1, 0) + n(1, 1))/n)$. Equation (8.23) is a Moran coefficient computed at distance h. These adjustment terms play the same role and are similar to trace($\mathbf{R}_x\mathbf{R}_y$) in (8.16). As before the adjustment is only needed if both X and Y are spatially correlated. Both of these adjustments can be adapted to the case of data on an irregular spatial framework.

Table 8.4 *Testing the spatial association of Maple and Hickory trees, Lansing Woods, Michigan*

Maple/Hickory	Present (1)	Absent (0)	Total
Present (1)	116	121	237
	(145.6)	(91.3)	
Absent (0)	238	101	339
	(208.3)	(130.6)	
Total	354	222	576

Notes: The first numbers in each cell denote observed counts ($n(s, t)$); numbers in brackets denote expected counts ($e(s, t)$) under the null hypothesis. $W = 26.10$ (strongly significant). Note the W statistic can also be computed as:

$$\Sigma_{s=0,1} \Sigma_{t=0,1} \left[(n(s, t) - e(s, t))^2 / e(s, t) \right]$$

Case (a): $\lambda_a = 2.656$. $W^* = 7.14$. (P-value: 0.008)
Case (b): $\lambda_b = 1.635$. $W^* = 9.90$ (P-value: 0.002)

The worked example by Cerioli uses data on the cross classification of 576 contiguous quadrats in Lansing Woods, Michigan where for each quadrat the presence or absence of hickory and maple trees is recorded (Upton and Fingleton, 1989). The cross-classified counts are shown in table 8.4 whilst figure 8.3 shows the semi-variograms for the two sets of data.

The value of W^* represents a severe deflation of the original value, W, but the null hypothesis is still rejected. There is evidence of negative association between the two species that is also reflected in the relationship between the observed and expected counts. The need to take into account spatial correlation and compute W^* is likely to be particularly important in those situations where there is strong spatial correlation in both of the categorical variables and/or the rejection of the null hypothesis using W is close to the chosen significance level.

Modelling spatial data

9

Models for the statistical analysis of spatial data

Data modelling enables the analyst to test hypotheses and to see to what extent a collected set of data support or refute a set of hypotheses. By specifying a model to represent the variation in data, rather than just testing a null hypothesis against a non-specific alternative (see chapters 7 and 8), the analyst is able to construct tests of hypothesis with greater statistical power. In the case of spatial data this includes modelling the *spatial* variation in the data.

Dobson (1999, pp. 10–11) describes a statistical model in information theory terms. Data represent measurements that consist of *signal* (the 'message' or 'process') distorted by *noise*. The variation in data is then composed of these two elements. The analyst constructs mathematical models that include both these components where the signal is regarded as deterministic and the noise is regarded as random. Combining these two components into a single model yields a probabilistic mathematical model called a statistical model. Another way she suggests of thinking of a statistical model is to consider the signal as a mathematical description of the important features of the data whilst the noise is the part of the data that is 'unexplained' by the signal component of the model. In descriptive spatial modelling the signal is a mathematical expression that describes spatial pattern or spatial structure, whilst in explanatory modelling it is a mathematical expression that incorporates predictors.

Fitting a statistical model to data allows inference to be carried out on potentially observable and other quantities (e.g. unobservable quantities like parameters) we want to learn about in the population. This may include performing (frequentist) tests of hypothesis, or constructing (Bayesian) probability or (frequentist) confidence intervals for unknown quantities of interest. Fitting a model consists first in specifying a model for the signal and noise including a probability distribution for describing the random variation. This process of *model specification* formalizes what the analyst either knows or is willing to assume about the population given the scientific problem and the data collection process. Frequentist analysis uses the likelihood function, the joint probability distribution of the data where the unknown quantities of interest are the parameters of the probability model, that is, they are assumed to have a fixed, but unknown value. Bayesian analysis starts with a joint probability distribution for all the observable *and* unobservable quantities and obtains the posterior probability distribution of the unobserved quantities of interest given the data (Gelman et al., 1995, p. 3).

After model specification the next step is to *estimate* or arrive at probability statements about the unknown quantities in the model using the data. The third step is to *assess* or *validate* the model. Model assessment may include partitioning the total variability in the data into the part that is attributable to the signal (explained variation) with the remainder attributable to the noise (unexplained variation). A good model, in statistical terms, is one which explains a large part of the variability.

Model assessment considers criteria other than variance explained. The principal of parsimony (Occam's Razor) is often invoked to justify a simpler model which provides an adequate description of the model over a more complicated model which explains more of the variability in the data. An information criterion such as Akaike's (1973) may be used to help the analyst choose between complex models which achieve high levels of explanation and simpler models which achieve lower but still adequate levels of explanation. Model assessment includes, where possible, implementing diagnostics that check statistical assumptions and identify data effects. A small number of data cases may have a disproportionate influence on the fit of a model and the analyst will want to consider these. Failure to satisfy a statistical assumption may or may not be serious depending on the statistical consequences of the failure in relation to the purpose of the modelling. Model assessment also includes examining the sensitivity of results to model assumptions.

Spatial data dependence in a variable can arise through the operation of certain types of spatial processes (see chapter 1). It can also be an inherited property of a variable as a result of its association with other variables that are spatially dependent. It can also arise from operations associated with data capture: the scale of the spatial unit in relation to the spatial scale of variability in the phenomena under study and forms of measurement error (chapters 2 and 3). Spatial data can acquire dependence if spatial interpolation methods are used to generate some data values (chapter 4). Statistical models for spatial data must account for the variation between observed quantities at different locations. Variation can be represented through the mean or the correlation structure or a combination of both. The various methods for representing spatial variation give the area of spatial statistics its distinctive flavour.

In some applications randomness is only associated with the imperfect process of data capture. The underlying process is deterministic but a statistical model is specified to accommodate the effects of omitted variables, measurement error or in the case of certain types of spatial data, aggregation effects. In other contexts the underlying process responsible for generating the phenomena under study is conceptualized as stochastic, as with the number of cases of a disease in a neighbourhood in a given period of time (see section 2.1.4). Further randomness may be introduced into the model associated, for example, with errors in counting the number of disease cases.

In undertaking statistical modelling the analyst concerned with a specific scientific question will have in mind the interpretation of the model, for example whether it has scientific meaning and enables important hypotheses to be tested. Whilst statistical criteria will be important, subject matter issues will be important. The interpretability of the model is a matter of special concern in non-experimental science where the analyst is using the data and statistical decision rules to arrive at an acceptable model. In the policy context the policy maker will want to know about the importance of variables whose levels can be varied through policy action.

This chapter proceeds by identifying classes of models for spatial data that correspond to types of modelling problems. The aim is to identify some of the practical issues encountered in modelling spatial data and suggest appropriate modelling approaches. We distinguish between statistical models that can be used to provide a *description* of the spatial variation in a variable and models that can be used to provide an *explanation* of the variability in a response variable in terms of other variables. This distinction is used to provide the main structure for this chapter with descriptive modelling discussed in section 9.1 and explanatory modelling in section 9.2.

In descriptive modelling data are recorded for only one variable (Z) together with their locations ($\mathbf{s}$) and a simple functional representation of the variation is sought with as few parameters as possible. In explanatory modelling data are recorded for many variables and the aim is to account for the variation in the response variable in terms of other variables called covariates, predictors or explanatory variables. In both these cases it will be necessary to distinguish between modelling where the spatial data are from a continuous surface of values and modelling where data refer to areas.

Cressie (1991, p. 8) identifies four different types of spatial models. The k observed quantities $\mathbf{z}(\mathbf{s}(i)) = (z_1(\mathbf{s}(i)), \ldots, z_k(\mathbf{s}(i)))_i$ are assumed to be realizations of random quantities at fixed spatial locations $\mathbf{s}(i)$ in two-dimensional space. The $\mathbf{s}(i)$ vary over the set D which is a subset of two-dimensional space. The observed values represent a realization of a (two-dimensional) multivariate random field

(Cressie, 1991, p. 8):

$$\{\mathbf{Z}(\mathbf{s}(i)) \mid \mathbf{s}(i) \in D\} \tag{9.1}$$

where $\mathbf{Z}(\mathbf{s}(i))$ is a vector of random variables representing a 'potential' datum on the k variables. This is referred to as the 'superpopulation' view of spatial data (see section 2.1.4), since potentially there are many different realizations of the random field. In the case of data from a continuous surface, what Cressie calls *geostatistical data*, D is a fixed area in continuous, two-dimensional, space. In the case of data referring to fixed points or areas (possibly represented as points), what Cressie calls *lattice data* and which is also sometimes referred to as *regional data*, D is a collection of points. As discussed in chapter 2, D is normally a graph $(D = \{\mathbf{s}(i), N(i)\})$ that also specifies the neighbours $(N(i))$ for each $\mathbf{s}(i)$. These are the data types considered here.

Cressie's (1991, p. 8) typology of spatial data types includes two more that are not treated here. The first is *point pattern data* where D is a point process in two-dimensional space, and associated with each $\mathbf{s}(i)$ is a vector of random variables $\mathbf{Z}(\mathbf{s}(i))$. The second is *object data* where D is a point process in two-dimensional space, and $\mathbf{Z}(\mathbf{s}(i))$ is a random set – an object with a spatial extent. Point pattern data are, superficially at least, similar to Cressie's geostatistical data and a type of lattice data. Object data are, superficially, similar to a type of lattice data. The differences lie in the specification of the process associated with D. Geostatistical and lattice data treat D as fixed, point pattern and object data treat D as a process.

9.1   Descriptive models

In this section we consider statistical models for representing spatial variation in a single variable and making population inferences. Spatial variation can be partitioned into large-scale and small-scale variation using statistical models. Unlike the conceptual models described in chapter 5, which component of a model refers to large-scale and which to small-scale variation is explicit. However it is partly the decision of the modeller based on their understanding of the problem and partly a statistical decision based on model fit and parsimony that determines for any given set of data the allocation of total spatial variation to these two main components. In real data sets there is likely to be a continuum of scales of variation present and this is also mirrored in the models which may capture a range of different scales of spatial variation depending on their specific form and parameter values.

9.1.1   Models for large-scale spatial variation

Broad or large-scale variation can be modelled through fitting smooth functions (trend surfaces) to Z. The following model for large-scale variation can be applied to both continuous and area data. Now $\mathbf{s}(i) = (s_1(i), s_2(i))$ is the co-ordinate position of the ith case on the surface or area. The general expression for a trend surface is written:

$$f(\mathbf{s}(i)) = \Sigma_{r+s \leq q} \theta_{r,s} \, s_1(i)^r s_2(i)^s \qquad (9.2)$$

where q, an integer, denotes the order of the surface. In the case of a flat surface $(q=0)$, the mean is constant, independent of i. A linear trend $(q=1)$ is a plane that tilts, whilst a quadratic trend $(q=2)$ has curvature (see, e.g., Haining, 1990, pp. 77–9). The low-order surfaces are:

flat: $\theta_{0,0}$

linear: $\theta_{0,0} + \theta_{1,0}s_1(i) + \theta_{0,1}s_2(i)$.

quadratic: $\theta_{0,0} + \theta_{1,0}s_1(i) + \theta_{0,1}s_2(i) + \theta_{2,0}s_1(i)^2$
$\qquad\qquad + \theta_{1,1}s_1(i)s_2(i) + \theta_{0,2}s_2(i)^2$

Since any fitted surface should not depend on the co-ordinate system, in practical application, only full-order surfaces (with a few exceptions) are allowed (Cliff and Kelly, 1977).

A model for Z that describes large-scale variation plus independent error can be written:

$$Z(\mathbf{s}(i)) = Z(i) = f(\mathbf{s}(i)) + e(\mathbf{s}(i)) \qquad (9.3)$$

where $e(\mathbf{s}(i))$ is independent and identically distributed with a mean of zero and variance σ^2. The mean or expected value of $Z(i)$ given by (9.3) is:

$$E[Z(i)] = \mu(i) = f(\mathbf{s}(i)) = \Sigma_{r+s \leq q} \theta_{r,s} \, s_1(i)^r s_2(i)^s$$

so that in this model large-scale variation is described through first-order properties. Note that if the trend surface is flat $\mu(i)$ is constant, independent of i, and so can be written as μ.

9.1.2   Models for small-scale spatial variation

(a)   Models for data from a surface

Models for small-scale spatial variation of surface data can be specified directly in terms of autocovariances. Weak stationarity about a constant mean implies:

$$E\left[Z(\mathbf{s}(i))\right] = \mu \quad \text{and} \quad \text{Cov}[Z(\mathbf{s}(i), Z(\mathbf{s}(j))] = C(\mathbf{s}(i) - \mathbf{s}(j)) \qquad (9.4)$$

for all $\mathbf{s}(i)$ and $\mathbf{s}(j)$ in D. The mean ($E[.]$) is assumed independent of location and the covariance between two values depends only on their relative locations. If the covariance ($\text{Cov}[.,.]$) depends only on the distance between the two points, $d(\mathbf{s}(i) - \mathbf{s}(j))$, and not direction, the process is called isotropic. In (9.4) there is no large-scale variation (the mean is a constant) and the covariance describes small-scale spatial variation.

The analyst often has data from only one realization out of many possible realizations of the random field but needs to be able to estimate the mean and covariances (from this sample of size one). The additional assumption that makes inference possible is that the random field is ergodic. The ergodic assumption allows means and covariances that are properties over the set of all possible realizations to be estimated by taking averages over the one available realization. If it can be assumed that the random field is normally distributed then if $C[d(\mathbf{s}(i) - \mathbf{s}(j))] \to 0$ as $d(\mathbf{s}(i) - \mathbf{s}(j)) \to \infty$ then the random field is ergodic (Cressie, 1991, p. 58). This assumption can be assessed against the data. If data are available for several time periods then it may be reasonable to view these as replications of the same field after removing time trends or other features of temporal variability.

For a function to be a valid covariance model it must be positive-definite – see Isaaks and Srivastava (1989, pp. 370–2) for a discussion of this property. Christakos (1984), and summarized in Cressie (1991, p. 86), identifies sufficient conditions for positive-definiteness of $C(h)$ where h denotes distance in two-dimensional space. These conditions include that $C(0)$ (the variance) must be non-negative; $C(h) = C(-h)$ where $-h$ denotes distance measured in the opposite direction; and $|C(h)| \leq C(0)$ where $|.|$ denotes the absolute value. Figure 9.1 shows examples of valid isotropic correlation models, $R(h) = C(h)/C(0)$.

Intrinsic stationarity is a more general assumption than weak stationarity and leads to modelling spatial variation using the semi-variogram, $\gamma(.)$ (for an example of a process that is intrinsically but not weakly stationary, see Cressie, 1991, p. 68). In the case of a random field that is intrinsically stationary:

$$E\left[Z(\mathbf{s}(i))\right] = \mu \quad \text{and} \quad \text{Var}[Z(\mathbf{s}(i) - Z(\mathbf{s}(j))] = 2\gamma(\mathbf{s}(i) - \mathbf{s}(j)) \qquad (9.5)$$

for all $\mathbf{s}(i)$ and $\mathbf{s}(j)$ in D. In a random field that satisfies (9.5) there is no large-scale variation and the semi-variogram describes the small-scale spatial variation. The semi-variogram is a function only of the difference $(\mathbf{s}(i) - \mathbf{s}(j))$ and

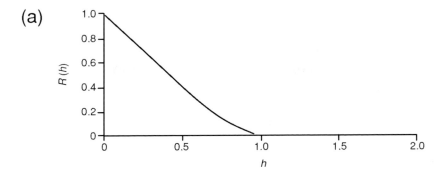

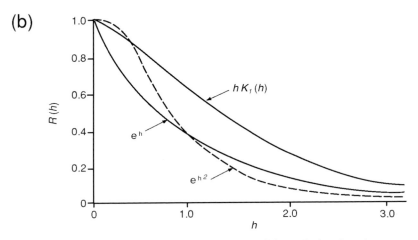

Figure 9.1 Examples of valid isotropic spatial correlation functions:
(a) spherical model; (b) negative exponential and bessel function models

is not a function of location. If it is not dependent on direction the semi-variogram is said to be isotropic and can be written $\gamma(h)$ where as before h denotes distance, otherwise it is called anisotropic.

In the case of an isotropic weakly stationary processes, the semi-variogram $\gamma(h) = (C(0) - C(h))$. This means that any permissable covariance function also can be used as a model for the semi-variogram of a weakly stationary process. The conditions for a valid semi-variogram for an intrinsically stationary random field (conditional negative-definiteness) are given by Cressie (1991, p. 87). Figure 9.2 shows examples of commonly used, valid semi-variograms for weakly stationary fields. The sill is the plateau a function reaches and which corresponds to $C(0)$. The range is the distance at which the plateau is reached. The range (r) defines the distance separation between observations for values in a random field to be uncorrelated $(R(r) = C(r)/C(0) = 0)$ because at this distance, $C(r) = 0$. For further discussion of semi-variogram models both isotropic

(a)

Exponential	$\gamma(h) = C_0 + C\left[1 - \exp\left(\frac{3h}{a}\right)\right]$	$C, a \geqslant 0$ Approaches asymptotically an upper limiting value.	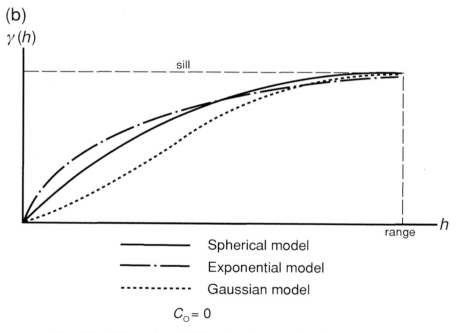
Gaussian	$\gamma(h) = C_0 + C\left[1 - \exp\left(-3\left(\frac{h}{a}\right)^2\right)\right]$	$C, a \geqslant 0$ Approaches asymptotically an upper limiting value.	
Hole	$\gamma(h) = C_0 + C\left[1 - a\sin(ah)/h\right]$	$C, a \geqslant 0$ Increasing to a plateau with oscillation.	
Spherical	$\gamma(0) = 0$ $\gamma(h) = C\left[\frac{3h}{2a} - \frac{1}{2}\left(\frac{h}{a}\right)^3\right] + C_0$ $\gamma(h) = C + C_0 \quad h > a$	$C, a \geqslant 0$ Stationary (transition) model.	

$C_0 > 0$ Implies a nugget effect.

(b)

$\gamma(h)$

——————— Spherical model

—— · —— Exponential model

· · · · · · · · · · Gaussian model

$C_0 = 0$

Figure 9.2 (a) Examples of valid semi-variograms. (b) Relationship between spherical, exponential and Gaussian models

and anisotropic, see Isaaks and Srivastava, 1989, pp. 372–86 and Webster and Oliver, 2001, pp. 109–27.

The functional form and parameter values of the covariance or semi-variogram functions determine the scale of spatial variation present in the random field. Functions that decline steeply to zero (in the case of the covariance function) or climb steeply to the sill (in the case of the semi-variogram function) depict random fields dominated by small-scale spatial variation. In situations where the function gradients extend over longer distances then these depict random fields with larger scales of spatial variation.

(b)　　Models for continuous-valued area data

In order to focus the treatment, in this section we concentrate on models for continuous-valued variables except for the purposes of introducing the concept of the spatial Markov property. The treatment is extended to discrete-valued variables afterwards.

The concepts of stationarity and isotropy as described above are not now usually invoked. Realizations only occur on the fixed points of D and cannot occur elsewhere so the covariance expression in (9.4) which depends on distance is not defined. The usual approach is to use the information on the spatial locations of sites and assumptions about the graph structure (the set of neighbours of each site) to build probability models for $\{\mathbf{Z}(\mathbf{s}(i)) : \mathbf{s}(i) \in D\}$. These models can be constructed to be stationary on large lattices but in practice this is not usually an important consideration.

A starting point for the construction of valid probability models that incorporate spatial dependence is to assume the Markov property. The 'spatial' Markov property is best approached by first considering the Markov property in time series modelling. Suppose the time series observations are discrete valued and refer to points in time $t = 1, 2, 3, \ldots$ If the notation $\mathrm{Prob}\{Z(t) = z(t) \mid Z(1) = z(1), \ldots, Z(t-1) = z(t-1)\}$ denotes the conditional probability that the value of the random variable Z at time t equals $z(t)$, given the values of Z at the all the preceding points in time $(1, \ldots, t-1)$, then the conditional probability approach to the definition of the Markov property is that:

$$\mathrm{Prob}\{Z(t) = z(t) \mid Z(1) = z(1), \ldots, Z(t-1) = z(t-1)\}$$
$$= \mathrm{Prob}\{Z(t) = z(t) \mid Z(t-1) = z(t-1)\} \tag{9.6}$$

The nature of the time series dependence is that only the most recent past determines the conditional probability of the present. The alternative, joint probability, approach to defining the Markov property for a temporal process is equivalent to (9.6) (Cressie, 1991, p. 403). The directional flow of time makes this a natural way to define dependence in time (figure 9.3(a)).

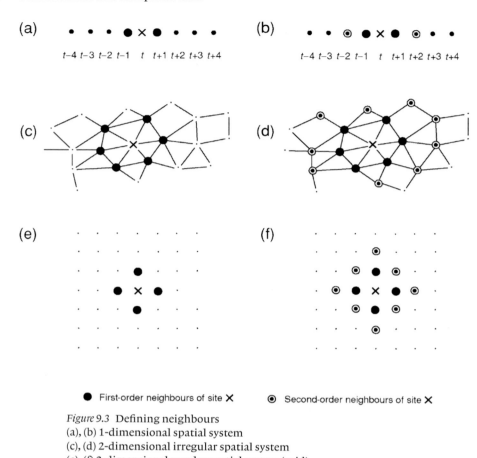

Figure 9.3 Defining neighbours
(a), (b) 1-dimensional spatial system
(c), (d) 2-dimensional irregular spatial system
(e), (f) 2-dimensional regular spatial system (grid)

The analogue to (9.6) for spatial data is based on invoking the graph structure on D. In figure 9.3(c) a distribution of sites is shown together with the set of neighbours to site i ($N(i)$). The spatial Markov property is defined as:

$$\text{Prob}\{Z(i) = z(i) \,|\, \{Z(j) = z(j)\}, j \in D, j \neq i\}$$
$$= \text{Prob}\{Z(i) = z(i) \,|\, \{Z(j) = z(j)\}, j \in N(i)\} \qquad (9.7)$$

In the case of a rectangular lattice, the neighbours to site i might be the four sites to the north, east, south and west (figure 9.3(e)).

In time series modelling the Markov property can be specified at different orders. For a model satisfying the second-order Markov property the right-hand side of (9.6) becomes:

$$\text{Prob}\{Z(t) = z(t) \,|\, Z(t-1) = z(t-1), Z(t-2) = z(t-2)\} \qquad (9.8)$$

A model satisfying (9.8) might have two different time dependency parameters to reflect an expectation that the present is affected differently by the immediate past $(t-1)$ than the more remote past $(t-2)$ (see figure 9.3(b)). In the case of a random field satisfying the second-order Markov property the right-hand side of (9.7) becomes:

$$\text{Prob}\{Z(i)=z(i)\,|\,\{Z(j)=z(j)\}, \{Z(k)=z(k)\}, j \in \mathrm{N}_1(i), k \in \mathrm{N}_2(i)\} \quad (9.9)$$

where $\mathrm{N}_1(i)$ denotes the set of first-order neighbours of site i and $\mathrm{N}_2(i)$ denotes the set of second-order neighbours of site i. Models satisfying (9.9) might have different parameters to reflect the expectation that site i is affected differently by its close neighbours than by its more remote neighbours (see figures 9.3(d) and 9.3(f)).

This is called the *conditional approach* to specifying random field models for regional data. For the analyst concerned to build spatial models that satisfy a spatial form of the Markov property then (9.7), (9.9) and higher-order generalizations provide the natural basis for doing so (Cressie, 1991, p. 404). However, the lack of a strict ordering in space, as compared to time, means there is no unilateral flow to dependency in space and this complicates parameter estimation. As a consequence the Markov property does not have the same 'natural' status in modelling area data (as a means of escaping from the independence assumption in model specification) that it has in time series modelling. Moreover, because there is no unilateral flow to dependency in space, joint and conditional approaches to specifying Markov random fields are not equivalent, except in special circumstances, whereas they are in the time series case (Brook, 1964). Kingman (1975) provides further discussion of the Markov property for spatial data.

A multivariate normal spatial model satisfying the first-order Markov property can be written as follows (Besag, 1974; Cressie, 1991, p. 407):

$$E\left[Z(i)=z(i)\,|\,\{Z(j)=z(j)\}_{j\in N(i)}\right]$$
$$= \mu(i) + \Sigma_{j\in N(i)}k(i,j)[Z(j)-\mu(j)] \quad i=1,\ldots,n \quad (9.10)$$

and

$$\text{Var}[Z(i)=z(i)\,|\,\{Z(j)=z(j)\}_{j\in N(i)}] = \sigma(i)^2 \quad i=1,\ldots,n$$

where $k(i,i)=0$ and $k(i,j)=0$ unless $j \in N(i)$. This is called the autonormal or *conditional autoregressive model* (CAR model). Unconditional properties of the model are, using matrix notation:

$$E[\mathbf{Z}] = \boldsymbol{\mu} \quad \text{and} \quad \text{Cov}[(\mathbf{Z}-\boldsymbol{\mu}), (\mathbf{Z}-\boldsymbol{\mu})^T] = (\mathbf{I}-\mathbf{K})^{-1}\mathbf{M} \quad (9.11)$$

$\mathbf{K}$ is the $n \times n$ matrix $\{k(i, j)\}$ and $\boldsymbol{\mu} = (\mu(1), \dots, \mu(n))^T$. $\mathbf{K}$ is specified exogenously. $\mathbf{M}$ is a diagonal matrix where $m(i, i) = \sigma(i)^2$. Because $(\mathbf{I} - \mathbf{K})^{-1}\mathbf{M}$ is a covariance matrix, it must be symmetric and positive definite. It follows that $k(i, j)\sigma(j)^2 = k(j, i)\sigma(i)^2$. Note that if the conditional variances are all equal then $k(i, j)$ must be the same as $k(j, i)$.

A simple, first-order, form of this model is to set $\mu(i) = \mu$, $k(i, j) = \tau w(i, j)$ where τ is a parameter and $\sigma(i)^2 = \sigma^2$. To ensure invertibility of $(\mathbf{I} - \tau\mathbf{W})$, then τ lies between $(1/\omega_{min})$ and $(1/\omega_{max})$ where ω_{min} and ω_{max} are the smallest and largest eigenvalues of the connectivity matrix $\mathbf{W}$. Haining (1990, p. 89) shows examples of where the (unconditional) variances are not the same for all sites. The model is not stationary on finite lattices.

For square lattices where each site has four neighbours $(w(i, j) = 0$ or $1)$, as lattice size increases $\omega_{min} \downarrow -4$ and $\omega_{max} \uparrow +4$ so that $|\tau| < 0.25$. As τ increases from 0 to its upper boundary, correlations increase between all pairs of sites. In some applications large values of the autoregressive parameter are needed to generate realistic spatial correlation levels and this has lead to the development of limiting forms of the CAR model referred to as *intrinsic autoregressions* (Besag and Kooperberg, 1995). This class of models is discussed in Congdon (2001, pp. 338–9) and will appear in section 9.1.4. The intrinsic CAR model is specified by an improper density that computes spatial averages. This model removes the practical problem of estimating the autoregressive parameter (τ in the above example) that arises in CAR models.

If the analyst of regional data does not attach importance to satisfying a Markov property another option is available called the *simultaneous approach* to random field model specification. Let $\mathbf{e}$ be independent normal $\mathrm{IN}(\mathbf{0}, \sigma^2\mathbf{I})$ and $e(i)$ is the variable associated with site $i(i = 1, \dots, n)$. Define the expression:

$$Z(i) = \mu(i) + \Sigma_{j \in N(i)} s(i, j)[Z(j) - \mu(j)] + e(i) \quad i = 1, \dots, n \tag{9.12}$$

where $s(i, i) = 0$. Although it is not a requirement that $s(i, j) = s(j, i)$ in practice this assumption is often made (Cressie, 1991, p. 410). In matrix terms (9.12) can be written as:

$$(\mathbf{I} - \mathbf{S})(\mathbf{Z} - \boldsymbol{\mu}) = \mathbf{e} \tag{9.13}$$

where $\mathbf{S}$ is the $n \times n$ matrix $\{s(i, j)\}$ and $\boldsymbol{\mu} = (\mu(1), \dots, \mu(n))^T$. $\mathbf{S}$ is specified exogenously. It follows that $E[\mathbf{Z}] = \boldsymbol{\mu}$ and $\mathrm{Cov}[(\mathbf{Z} - \boldsymbol{\mu}), (\mathbf{Z} - \boldsymbol{\mu})^T] = \sigma^2 (\mathbf{I} - \mathbf{S})^{-1} \times (\mathbf{I} - \mathbf{S}^T)^{-1}$. $\mathbf{Z}$ is multivariate normal. More generally if $\mathbf{e}$ has a variance–covariance matrix Λ which is diagonal (thereby still ensuring independence) but where the elements are not all necessarily identical then:

$$\mathrm{Cov}[(\mathbf{Z} - \boldsymbol{\mu}), (\mathbf{Z} - \boldsymbol{\mu})^T] = (\mathbf{I} - \mathbf{S})^{-1}\Lambda(\mathbf{I} - \mathbf{S}^T)^{-1} \tag{9.14}$$

Model (9.14) is called a *simultaneous autoregressive model* (SAR model). A simple form of (9.14) is obtained by setting $\mu(i) = \mu, s(i, j) = \rho w(i, j)$ where ρ is a parameter and $w(i, j) = w(j, i)$. The constraints on ρ are set by the largest and smallest eigenvalues of $\mathbf{W}$ just as in the case of the simple CAR model. Again unconditional variances are not the same for all sites and the model is not stationary on finite lattices (Haining, 1990, p. 83). As ρ increases from 0 to its upper boundary the correlation between pairs of sites increases. The connectivity matrix $\mathbf{W}$ does not need to be symmetric so row sums can be standardized to 1 and in this case the constraint is $|\rho| < 1.0$. There is further discussion of properties in Kelejian and Robinson (1995).

Superficially this simple form of the SAR model looks to be the spatial analogue of the simplest time series autoregressive model:

$$Z(t) = \mu + \rho[Z(t-1) - \mu] + e(t) \quad t = 2, \ldots, T \tag{9.15}$$

where $Z(1)$ denotes the initial value (boundary) of the time series.

In social science applications, in a clear majority of cases, the simultaneous spatial autoregression model is the preferred representation for a Gaussian spatial model when making the simplest departure from the assumption of independence. This is probably because the dependency structure is specified directly in terms of interactions between variables and model specification at least in the social sciences, is usually undertaken in this way. However Cressie (1991, p. 408) comments that the single parameter form of (9.10) has a stronger claim to being called the spatial analogue of (9.15). This is because the $\{e(t)\}$ are independent of the $\{Z(t)\}$ in (9.15) and the pseudo-errors, $(\mathbf{I} - \mathbf{K})(\mathbf{Z} - \boldsymbol{\mu})$, are independent of $\mathbf{Z}$ in (9.10) and hence represent true innovations or random shocks. By contrast the errors $\{e(i)\}$ in (9.12) are not independent of $\{Z(i)\}$ and hence not true innovations. The conditional expectation of the single parameter form of (9.12) involves the neighbours of the nearest neighbours which also suggests that (9.12) does not in fact represent the simplest and most natural departure from the independence assumption (Cressie, 1991, pp. 409–10). The equivalence of the two forms of spatial model are established if and only if their variance matrices are equal, that is:

$$(\mathbf{I} - \mathbf{S})^{-1}\boldsymbol{\Lambda}(\mathbf{I} - \mathbf{S}^{T})^{-1} = (\mathbf{I} - \mathbf{K})^{-1}\mathbf{M}$$

Because $\mathbf{M}$ is diagonal, any simultaneous autoregressive model can be represented as a conditional autoregressive model (although not necessarily as parsimoniously) by setting $\mathbf{K} = \mathbf{S} + \mathbf{S}^{T} - \mathbf{S}\mathbf{S}^{T}$ but the reverse is not possible except by introducing rather artificial spatial dependency assumptions on the $\{e(i)\}$ in (9.12) (see Martin, 1987).

Other models can be generated using this simultaneous approach. The moving average model introduced in chapter 4 for modelling errors in spatial databases is defined as:

$$Z(i) = \mu(i) + \Sigma_{j \in N(i)} t(i, j) e(j) + e(i) \quad i = 1, \ldots, n \tag{9.16}$$

and equation (4.1) follows by setting $t(i, j) = \nu w(i, j)$ where ν is a parameter. Whittle (1954) defines an 'errors in variables' model that superimposes independent normal errors on to a simultaneous autoregressive model. This creates a discontinuity in the correlation structure at the origin ($h = 0$) and may be used to model a random field with superimposed site-specific spatially random measurement error (see also Besag, 1977). This is similar to the introduction of nugget variance in geostatistics (Isaaks and Srivastava, 1989, p. 3). Kelejian and Robinson (1995) propose a similar model for area data.

In all the models in this section the specification of the set of neighbours $\{N(i)\}_i$ will affect model properties. For a given set of parameters, enlarging the neighbourhood set and extending it over longer distances strengthens correlation in the random field over longer distances. As connectedness is increased and intensified between subsets of adjacent neighbours the correlation structure will usually strengthen between such highly interconnected sites (Haining, 1990, pp. 110–12). This and other properties can be explored through numerical evaluation of the covariance matrix – see for example (9.11) and (9.14). In practical situations, the aim of the analyst is to model spatial variation using as parsimonious a model as possible and making simple assumptions about the neighbourhood structure rather than trying to model the correlation structure.

Assumptions made about the neighbourhood structure have implications for other model properties including means and variances and conditional means and variances. For these reasons the exogenously determined weights $\{w(i, j)\}$ in the neighbourhood structure may be chosen to meet criteria that are in addition to or even distinct from the dependency structure in the data. Unconditional variances for (9.10) and (9.12) are given by the diagonal elements of (9.11) and (9.14) respectively. Conditional means and variances for the conditional autoregressive model are specified by (9.10). The conditional variances, $\{\sigma^2(i)\}$ will all be identical if $k(i, j) = k(j, i)$. This will arise if an inverse distance or binary connectivity specification is adopted for $\{w(i, j)\}$. However if modelling areas and a shared common border length definition is adopted (see section 2.4) then we might specify:

$$k(i, j) = (l_{i,j} / l_i) \phi$$

Here ϕ is a parameter and $l_{i,j}$ is the common border length between areas i and j

and l_i is the length of the border of i. Now the conditional variance $\sigma(i)^2 = \sigma^2/l_i$ and the conditional variances are smaller the larger the areal unit. Besag (1975) suggests this model property may be desirable where the variable Z is a rate or density (see sections 4.2.1 and 6.2.1). Cressie and Chan (1989) give an example where:

$$k(i, j) = (n_j/n_i)^{1/2} d_{i,j}^{-\alpha} \phi$$

The term n_j is the population in area j, $d_{i,j}$ is the distance between areas i and j and α is an additional parameter. Specifying weights in terms of population ratios has a rationale in certain models of spatial interaction but more importantly it follows that the conditional variance for site i is σ^2/n_i so that this decreases as population size increases – again a desirable property if Z is a rate.

The following result can be used to explore these properties. If $\mathbf{V}$ denotes the unconditional variance–covariance matrix of a normal random field and if $\mathbf{V}^{-1} = \{v^{i,j}\}$ then:

$$E\left[Z(i) = z(i) \mid \{Z(j) = z(j)\}_{j \in N(i)}\right]$$
$$= \mu(i) - v^{i,i} \Sigma_{j \neq i} v^{i,j} [Z(j) - \mu(j)] \quad i = 1, \ldots, n \tag{9.17}$$

and:

$$\mathrm{Var}\left[Z(i) = z(i) \mid \{Z(j) = z(j)\}_{j \in N(i)}\right] = (v^{i,i})^{-1} \quad i = 1, \ldots, n$$

The result (9.17) can be used to show that the single parameter version of (9.12), in terms of the Markov property, includes the 'neighbours of the neighbours' in its conditional expectation. This follows by noting that $\mathbf{V} = \sigma^2[(\mathbf{I} - \rho\mathbf{W})^{-1} \times (\mathbf{I} - \rho\mathbf{W}^T)^{-1}]$ and that in the expansion of $\mathbf{V}^{-1}$ the weights matrix $\mathbf{W}$ occurs in the expansion as a product, $\mathbf{W}\mathbf{W}^T$, or as $\mathbf{W}^2$ if $\mathbf{W}$ is symmetric.

As noted in the case of surface models the choice of model and parameter values determine the scale of variation present in the random field. For illustration purposes, figure 9.4 shows a transect through the correlation structure for three models on a large rectangular lattice of sites and where the neighbours, $N(i)$, of site i are the four sites to the north, east, south and west. The three models are: (1) the single parameter conditional autoregressive model (equation 9.10 with a constant mean and $k(i, j) = \tau w(i, j)$); (2) the single parameter simultaneous autoregressive scheme (equation (9.12) with a constant mean and $s(i, j) = \rho w(i, j)$) and (3) the single parameter moving average model (equation (4.1)). Three parameter values are selected, the largest value (0.24) is close to the permitted maximum value for all three models on a rectangular lattice. Of the models displayed in figure 9.4, the simultaneous autoregression generates a random field with the largest scale of spatial correlation. However as noted

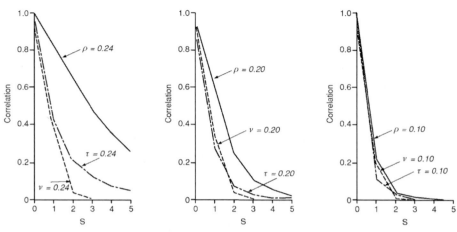

Figure 9.4 Spatial correlation functions for three spatial models
First-order simultaneous autoregression $(\rho) - (9.12)$
First-order conditional autoregression $(\tau) - (9.10)$
First-order moving average model $(\nu) - (9.16)$

above, in social science applications it is the interaction structure amongst the variables rather than the correlation structure that is of most interest for the purpose of model specification.

(c)   Models for discrete-valued area data
    The conditional approach to model specification leads to the construction of models for discrete-valued data (Besag, 1974). These include a spatial version of the logistic model for binary data (the autologistic model) and spatial versions of the binomial and Poisson models for count data (the auto-binomial and auto-Poisson models). The non-spatial versions of these models are special cases of these spatial versions.
    The logistic model is used to model presence or absence of an event $(Z(\mathbf{s}(i)) = 0$ or $1)$, for example whether an enumeration district is a crime hot spot or not (Craglia et al., 2001). The binomial or Poisson probability models are used to model counts $(Z(\mathbf{s}(i)) = 0, 1, 2, \ldots)$. The Poisson is an approximation to the binomial model where p (the probability of an event) is close to 0 and the number of 'trials', n, is large. The Poisson model is often used as a model for rare diseases (Mollie, 1996). Disease rates are often analysed as observed counts divided by expected counts and the numerator treated as a Poisson variable. In the case of common diseases, or crimes, then the binomial is a more appropriate model (e.g. Cressie, 1991, p. 431; Knorr-Held and Besag, 1998; see also Webster et al., 1994). The spatial (auto) versions of these models specify conditional probabilities given the value of the random variable (Z) at the neighbouring sites.

In the case of the autologistic model for $z(i) = 0$ or 1:

$$\text{Prob}\{Z(i) = z(i) \mid \{Z(j)\}\, j \in N(i)\}$$
$$= \exp[z(i)(\alpha(i) + \Sigma_{j\in N(i)}b(i, j)z(j))]/[1 + \exp[(\alpha(i)$$
$$+ \Sigma_{j\in N(i)}\, b(i, j)z(j))]] \qquad (9.18)$$

So in (9.18) the conditional probability that $Z(i)$ equals 0 or 1 depends on the value of Z at the neighbouring sites. The two conditional probabilities are obtained by substituting $z(i) = 0$ and $z(i) = 1$ into the right-hand side of (9.18).

There are two sets of parameters, one denoting intra-site or site-specific effects $\{\alpha(i)\}$, the other inter-site or between site interaction effects $\{b(i, j)\}$. A condition of the model is that $b(i, j) = b(j, i)$. Here, as in the other models of this section, non-spatial versions can be constructed by setting $b(i, j) = 0$ for all i and j.

Autologistic models for infinitely large lattices display 'long-range dependence' when the spatial dependence parameter exceeds a particular critical value. This means that whilst spatial correlation decreases with distance it never becomes 0 no matter how far apart two sites are. This property has made the autologistic model useful for modelling certain types of processes that display 'phase transitions' (see Gumpertz et al., 1997, p. 135; Haining, 1985). Cross and Jain (1983) provide realizations of (9.18) for the case $\alpha(i) = \alpha$ and $b(i, j) = \beta w(i, j)$ where β is a spatial dependence parameter. Their maps show how patches of 0s and 1s emerge on the map and increase as β increases.

The autobinomial model is defined $(Z(i) = 0, 1, 2, \ldots, c(i))$:

$$\text{Prob}\{Z(i) = z(i) \mid \{Z(j)\}\, j \in N(i)\}$$
$$= (c(i)!/z(i)!(c(i) - z(i))!)\,\theta(i)^{z(i)}(1 - \theta(i))^{c(i)-z(i)}$$
$$\theta(i) = \exp[\alpha(i) + \Sigma_{j\in N(i)}b(i, j)z(j)]/[1 + \exp[\alpha(i) \qquad (9.19)$$
$$+ \Sigma_{j\in N(i)}b(i, j)z(j)]]$$

which for $c(i) = 1$, is the autologistic model. A requirement of the model is that $b(i, j) = b(j, i)$.

The parameters of the autobinomial model correspond to those in the binomial model in the sense that there are two main sets: $c(i)$, the number of 'trials' at site i (a parameter that is assumed known) and $\theta(i)$, the probability of a 'success' at site i, which is usually the parameter of interest. In a disease context, $c(i)$ might be the population at risk in area i and $\theta(i)$ the probability that any individual in i has the particular disease. $\theta(i)$ depends on the parameters embedded within $\{\alpha(i)\}$ and $\{b(i, j)\}$. The parameter set $\{\alpha(i)\}$ reflect site-specific effects whilst the parameter set in $\{b(i, j)\}$ capture spatial dependencies between the $\{Z(i)\}$. Cross and Jain (1983) provide realizations of (9.19) for the case $\alpha(i) = \alpha$ and $b(i, j) = \beta w(i, j)$ where β is a spatial dependence parameter.

The auto-Poisson model is defined $(Z(i) = 0, 1, 2, \ldots)$:

$$\text{Prob}\{Z(i) = z(i) \mid \{Z(j)\}\, j \in N(i)\} = \exp[-\lambda(i)]\, \lambda(i)^{z(i)}/z(i)!$$
$$\lambda(i) = \exp[\alpha(i) + \Sigma_{j \in N(i)} b(i, j)\, z(j)] \qquad (9.20)$$

Spatial dependency is introduced through the intensity parameter $\lambda(i)$. Note again the presence of the site-specific term $\alpha(i)$. As with the previous models, $b(i, j) = b(j, i)$ but now there is an additional requirement that $b(i, j) \leq 0$ for all i and j. This implies that the auto-Poisson can only model competitive rather than co-operative neighbourhood processes, that is it can only model negative spatial dependence. For example, if $b(i, j) = \beta w(i, j)$ then $\beta \leq 0$. Kaiser and Cressie (1997) show that this negativity constraint can be removed by Winsorizing the distribution to a finite set of integers $Z(i) = 0, 1, 2, \ldots, R$. If R is much larger than any of the $\{\lambda(i)\}$, say $R \geq 3[\max_i\{\lambda(i)\}]$, then the effect of Winsorizing will be small. The negativity condition also applies to the spatial version of the negative binomial model.

The reader who wishes to follow up the supporting statistical theory that ensures that the conditional probabilities are specified consistently so that joint probabilities can be calculated (and which therefore ensure likelihood inference) should consult Besag (1974) or Cressie (1991, pp. 410–23).

9.1.3   Models with several scales of spatial variation

The models in 9.1.2(a) and (b) can be extended to include large-scale variation through the mean function which is allowed to vary spatially, $\mu(i)$, using (9.2). In making this substitution the large scales of spatial variation are described through the mean using a specified order of trend surface. Smaller scales of spatial variation are described through the covariance. So, for example, in (9.12) setting $\mu(.) = \theta_{0,0} + \theta_{1,0}s_1(.) + \theta_{0,1}s_2(.)$ and $s(i, j) = \rho w(\mathbf{s}(i), \mathbf{s}(j))$, $Z(i)$ is a normal first-order simultaneous spatial autoregression with a linear trend. In the case of the models in section 9.1.2(c) then (9.2) may be substituted as a model for $\alpha(i)$.

Large scales of variation, if they were modelled through the mean would require a trend surface that is not first or second order, may be better represented using the models of section 9.1.2(a) and (b). *How* different scales of variation are represented may depend on the scientific context. Cressie (1991, p. 25) writes: 'what is one person's (spatial) covariance structure may be another person's mean structure'. There may also be a trade off between model fit and parsimony (as measured by the number of parameters).

Regions (see chapter 5) represent another possible form of large-scale variation in a data set. If K regions can be defined then different descriptive models can be fit to each of the K regional subsets. One formal representation of regional variation is through the use of $K - 1$ dummy or indicator variables $(D_1(.), \ldots, D_{K-1}(.))$ – with region K acting as the reference region. If a particular site i, is in the kth region then $D_1(i) = 0, \ldots, D_{k-1}(i) = 0, D_k(i) = 1, D_{k+1}(i) = 0, \ldots, D_{K-1}(i) = 0$. If site i is in region K then all the $K - 1$ dummy variables have the value 0. Including these $K - 1$ additional variables in (9.2) allows $\theta_{0,0}$ to vary between regions. The remaining parameters can be allowed to vary by forming $K - 1$ products for each term in (9.2) (for example $D_k(i)s_1(i)^r$) and including these in the model. As a simple example, a two-region linear trend model corresponding to (9.3), with both the intercept ($\theta_{0,0}$) and gradient parameters ($\theta_{1,0}, \theta_{0,1}$) varying between the two regions, would be written:

$$Z(\mathbf{s}(i)) = \theta_{0,0} + \theta_{0,0;D} D(i) + \theta_{1,0} s_1(i) + \theta_{1,0;D}[D(i)s_1(i)]$$
$$+ \theta_{0,1} s_2(i) + \theta_{0,1;D}[D(i)s_2(i)] + e(\mathbf{s}(i)) \tag{9.21}$$

where $D(i) = 1$ if i is in region 1 and is 0 if i is in region 2. $[D(i)s_1(i)]$, for example, is equal to $s_1(i)$ if i is in region 1 and 0 if i is in region 2. So in region 1 the intercept parameter is ($\theta_{0,0} + \theta_{0,0;D}$) whilst in region 2 it is ($\theta_{0,0}$), and so on for the other parameters.

This model allows the analyst to include regional variation into a descriptive model and is an adaptation of the conventional use of dummy variables. Unless the data set is large, the number of regions will probably need to be relatively few and the order of surface relatively low. The model will tend to produce discontinuities in the surface at the regional boundaries. An interesting feature of (9.21) is that it is a simple example of a model where the *parameters* specifying the mean vary spatially. However it can be difficult to be objective in the specification of regions. The next section and section 9.2.4 consider other ways to model location effects.

9.1.4   Hierarchical Bayesian models

This section considers a class of models of particular interest in modelling spatial variation in discrete valued data. Observed outcomes are modelled so that they are conditional on a set of parameters which are themselves given a probabilistic specification in terms of other parameters – hence the term hierarchical. In these models spatial dependence is introduced through the specification of the parameter of interest. Unlike the models of section 9.1.2(c) spatial dependence is not defined directly with respect to the discrete-valued observations, nor is there a spatial parameter to estimate that corresponds to $\{b(i, j)\}$.

One important area of application of these models is in improving the precision of the estimator of parameters that are of the same type – for example relative risks across a study region divided into subareas. The estimate for the parameter of interest for any given subarea in the study region is obtained by 'borrowing' strength from other subareas in the study region.

Let $Z(i) = O(i)$ denote the number of deaths observed in area i during a specified period of time of a rare but non-infectious disease. It is assumed that $O(i)$ is independent and identically Poisson distributed with intensity parameter $\lambda(i) = E(i)r(i)$. Now $E(i)$ denotes the expected number of deaths from the disease in area i and $r(i)$ is the positive area-specific relative risk of dying from the disease in area i. Then:

$$\text{Prob}\{O(i) \mid r(i)\} = \exp\left[-E(i)r(i)\right](E(i)r(i))^{O(i)}/O(i)! \tag{9.22}$$

It is the variation in relative risk that is of interest. From (9.22) it can be shown that the standardized mortality ratio (SMR) for area i, $O(i)/E(i)$, where $O(i)$ is now the observed number of deaths, is the maximum likelihood estimator of $r(i)$.

The SMR for area i is an unbiased estimator of $r(i)$ but, because it treats each area independently, its precision depends on the size of the population at risk in i. This makes the interpretation of disease maps difficult when different subareas have very different population sizes (see section 6.2.1). For the purpose of mapping $\{r(i)\}$, a better estimator would be one with more uniform levels of precision across the set of areas – and preferably higher levels of precision.

Non-parametric approaches to this problem have been described in section 7.1.3. The approach here is to assume the $\{r(i)\}$ are random variables and in particular to assume that each $r(i)$ is drawn from a probability distribution called, in Bayesian terminology, the *prior* distribution. Two types of random effects models are of interest here: one where the random variation of the $\{r(i)\}$ is spatially unstructured, the other where it is spatially structured (Mollie, 1996). Both types of model have attracted interest because they can be used to 'stabilize' maps of relative risk by 'borrowing strength' from estimates elsewhere on the map. This also has the effect of improving estimator precision. This general approach to estimating parameter values on a map has arisen in previous sections, such as those dealing with map interpolation and non-parametric smoothing. The difference is that here the information from other subareas on the map is formalized through the specification of the prior distribution.

Let the notation $P(E(i)r(i))$ denote the distribution of $O(i)$ in (9.22). In the case of *unstructured* parameter variation, the $\{r(i)\}$ are *independent* drawings from the same probability distribution. For example, suppose the $\{r(i)\}$ are assumed gamma distributed $(G(v, \alpha))$ where v is the shape parameter and α is the scale

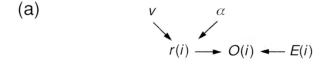

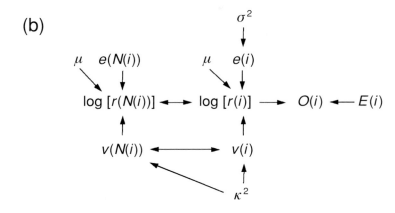

N(i) denotes the neighbours of site *i*

Figure 9.5   (a) Representation of the Poisson–gamma model for relative risk (9.23)
(b) Representation of the Poisson–log normal model for relative risk with spatially
structured (*v*) and unstructured (*e*) random effects (9.27; 9.28)

parameter. The gamma distribution in this case is the prior distribution and
its parameters *v* and α are called, in Bayesian terminology, *hyperparameters*. The
quantity v/α is the mean of the gamma distribution of relative risks from which
any particular $r(i)$ has been drawn, whilst v/α^2 is the variance of the distribution of relative risks. So for fixed *v*, as α increases above 1.0, the variation in
risk across the set of areas decreases. The heterogeneity of relative risk is unstructured because the $\{r(i)\}$ are independent of one another. The model can
be summarized as follows, using $\sim$ (to denote 'is distributed'):

$$O(i) \sim P(E(i)r(i));$$
$$r(i) \sim G(v, \alpha) \tag{9.23}$$

Figure 9.5(a) provides a pictorial representation of the model

Another model with unstructured random effects is to assume that the
logarithm of the relative risks, $\log(r(i)) = \mu + e(i)$ where $e(i)$ is independently
distributed $N((0, \sigma^2)$. This model is summarized:

$$O(i) \sim P(E(i)r(i));$$
$$\log(r(i)) \sim N(\mu, \sigma^2) \tag{9.24}$$

and each $\log(r(i))$ is an independent drawing from the $N(\mu, \sigma^2)$ distribution with μ and σ^2 fixed (Clayton and Kaldor, 1987). These models will be discussed further in chapter 10.

Spatially structured random effects can be introduced into (9.24). The $\{\log[r(i)]\}$ in (9.24) can be assumed to be drawn from a Gaussian Markov random field prior. For this class of model, 'the conditional distribution of the relative risk in area i, given values for the relative risks in all other areas $j \neq i$, depends on the relative risk values in the neighbouring areas ($N(i)$) of area i. Thus in this model relative risks have a locally dependent prior probability structure' (Mollie, 1996, p. 365). A spatially structured model would be appropriate if the underlying but unmeasured predictors that are believed to be responsible for the variation in relative risk are themselves spatially structured. Note that spatial dependence is specified on the parameters of interest, $\{\log[r(i)]\}$, not the observed counts, $\{O(i)\}$, although the latter will inherit spatial dependence from their dependence on the $\{\log[r(i)]\}$.

A model for pure spatially structured variation of relative risk is provided by the *intrinsic* Gaussian autoregression, a limiting form of a conditional autoregressive model (see section 9.1.2(b)), with a single dispersion parameter κ^2 (Besag et al., 1991). In this model if a binary connectivity or weights matrix ($\mathbf{W}$) is assumed ($w(i, j) = 1$ if i and j are neighbours, 0 otherwise) then:

$$E\left[\log(r(i)) \mid \{\log(r(j))\}, j \in N(i), \kappa^2\right] = \Sigma_{j=1,\dots,n} w^*(i, j) \log(r(j)) \quad (9.25)$$

$$\mathrm{Var}[\log(r(i)) \mid \{\log(r(j))\}, j \in N(i), \kappa^2] = \kappa^2 / \Sigma_{j=1,\dots,n} w(i, j) \quad (9.26)$$

where $\mathbf{W}^* \{w^*(i, j)\}$ denotes the row standardized form of $\mathbf{W}$ (see section 2.4). So the conditional expected value of the log relative risk of area i is the average of the log relative risks in the neighbouring areas. The conditional variance is inversely proportional to the number of neighbouring areas. Note that there is no *spatial* parameter to estimate in this model.

The intrinsic autoregression implies strong spatial dependence and rates will be smoothed to the neighbourhood averages. Besag et al. (1991) propose a model for the log relative risks that combines independence and local spatial dependence called a *convolution* Gaussian model. This is the sum of two components that are mutually independent to which can be added a mean effect μ so that:

$$\log(r(i)) = \mu + v(i) + e(i) \quad (9.27)$$

The term $e(i)$ is an independent normal variable $N(0, \sigma^2)$ and captures unstructured heterogeneity. The term $v(i)$ is an intrinsic Gaussian autoregression defined by (9.25 and 9.26). The parameters κ^2 and σ^2 specify the variability in $v(i)$ and $e(i)$ respectively. For identifiability of the model, that is to decompose the random effects into these two elements, there is the requirement that $\Sigma_i v(i) = \Sigma_i e(i) = 0$.

The conditional variance of $\log(r(i))$ given all the other log relative risks is:

$$\text{Var}[\log(r(i)) \mid \{\log(r(j))\}, j \in N(i), \sigma^2, \kappa^2] = (\kappa^2 / \Sigma_{j=1,\ldots,n} w(i,j)) + \sigma^2$$

$$(9.28)$$

If κ^2 / σ^2 is close to the average number of neighbours in the region $((1/n)\Sigma_{i=1,\ldots,n}\Sigma_{j=1,\ldots,n} w(i,j))$ both components of variation (spatially structured and unstructured) have the same importance. If κ^2 / σ^2 is small then unstructured variation dominates and if κ^2 / σ^2 is large then spatially structured variation dominates (Mollie, 1996, p. 366). In the case where spatially structured heterogeneity dominates then relative risk rates for areas with small populations are shifted towards the average rate for the areas that are the geographical neighbours. However, depending on the relative strength of the two components individual rates will be smoothed either towards the overall average or the neighbourhood average (see for example Congdon, 2001, p. 339). This model is summarized:

$$O(i) \sim P(E(i)r(i));$$
$$\log(r(i)) = \mu + v(i) + e(i)$$
$$e(i) \sim N(0, \sigma^2)$$
$$v(i) \mid v(j); j \in N(i) \sim N(\Sigma_{j=1,\ldots,n} w^*(i,j)v(j), \kappa^2 / \Sigma_{j=1,\ldots,n} w(i,j))$$

Figure 9.5(b) provides a pictorial representation of this model which performed well in comparative trials reported by Lawson et al. (2000).

Fully Bayesian models specify distributions for the parameters of the prior distribution. These distributions are referred to as *hyperprior* distributions. Empirical Bayesian analysis estimates the parameters of the prior distribution (v and α in the case of (9.23)) treating these quantities as fixed, and hence does not take into account any uncertainly associated with the values of these parameters.

For further discussion of the models of this section see Gelman et al. (1995, pp. 119–23) and Congdon (2001, 2002). There will be further examples of Bayesian models for spatial data that have been used in disease mapping in chapter 10.

9.2   Explanatory models

This section looks at models in which the variation in a response variable is accounted for (explained) in terms of one or more predictors. To distinguish the two types of variables, the response variable is denoted Y and the predictors as $X_1, \ldots, X_k$. Descriptive models of the type considered in section 9.1 become explanatory models when predictors or covariates are introduced into the model specification. The review of models is in two parts, first for continuous-valued variables (on surfaces or for areas) and then for discrete-valued area data.

9.2.1   Models for continuous-valued response variables: normal regression models

In the case of linear regression with normally distributed responses the known predictors for the response $Y(i)$ enter into the specification of the mean function or expected value of $Y(i)$, $\mu(i)$. For example:

$$\mu(i) = \beta_0 + \beta_1 X_1(i) + \beta_2 X_2(i) + \cdots + \beta_k X_k(i) \quad i = 1, \ldots, n$$

where β_0 denotes the intercept parameter and $\beta_1, \beta_2, \ldots, \beta_k$ denote the regression parameters of the model which are the parameters of interest rather than $\mu(i)$ itself. The main purpose of this section is to examine different types of explanatory model specification that can be handled by $\mu(i)$. The analysis of spatial variation in a response variable may need to reflect the fact that spatial units (areas or points) interact in some way. The effect of the level of a predictor variable at location i need not be restricted to the level of the response at location i. The reasoning lying behind this observation has been explored in some detail in chapter 1. The normal linear model is used for illustrative purposes.

The normal linear regression model is specified:

$$Y(i) = \mu(i) + e(i) \quad i = 1, \ldots, n$$
$$\mu(i) = \beta_0 + \beta_1 X_1(i) + \beta_2 X_2(i) + \cdots + \beta_k X_k(i) \tag{9.29}$$

where n denotes the number of points or areas, values for the predictors are treated as fixed and $\{e(i)\}$ are independent $N(0, \sigma^2)$. This model will be appropriate as a model for spatial variation if predictor effects on Y are purely 'vertical' (the value of the response at location i is only a function of predictor levels in i) and the errors in the model $\{e(i)\}$ can be assumed independent. The latter implies there is no spatially correlated measurement error and any unmeasured predictors excluded from the model specification are also spatially uncorrelated. If unmeasured predictors are spatially correlated this property will be inherited by the errors. The assumption of constant variance (homoscedasticity) is likely to require checking, particularly if the response variable is a rate (see, e.g., the test due to Breusch and Pagan (1979) for normal errors, the test of Koenker and Bassett (1982) for non-normal errors or the more general test of White (1980)). In this case where areas vary in size then non-constant variance (heteroscedasticity) in the errors may need to be modelled by specifying $\{e(i)\}$ as independent $N(0, \sigma^2(i))$. This problem and some suggestions for the form of $\sigma^2(i)$ have been suggested in section 4.2.1.

The spatial distribution of the response variable may be spatially correlated. This does not invalidate the model defined by (9.29). Spatial correlation in the response may be accounted for by the spatial correlation present in one or more of the predictors. Put slightly differently, the spatial correlation in the response variable has been inherited from one or more of the predictors. One of the important criteria for achieving a successful fit is that the model residuals are not spatially autocorrelated (see chapter 11).

There may be spatially correlated measurement error. There may be many unmeasured predictors that the analyst believes are likely to be spatially correlated but for which no data are available, or it is impractical to attempt to incorporate them within the model. In either case (9.29) may be re-specified as:

$$Y(i) = \mu(i) + u(i) \quad i = 1, \ldots, n$$
$$\mu(i) = \beta_0 + \beta_1 X_1(i) + \beta_2 X_2(i) + \cdots + \beta_k X_k(i) \tag{9.30}$$

where now $u(i)$ is a spatial error model and $\mathbf{u} = (u(1), \ldots, u(n))^T$ is MVN$(\mathbf{0}, \mathbf{V})$ where $\mathbf{V}$ models the small-scale spatial variation in the errors as described in section 9.1.2(a) and (b). If the analyst suspects the presence of local spatial structure in the measurement error a spatial moving average model may be appropriate for $\mathbf{V}$ (see 4.1). If there are unmeasured predictors, the form of $\mathbf{V}$ might be influenced by whatever is known about their spatial structure. In social science applications, where the spatial units are irregular, $\mathbf{V}$ is often modelled using a simultaneous spatial autoregression. In other field of application the residuals, as estimates of the errors, are used to specify $\mathbf{V}$.

The relationship between predictors and response need not be purely 'vertical' but may have a 'horizontal' component. This means that the value of the response variable at location i may be a function of levels of one or more predictors at, for example, the set of locations that represent neighbours of i, $N(i)$. Model (9.29) may be re-specified to model a neighbourhood context effect operating through, for example, X_1:

$$Y(i) = \mu(i) + e(i) \quad i = 1, \ldots, n$$
$$\mu(i) = \beta_0 + \beta_1 X_1(i) + \beta_{1;N} \Sigma_{j \in N(i)} w(i, j) X_1(j) + \beta_2 X_2(i) + \cdots + \beta_k X_k(i) \tag{9.31}$$

where $\beta_{1;N}$ is the parameter associated with the neighbourhood effect. The spatial scale of the neighbourhood effect is handled through the specification of the non-zero elements of the weights matrix. It is often appropriate to row-standardize the weights matrix so that the neighbourhood effect represents a neighbourhood average. This 'horizontal' effect of predictor X_1 is modelled as with the other predictors through $\mu(i)$. Model (9.31) is sometimes referred to as the linear regression model with spatially lagged predictor (or independent) variables. If $Y(i)$ measures, say, the incidence of a respiratory condition, according to home address, by small areas in a city and X_1 is a measure of air quality, the introduction of a horizontal effect on X_1 into the specification may be to reflect the fact that people move about within the city and are thus exposed to average levels of pollution over an area wider than just the area where they live.

Another type of horizontal effect may be introduced through the response variable itself. Thus (9.29) may be re-specified as:

$$Y(i) = \beta_0 + \beta_1 X_1(i) + \beta_2 X_2(i) + \cdots + \beta_k X_k(i)$$
$$+ \rho \Sigma_{j \in N(i)} w(i, j) Y(j) + e(i) \quad i = 1, \ldots, n \quad (9.32)$$

where ρ is the parameter associated with the interaction effect. Model (9.32) is sometimes referred to as the linear regression model with a spatially lagged response (or dependent) variable. If $Y(i)$ measures the price of a commodity at location i then this price might be a function of price levels at neighbouring, competitor sites. If $Y(i)$ measures the incidence rate of an infectious disease in area i at the end of an outbreak the incidence rate may be a function of incidence rates in neighbouring areas.

At first sight it may appear that the horizontal effect in (9.32) is modelled through $\mu(i)$ as in (9.31). In fact (9.32) has the form of a simultaneous autoregression which is evident when comparison is made with (9.12):

$$Y(i) = \mu(i) + \rho \Sigma_{j \in N(i)} w(i, j) Y(j) + e(i) \quad i = 1, \ldots, n$$
$$\mu(i) = \beta_0 + \beta_1 X_1(i) + \beta_2 X_2(i) + \cdots + \beta_k X_k(i)$$

The conditional autoregressive model (9.10) can also be interpreted as a model where there is spatial interaction amongst the responses and can be extended in the same way to include predictors in $\mu(i)$. The interaction parameters in these two forms of autoregressive model are both measures of spatial dependence although they differ in statistical interpretation (Cressie, 1991, p. 408).

Writing (9.32) in matrix notation and performing a sequence of matrix operations then:

$$\mathbf{Y} = \rho\mathbf{WY} + \mathbf{X}\boldsymbol{\beta} + \mathbf{e}$$

$$\mathbf{Y} - \rho\mathbf{WY} = \mathbf{X}\boldsymbol{\beta} + \mathbf{e}$$

$$(\mathbf{I} - \rho\mathbf{W})\mathbf{Y} = \mathbf{X}\boldsymbol{\beta} + \mathbf{e}$$

$$\mathbf{Y} = (\mathbf{I} - \rho\mathbf{W})^{-1}\mathbf{X}\boldsymbol{\beta} + (\mathbf{I} - \rho\mathbf{W})^{-1}\mathbf{e}$$

$$\equiv (\mathbf{I} - \rho\mathbf{W})^{-1}\mathbf{X}\boldsymbol{\beta} + \mathbf{u} \tag{9.33}$$

If $\mathbf{e} = (e(1), \ldots, e(n))^T$ is $N(\mathbf{0}, \boldsymbol{\Lambda})$ then $\mathbf{u}$ is $\mathrm{MVN}(\mathbf{0}, (\mathbf{I} - \rho\mathbf{W})^{-1}\boldsymbol{\Lambda}(\mathbf{I} - \rho\mathbf{W}^T)^{-1})$. This simplifies if, as in (9.29), $\boldsymbol{\Lambda} = \sigma^2\mathbf{I}$. So the regression model with lagged response variable is formally the same as (9.33), a model which has a particular spatial correlation structure in the errors and a particular structure of horizontal effects in the predictors.

Models (9.31) and (9.32) need to be interpreted keeping in mind that the other covariate effects are measured *given* the state of $\sum_{j\in N(i)} w(i, j)X_1(j)$ in (9.31) or *given* the state of $\sum_{j\in N(i)} w(i, j)Y(j)$ in (9.32). The relationship between different types of models with spatial lag structures and their interpretation will be discussed further in chapter 11.

In the preceding models the mean of the normal distribution varies spatially as a consequence of spatial variation in the level of the predictors. In some circumstances the analyst may believe there is additional variation associated with spatial variation in the parameters of the mean – the intercept (β_0) and regression parameters ($\beta_1, \ldots, \beta_k$) in (9.29). Regional effects similar to those defined in (9.21) can be included in (9.29). Using the same notation on the dummy variable D as employed for (9.21) and assuming a two-region model (region 2 being the reference region) with regional effects on the intercept and regression parameters on both of two predictor variables then:

$$Y(i) = \mu(i) + e(i). \quad i = 1, \ldots, n$$

$$\mu(i) = \beta_0 + \beta_{0;D} D(i) + \beta_1 X_1(i) + \beta_{1;D}[D(i)X_1(i)] \tag{9.34}$$

$$+ \beta_2 X_2(i) + \beta_{2;D}[D(i)X_2(i)]$$

In region 1 the intercept term is $(\beta_0 + \beta_{0;D})$ and the regression coefficients on X_1 and X_2 are $(\beta_1 + \beta_{1;D})$ and $(\beta_2 + \beta_{2;D})$ respectively. For region 2 the three parameters are: β_0, β_1 and β_2 respectively. The application of a model like (9.34) depends on being able to specify the regional partition as part of model specification. The Chow test (Chow, 1960) is used to test whether the parameters are significantly different for the two regions.

Other models for spatial variation in parameter values have been proposed. Returning to (9.29) let:

$$\beta_j = \beta_j(i) = \beta_{j;0} + \beta_{j;1}s_1(i) + \beta_{j;2}s_2(i) \quad j = 0, \ldots, k \tag{9.35}$$

where as in (9.21) the terms $s_1(i)$ and $s_2(i)$ denote the spatial co-ordinates of the ith area or site. For a model with two predictors corresponding to (9.34) the expanded form of the regression model (9.29) can be written:

$$Y(i) = \mu(i) + e(i) \quad i = 1, \ldots, n$$
$$\mu(i) = \beta_{0;0} + \beta_{0;1}s_1(i) + \beta_{0;2}s_2(i) + \beta_{1;0}X_1(i) + \beta_{1;1}[s_1(i)X_1(i)]$$
$$+ \beta_{1;2}[s_2(i)X_1(i)] + \beta_{2;0}X_2(i) + \beta_{2;1}[s_1(i)X_2(i)] + \beta_{2;2}[s_2(i)X_2(i)]$$
$$\tag{9.36}$$

This expansion of the regression model allows parameters to vary continuously over the study area as a function of location rather than discontinuously in terms of pre-defined regions. The linear trend surface assumed in (9.35) models linear trend parameter variation over the study region. Higher-order, non-linear, trend surface terms can be used in (9.35) to allow more complex spatial variation of parameter values. This is referred to as an 'expansion' model and the form of (9.36) as a 'spatial expansion' model (Jones and Casetti, 1992). Predictors can be used rather than co-ordinate locations in (9.35).

Section 9.2.2 considers a more general class of models to those illustrated by (9.34) and (9.36). Figure 9.6 provides a picture of some of the models reviewed in this section.

9.2.2   Models for discrete-valued area data: generalized linear models

The models for discrete-valued area data discussed in section 9.1.2(c) can be generalized to include predictors. Consider first the non-spatial versions of these models where $b(i, j) = 0$ for all i and j. The logistic, binomial and Poisson are all members of the class of generalized linear models as is the normal model with independent errors (9.29) (see, e.g., Dobson, 1999, pp. 30–1).

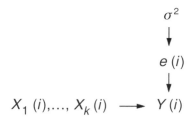

Figure 9.6 Representations for different explanatory models for continuous-valued response variables
(a) Normal linear regression (9.29)

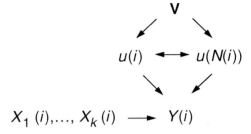

$$u(i) \longleftrightarrow u(N(i))$$

$$X_1(i),\ldots, X_k(i) \longrightarrow Y(i)$$

Figure 9.6(b) Normal linear regression with spatially correlated errors (9.30)

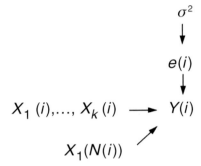

Figure 9.6(c) Normal linear regression with a spatially lagged covariate X_1 (9.31)

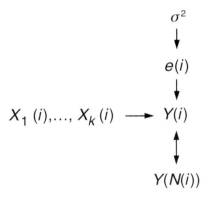

Figure 9.6(d) Normal linear regression with the response appearing in a spatially lagged form as an explanatory variable (9.32)

As before let $\mu(i)$ denote the expected value of the response $Y(i)$ then for the three models (note $c(i)$ the number of trials in the binomial model is assumed known):

Logistic: $\mu(i) = \text{Prob}\{Y(i) = 1\} = \theta(i) = \exp(\alpha(i))/(1 + \exp(\alpha(i)))$
Binomial: $\mu(i)/c(i) = \theta(i) = \exp(\alpha(i))/(1 + \exp(\alpha(i)))$
Poisson: $\mu(i) = \lambda(i) = \exp(\alpha(i))$

As in section 9.2.1, the parameter $\mu(i)$ is not of direct interest because there are as many of these as observations or cases (n). For a generalized linear model it is possible to write down a linear combination of a smaller set of parameters $\beta_0, \beta_1, \beta_2, \ldots, \beta_k$ which is equal to some function of $\mu(i)$. This function, which includes the predictor variables, is called the link function $g(\mu(i))$ where:

$$g(\mu(i)) = \beta_0 + \beta_1 X_1(i) + \beta_2 X_2(i) + \cdots + \beta_k X_k(i) \quad i = 1, \ldots, n \quad (9.37)$$

The link functions for the three discrete-valued models are:

Logistic: $g(\theta(i)) = \text{logit}[\theta(i)] = \log[\theta(i)/(1 - \theta(i))] = \alpha(i)$
Binomial: $g(\theta(i)) = \text{logit}[\theta(i)] = \log[\theta(i)/(1 - \theta(i))] = \alpha(i)$
Poisson: $g(\lambda(i)) = \log[\lambda(i)] = \alpha(i)$ (9.38)

For the Poisson model the link function is the log function. For the binomial and logistic models the link function is the logit function and $\alpha(i)$ is called the log *odds*. In all cases the link function is defined in terms of the set of predictor variables and a set of $(k + 1)$ parameters by setting $\alpha(i)$ equal to the right-hand side of (9.37). These transformations ensure that the parameters are positive valued as required. Note that for the normal model $g(\mu(i)) = \mu(i)$, as can be seen from (9.29).

Consider the case of modelling the variation in the observed number of cases of a disease $O(i)$ across a group of areas $i = 1, \ldots, n$. As noted above $O(i)$ is Poisson ($\lambda(i)$) where $\lambda(i) = E(i)r(i)$. $E(i)$, the expected number of cases of the disease, is assumed to be known and $r(i)$ is the relative risk – the parameter of interest. The specification of the Poisson model is from (9.37) and (9.38) where $\mathbf{E}[.]$ denotes expectation:

$$
\begin{aligned}
\log(\mathbf{E}[O(i)]) = \log(\lambda(i)) &= \log(E(i)) + \log[r(i)] \quad i = 1, \ldots, n \\
&= \log(E(i)) + \beta_0 + \beta_1 X_1(i) + \beta_2 X_2(i) + \cdots + \beta_k X_k(i)
\end{aligned}
$$
(9.39)

where $\log(E(i))$ is called an offset variable, its (implicit) parameter set to 1. The parameter β_0 is needed if $\{\log(E(i))\}$ are centred, if $\Sigma_i O(i) \neq \Sigma_i E(i)$ or if the co-variates are not centred. The Poisson model has the property that the mean and variance must be the same. However, because of spatial heterogeneity across the areas and the effects of missing or unmeasured predictors influencing the relationship, variation in $O(i)$ is likely to exceed Poisson variation. The presence of overdispersion should be investigated and if found its effects allowed for in interpreting statistical significance. The presence of overdispersion means that real type I error rates are larger than those calculated from generalized linear model theory and reported by conventional packages. Some software (e.g. STATA) allow for overdispersion in the fitting procedure. Bayesian modelling (see sections 9.2.3 and 11.2.4 and 11.2.5) allows overdispersion to be explicitly modelled through the random effects and decomposed into spatially structured and unstructured components.

With the availability of software to fit generalized linear models which use iterative re-weighted least squares, (9.39) is now one of the preferred methods of modelling disease counts across areas. This strategy is preferred, for example, to modelling $\log[O(i)/E(i)]$ as a normal linear regression and fitting by weighted least squares where $E(i)$ or $O(i)$ are used as the weights (Pocock et al., 1981; Haining, 1991(b)). For examples comparing the two approaches see Haining (1990, pp. 365–72) or Bailey and Gatrell (1995, pp., 311–13). Clayton et al. (1993) provide a brief review. Richardson et al. (1995) provide an illustration of the use of Poisson regression where a linear spatial trend is added to the set of covariates in (9.39).

Parameter estimates in the three models are interpreted as follows. In (9.39) the estimate of $\exp(\beta_j)$ is the estimate of the change in relative risk for a unit increase in X_j. Suppose deprivation is categorized into just the top (most affluent) 50% and the bottom 50% and the top category is the reference category so $X_1(j) = 1$ if area j is a deprived area, 0 otherwise. Now $\exp(\beta_1)$ measures the change in relative risk from being in a deprived population relative to an affluent population. Effects can be combined. If $\exp(\beta_2)$ is the relative risk for an urban/rural categorization (with rural as the reference group) then the product, $\exp(\beta_1)\exp(\beta_2)$, is the change in relative risk from being *both* deprived *and* living in an urban environment.

In the case of the logistic and binomial models, $\exp(\beta_j)$ measures the *odds ratio* corresponding to the covariate X_j. This measures the ratio of the odds for a unit increase in X_j:

$$\exp(\beta_j) = [\theta(i)/(1 - \theta(i))]/[\theta^*(i)/(1 - \theta^*(i))]$$

$\theta(i)$ is the probability $Y(i) = 1$ when $X_j = x_j + 1$ and $\theta^*(i)$ is the probability $Y(i) = 1$ when $X_j = x_j$. If $\beta_j = 0.2231$ then $\exp(\beta_j) = 1.25$ and the odds of $Y = 1$ to $Y = 0$ increases 25% for every unit increase in X_j.

As with the normal regression model (9.32), the full 'auto' versions (9.18), (9.19) and (9.20) need to be considered where spatial interaction underlies values of the response variable at different locations over and above effects due to covariates. Such spatial interaction may be a consequence of processes such as those discussed in chapter 1.

Some 'horizontal' spatial relationships can be accommodated within the non-spatial versions ($b(i, j) = 0$). For example, modelling the relationship between the response and predictors equivalent to (9.31) – the spatially lagged predictor variables model – can be accommodated within the standard specification by setting:

$$\alpha(i) = \beta_0 + \beta_1 X_1(i) + \beta_{1;N} \sum_{j \in N(i)} w(i, j) X_1(j) + \beta_2 X_2(i) + \cdots + \beta_k X_k(i)$$

(9.40)

and similarly for versions of (9.34) and (9.36).

9.2.3   Hierarchical models

(a)   Adding covariates to hierarchical Bayesian models

The models of section 9.1.4 can be generalized to accommodate predictors. The approach amounts to combining the model structures described in section 9.1.4 with linear functions of the predictors. In the case of (9.24) with spatially unstructured random effects and in the model with mixed unstructured and spatially structured random effects (9.27), predictors are accommodated by letting μ vary across the areas $\{\mu(i)\}$ and defining:

$$\mu(i) = \beta_0 + \beta_1 X_1(i) + \beta_2 X_2(i) + \cdots + \beta_k X_k(i) \quad i = 1, \ldots, n$$

In the case of (9.27), area-specific random effects are decomposed into spatially structured $(v(i))$ and spatially unstructured $(e(i))$ components. These components explicitly model the extra-Poisson variation in the observed counts not explained by the covariates. As noted in section 9.1.4, in a fully Bayesian approach the parameters $\beta_0, \beta_1, \beta_2, \ldots, \beta_k$ are also drawn from a probability distribution as are the parameters σ^2 and κ^2. Different choices can be made for the priors for these parameters and Bernardinelli et al. (1995) consider the implications of different choices. Richardson et al. (1995) provide an example.

This type of modelling can be extended to the situation where the populations in subareas are stratified say into age–sex cohorts or smokers/non-smokers (Clayton and Bernardinelli 1992; Clayton et al., 1993). Individual characteristics, relating to lifestyle or diet for example will probably act as confounders in analysing the relationship between environmental characteristics (measured at the area level) and health outcomes, so it is often important to obtain such data and to make use of it in modelling. The model may now take the form:

$$\log[r(i, j)] = \mu(i, j) + v(i) + e(i)$$

Here $r(i, j)$ is the relative risk for strata j in area i. Again spatially structured and unstructured random effects are included. The mean $\mu(i, j)$ comprises two types of covariates – k area-level covariates (X) measured only at the area level and m strata-level covariates (W) measured at each stratum for each area:

$$\mu(i, j) = \beta_0 + \beta_1 X_1(i) + \cdots + \beta_k X_k(i) + \gamma_1 W_1(i, j) + \cdots + \gamma_m W_m(i, j)$$

All strata in the same area experience the same area-level effects and share the same random effects. Further terms may be added to test for interaction effects between area-level characteristics and strata – for example interaction between area deprivation and age in modelling health outcomes since deprivation effects may only impact on health outcomes for certain (younger) age cohorts. These models can be fitted by Markov Chain Monte Carlo which can be implemented in WinBUGS (see appendix I).

Where spatial correlation is present in the extra-Poisson variation this modelling strategy seems now to be preferred to modelling $\log[O(i)/E(i)]$ as a normal log linear regression with spatially correlated errors and heteroscedastic errors (Cook et al., 1983; Haining, 1991(b)). Note however that whereas in this earlier approach there is a spatial parameter which quantifies the strength of nearest neighbour correlation and which is estimated, there is no such parameter in $\{v(i)\}$.

(b)   Modelling spatial context: multi-level models

Multi-level models are used when data have a hierarchical structure. In spatial data analysis the term 'levels' often refers to the influence of spatial contexts or higher spatial scales on outcomes defined at a lower level. These lower levels might be the individual or some small spatial aggregate such as an enumeration district.

An individual's response to a question about their perception of crime may reflect individual characteristics (age, gender, marital status) but also contextual influences relating to the neighbourhood where they live. Individuals might be expected to share common neighbourhood effects with other individuals in the same neighbourhood. There may be higher-level effects such as an urban-level effect. So if the sample extended over several urban areas, individuals might share some common effects with other individuals in different neighbourhoods in the same urban area but not with individuals in different urban areas. Whether a house is burgled or not may reflect individual characteristics of the house and attributes of the householders lifestyle but also contextual influences relating to the neighbourhood where the house is located and perhaps certain characteristics of the city such as the level of deprivation or affluence. In disease outcome modelling, data may be available at the individual level. The analyst wants to test if an area-level environmental factor is significant after controlling for individual-level confounders such as age, sex and occupation (Pocock et al., 1982). Models for these types of problems are often referred to, generically, as multi-level models or hierarchical linear models.

Multi-level modelling can also be used to model an ecological response at one spatial level in terms of attributes at that spatial level and attributes at higher spatial levels consisting of groupings of the lower-level spatial units. Aggregate contextual effects have been implicit in some of the models defined earlier (e.g. equations (9.31) and (9.34)). The spatial structure here, however, is somewhat different. Enumeration district-level data may be available and also data at higher levels of spatial aggregation such as area groupings of enumeration districts, city-level data and regional-level data. Modelling participation rates in a healthy lifestyles campaign may consider the influence of neighbourhood characteristics (since inter-personal contact might influence participation rates) city-level characteristics (since the priority given by each city to promotion campaigns might produce city to city differences) and so on. Enumeration district-level variation in crime rates across a county may be a response to enumeration district-level characteristics, ward-level characteristics and city-level or urban/rural-level characteristics. The conceptualization is that spatial setting makes a difference to outcome where setting can extend to several different spatial scales. Two areas with similar deprivation characteristics might differ markedly in terms of crime rates or health outcomes. These may reflect the wider spatial context within which they are located – for example whether they are in a region with many other deprived areas or whether they are in a region where most of the other areas are relatively affluent.

Where there are different spatial scales influencing the outcome of a response variable then simply using the ordinary regression model (9.29) will not be appropriate. This is because (9.29) ignores the spatial structure in the data. It treats the data as independent observations rather than as groups of related observations. The (spatial) index carries information that is relevant to modelling the outcome in the response so that the indexes $(1, \ldots, n)$ associated with the response variable $(y_1, \ldots, y_n)$ are not *exchangeable*. This is because of the known grouping that exists across the observations in terms of the higher levels of the model (Gelman et al., 1995, p. 373).

An example of a Gaussian multi-level model is given by the following:

$$
\begin{aligned}
Y(i, j) &= \mu(i, j) + e(i, j) \\
\mu(i, j) &= \eta_0 + \eta_1 X(i, j)
\end{aligned}
\tag{9.41}
$$

where $Y(i, j)$ is the response in (level 1) area i which is a member of (level 2) area grouping j. $X(i, j)$ is a level 1 ecological predictor. $e(i, j)$ is an independent normal error term. Now assume η_0 is normally distributed with mean β_0 and variance $\sigma_{0,j}^2$ whilst η_1 is normally distributed with mean β_1 and variance $\sigma_{1,j}^2$. The intercept and slope parameters in the regression model are drawn from prior distributions which differentiate between the groupings of the level 1 areas at the higher level. This model allows between-group variation in both the intercept and the slope coefficients in the relationship between the response Y and the level 1 predictor X.

Model (9.41) can be extended to include predictor variables at level 2. The model is being extended in the sense of finding predictors to explain the between-group-level parameter differences. Let $V(j)$ denote a level 2 ecological predictor then (9.41) becomes:

$$Y(i, j) = \mu(i, j) + e(i, j)$$
$$\mu(i, j) = \eta_0 + \eta_1 X(i, j) + \delta_1 V(j) + \delta_2 V(j)X(i, j) \qquad (9.42)$$

The coefficient δ_1 estimates the change in the response for a unit change in the value of the predictor at level 2. The coefficient δ_2 estimates the interaction effect and the change in the response for a simultaneous one unit change in the level 1 and level 2 predictors.

This type of modelling requires that the analyst specifies the groupings in advance. In many applications such as the analysis of educational attainment there is a well-defined hierarchical structure – pupil, class and school (Aitkin and Longford, 1986). Such a hierarchical structure may be found in some spatial applications as, for example, when the hierarchies refer to census and administrative tiers (see, e.g., Congdon, 2001, pp. 372–4). But some important spatial structures that underpin explanation in the social sciences like 'neighbourhoods' and 'social areas' are not clearly bounded, are overlapping rather than discrete at any level and non-nested between levels furthermore they may not be homogeneous environments or contexts – see, for example, Sampson, Raudenbush and Earls' (1997) study of violent crime in Chicago. Although multi-level modelling will handle some forms of spatial structure the types of models specified by (9.41) and (9.42) will need modification to handle the less well-defined and internally heterogeneous hierarchies encountered in areas of spatial modelling. This may mean the need for spatial error models to replace $e(i, j)$ that capture this aspect of spatial structure. The convolution Gaussian model that combines unstructured variation with the spatially structured intrinsic autoregression (see (9.27) and (9.28)) may be one way to handle this type of concern.

Jones et al. (1998) analyse geographical variations in British voting patterns using a multi-level model with lower level units nesting within two or more higher level units. This replaces the strict hierarchical structure with a cross classified structure (Goldstein, 1994; Hill and Goldstein, 1998). Individuals (level 1) are members of constituencies (level 2) which are grouped into geographical regions (level 3) and functional regions (also at level 3). The spatial units are still, however, well defined. There is further discussion of multi-level modelling with spatial data in section 11.2.3.

10

Statistical modelling of spatial variation: descriptive modelling

This is the first of two chapters dealing with some practical aspects together with examples of the statistical modelling of spatial data. This chapter considers models for describing spatial variation ('descriptive' modelling). The purpose of descriptive modelling is to find a representation of the variation of a response variable where covariates are not present in the model. The data, then, consist only of measurements on the variable of interest together with data on the location of each observation.

In 10.1 approaches to modelling spatial variation are discussed where parameters are assumed to be fixed values and estimated by likelihood methods. Section 10.1.1 reviews models for continuous-valued variables. Trend surface and covariance and semi-variogram modelling are covered. Section 10.1.2 examines models for discrete-valued variables. Section 10.2 reviews some special problems that arise in modelling spatial variation. Section 10.3 examines some Bayesian approaches to mapping spatial variation and provides an illustration using lung cancer data recorded at the ward level for Cambridgeshire.

Spatial variation may be captured directly through the correlation structure or through the specification of a spatial response function that refers to the mean of the process. Spatial variation may be modelled directly in terms of the observations or in terms of some unobserved parameter of interest that the observations depend on. The models of section 10.1 include one or more 'spatial' parameters that describe and quantify the structure of spatial variation. The models of section 10.3 have no such spatial parameters and spatial variation is represented through spatial averaging.

10.1   Models for representing spatial variation

This section is divided into statistical models for continuous-valued variables and statistical models for discrete-valued variables. Observations on continuous-valued variables may refer to samples from continuous space or to areas but in the second case there may be additional modelling issues to recognize. For example attribute variance may not be constant across all the areas and this has implications for parameter estimation.

10.1.1   Models for continuous-valued variables

(a)   Trend surface models with independent errors
A trend surface model for a response variable (Y) is defined:

$$\mathbf{y} = \mathbf{A}\boldsymbol{\theta} + \mathbf{e} \tag{10.1}$$

where $\mathbf{y}$ is an $n \times 1$ data vector corresponding to n sample sites, $\mathbf{A}$ is the matrix of location co-ordinates for the n sites and $\boldsymbol{\theta}$ is the vector of trend surface parameters. The size of $\mathbf{A}$ and $\boldsymbol{\theta}$ depend on the order of the trend surface (see section 9.1.1). For a second-order trend surface the ith row of $\mathbf{A}$ is:

$$(1, s_1(i), s_2(i), s_1(i)^2, s_2(i)^2, s_1(i)s_2(i))$$

where $(s_1(i), s_2(i))$ denotes the co-ordinate position of the ith point. The vector $\boldsymbol{\theta}^T = (\theta_{0,0}\,\theta_{1,0}\,\theta_{0,1}\,\theta_{2,0}\,\theta_{0,2}\,\theta_{1,1})$. The vector $\mathbf{A}\boldsymbol{\theta}$ represents the signal whilst the vector $\mathbf{e}$ represents the noise – an unobserved vector of independent normal random variables with mean, $E[\mathbf{e}]$, equal to 0 and variance matrix, $E[\mathbf{e}\mathbf{e}^T]$, equal to $\sigma^2\mathbf{I}$. Note that the signal is the expected value of $\mathbf{Y}$, since $E[\mathbf{Y}] = \mathbf{A}\boldsymbol{\theta}$.

Model (10.1) can be fitted by ordinary least squares using conventional software as a special case of multiple regression. However special precautions have to be taken since polynomial regression is an ill-conditioned least squares problem – a problem accentuated if datapoints are clustered. Even for a cubic order surface $\mathbf{A}^T\mathbf{A}$ can be nearly singular so that matrix inversion to obtain the estimate of $\boldsymbol{\theta}$ given by:

$$(\mathbf{A}^T\mathbf{A})^{-1}(\mathbf{A}^T\mathbf{y}) \tag{10.2}$$

can lead to unreliable estimates (Ripley, 1981, pp. 30–4; Upton and Fingleton, 1985, pp. 324). Scaling the study area to the unit square reduces this problem.

There is a further problem – order selection. Residuals from a trend surface (the estimate of the noise in the data) tend to be spatially autocorrelated especially if the order of the fitted model is too low. If the analyst selects the order of model by first fitting low-order models and then progressively higher-order ones, stopping when no higher-order terms are statistically significant, the effect is usually to fit a surface of too high an order. A similar effect is noted in multiple regression when spatial autocorrelation is present in the errors.

Variables may be retained in the model as statistically significant because one of the effects of autocorrelated residuals is to lead to underestimation of the sampling variance of parameter estimates. In the terminology of chapter 8 this is because the amount of information carried in the errors is less than standard statistical theory suggests. It is often safer to fit a model like (10.1) but assuming spatially autocorrelated rather than independent error. This approach is discussed below.

Frequency distributions of rates, such as relative risk rates of disease for a set of contiguous areas, are sometimes strongly skewed (e.g. see section 9.1.4 for a justification for assuming a gamma distribution for relative risks). If this is the case, rather than normalizing the distribution through a log transformation of the relative risk an alternative is to use generalized linear model fitting, providing the assumption of independence holds. Kafadar (1996) provides an example, characterizing geographical trends in disease data for several areas of the USA using a gamma model.

(b)   Semi-variogram and covariance models

Suppose the random variation itself contains a signal, that is there is a mathematical expression underlying the variation about the mean. In the context of spatial analysis this component of the signal describes how values at different locations in space are stochastically related – typically with near values tending to be more similar than values that are far apart. This spatial structure may be in addition to trend as described in section 10.1.1(a) but the issue of fitting models that contain signal in both the mean and the variation about the mean will be examined in section 10.1.1(c). So for the purposes of what follows it is assumed that either the component of the signal represented by the mean is constant or it is known and has been removed.

The semi-variogram or covariance plots can be used to represent spatial variation about the mean (second-order stochastic variation). The empirical problem is to fit a permissible model to describe this variation. Permissible models were discussed in section 9.1.2(a). Representing spatial variation non-parametically using estimated covariances or semi-variances can be problematic because the necessary conditions (e.g. positive definiteness in the case of covariances) are not necessarily satisfied.

Webster and Oliver (2001, p. 127) note that 'choosing models and fitting them to data remain among the most controversial topics in geostatistics'. There are two ways of fitting models to sample values. First, fitting a model by eye to plots using visual and graphical methods and, second, fitting by statistical methods. The latter divide into methods that fit on the semi-variogram or covariance plots and those such as maximum likelihood and restricted maximum likelihood that fit on the sample data and which assume the data are normally distributed (Cressie, 1991, pp. 91–4). We will only describe methods of fitting models directly to the semi-variogram plot.

Webster and Oliver (2001, p. 128) describe the reasons why fitting models to semi-variogram plots is difficult but their comments apply equally to fitting models to covariance plots when the process is second-order stationary. First, the precision of each estimate (at different distances) varies because of sample size differences. Second, the variation may not be the same in all directions (anisotropic). Third, the empirical plot may show fluctutation between adjacent estimates and estimates are correlated. Fourth, most models are nonlinear in one or more parameters. The first three reasons complicate fitting by eye whilst the third and fourth reasons complicate statistical approaches to fitting. They recommend a combination of fitting by eye and statistical fitting. First plot the empirical semi-variogram. Second, using visualization methods choose models that capture the main features in the plot that are important for the application. Third, fit each model by weighted least squares and finally plot the fitted models against the empirical semi-variogram and if more than one seem to fit well choose the model with the smallest mean of the squared residuals (MSR). This sum of squares criterion might weight those parts of the plot that are most important for the application. In kriging, for example, it is goodness of fit over short distances rather than over the whole semi-variogram that will be of most importance.

Since semi-variogram estimates are not of equal precision and neighbouring semi-variogram estimates are correlated, ordinary least squares which gives equal weight to all the empirical semi-variogram values is not appropriate for the fitting stage (stage three above). Generalized least squares might be used to cope with both of the difficulties cited but the form of the covariance matrix with diagonal and off-diagonal terms makes implementation difficult (Cressie, 1985). For this reason weighted least squares is usually used. Weights can be chosen to reflect the diagonal elements of the covariance matrix, although this may make model comparison difficult because the weightings will not be the same for different models (see Cressie, 1985; McBratney and Webster, 1986). A simple and commonly used weighting is to set weights to be proportional to the number of pairs used in calculating each semi-variogram estimate. Since most pairs are used in calculating the semi-variogram at intermediate distances (fewer at shorter and longer distances) this form of weighting may be somewhat less satisfactory when, as in kriging, the analyst is most concerned about the fit at shorter distances. Webster and Oliver (2001, p. 130) list programs for fitting by weighted least squares (see also appendix I).

The principal of parsimony can be made operational and incorporated into the decision process using Akaike's information criterion (AIC) (Akaike, 1973):

$$n \ln(\text{MSR}) + 2p \tag{10.3}$$

where n is the number of points on the empirical semi-variogram plot and p is the number of parameters in the model. The information criterion leads to the selection of the model with the smallest value of (10.3), with p introducing a penalty term for models with more parameters. For further discussion see Webster and Oliver (2001).

Isolated atypical values or patches of atypical values may affect the estimate of the semi-variogram. Cressie (1984, 1991, pp. 40–6) describes the use of the square-root differences cloud for identifying atypical values (see also section 6.3) and providing a *resistant* estimator for the semi-variogram. The pocket plot (see section 6.3) can be used to detect patches of atypical values. Cressie and Hawkins (1980) propose a *robust* approach to estimation of the semi-variogram (where there are small amounts of independent contamination of normally distributed data) based on thinking of the estimation of the semi-variogram as a problem of robust estimation of the location (mean) of square-root differences (see Cressie, 1991, pp. 40, 74–9 for a review).

Since much semi-variogram modelling is undertaken in geostatistics for the purposes of kriging the goodness of fit of a model may be better assessed against how well it predicts, that is by cross-validation. For each possible model, each data value is itself kriged using the model and the other data values. The kriging variance is also computed (see section 4.4.2(v)). The known sample values and their kriged values are then compared. Webster and Oliver (2001, pp. 189–92) discuss different statistical diagnostics (mean error; mean squared error; mean squared deviation error which is the average of the squared deviation divided by the associated kriging variance) and a graphical diagnostic (plot of estimated against true values) and give an example. The mean error and the mean squared error should both be small, the mean squared deviation error should be close to 1 if the model is a good one.

The reader is referred to specialist monographs such as Webster and Oliver (2001) for examples of semi-variogram modelling. Webster et al. (1994) fit semi-variogram models to obtain maps of relative disease risk by kriging the surface. The relative risk on any part of the map is a weighted function of the observed rates and is therefore based on co-kriging. Local values of the observed rates are used to provide relative risk estimates and the method allows for spatial clustering and spatial autocorrelation in the rate map (see section 4.4.2 (v)). Their approach is similar to the Bayesian methodology described in section 9.1.4 which uses local spatial smoothing (9.27 and 9.28).

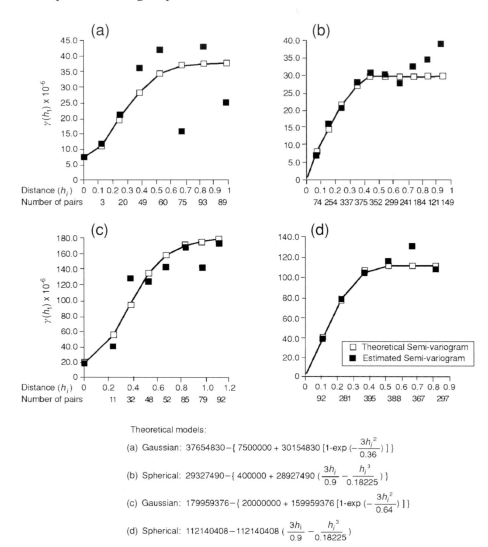

Theoretical models:

(a) Gaussian: $37654830 - \{\, 7500000 + 30154830 \,[1\text{-exp}\,(-\dfrac{3h_i^2}{0.36})\,]\,\}$

(b) Spherical: $29327490 - \{\, 400000 + 28927490 \,(\dfrac{3h_i}{0.9} - \dfrac{h_i^3}{0.18225})\,\}$

(c) Gaussian: $179959376 - \{\, 20000000 + 159959376 \,[1\text{-exp}\,(-\dfrac{3h_i^2}{0.64})\,]\,\}$

(d) Spherical: $112140408 - 112140408 \,(\dfrac{3h_i}{0.9} - \dfrac{h_i^3}{0.18225})$

Figure 10.1 Four semi-variogram fits to mean household income data by census tract. Gaussian models are fitted to low population density tracts (a) 1980 and (c) 1990; spherical models are fitted to high population density tracts (b) 1980 and (d) 1990. Syracuse, New York (Griffith et al., 1994)

Figure 10.1 shows examples of semi-variograms fitted by visual inspection to 1980 and 1990 mean household income data by census tract for Syracuse, New York (Griffith et al., 1994). Tracts were classified into high and low population density tracts to reflect that low-income households tend to live at higher density. Since the models were needed for interpolation the emphasis was on getting good fits at short distances. Figure 10.1 shows the fitted models for the four data sets with the spherical model chosen for high population density tracts and the Gaussian model for low population density tracts.

The ultimate objective of this modelling is often as a required input into implementing kriging either for spatial interpolation or smoothing (see sections 4.4.2 (v) and 7.1.3). The choice of model will affect kriging predictions and the fact that the model for Σ is not known, but has to be estimated from the sample data, means that the usual calculations (e.g. (4.38) and (4.41)) underestimate the prediction error. There is the further uncertainty as to whether the model provides a good representation across all parts of the map.

If universal kriging is used then there is also uncertainty associated with the order of the surface and the estimation of the parameters. Where interpolation or smoothing is local, ordinary kriging is often preferred to universal kriging. This is because inside the small window, where the interpolation is needed, it is often reasonable to assume the surface mean is constant.

(c)   Trend surface models with spatially correlated errors
We now consider the model defined by:

$$\mathbf{y} = \mathbf{A\theta} + \mathbf{u} \qquad\qquad (10.4)$$

where the terms are identical to (10.3) except that now the error term $\mathbf{u}$ is not independent N(0, $\sigma^2\mathbf{I}$) but instead MVN($\mathbf{0}$, Σ) where Σ denotes the variance–covariance matrix of the errors. Alternatively the spatial structure in the errors may be modelled using the semi-variogram. The aim is to fit (10.4) which involves selecting an appropriate order of trend surface and estimating parameters and also selecting a model for the errors and estimating the parameters associated with that part of the model. There can be a choice between fitting a low-order trend surface with one form of error model and fitting a higher-order trend surface with a different error model. Expressed differently, the analyst often has a choice as to the amount of the total variation that is accounted for by the trend and by the stochastic elements of the signal. Both types of model can give similar fits and the final decision may have to be based on what seems most appropriate in the scientific context.

If data values refer to areas their location is usually given by selecting a representative point in each area (such as the centroid). The model for the errors is then defined by a spatial autoregressive model (see section 9.1.2(b)) rather than a covariance or semi-variogram function. If data values refer to a variable with unequal variances across the areas then the estimation procedure must take this into account. This can often arise where the area values are population averages or densities and where the areas differ in terms of population or area size.

Iterative maximum likelihood fitting is recommended. Trend parameters are estimated by downweighting data values that are associated with large error variances and clusters of data values that are positively spatially autocorrelated. In the case of normal variables, the log likelihood function to be maximized is:

$$-(1/2)[\ln|\Sigma| + (\mathbf{y} - \mathbf{A}\boldsymbol{\theta})^T \Sigma (\mathbf{y} - \mathbf{A}\boldsymbol{\theta})] \tag{10.5}$$

and at the first iteration an initial estimate of $\boldsymbol{\theta}$ is obtained using (10.2). These values are substituted into (10.5) so an initial estimate of Σ can be obtained by maximizing (10.5). At this stage, the residuals from the trend surface can be used to specify the model for the errors. At the second iteration $\boldsymbol{\theta}$ is re-estimated:

$$(\mathbf{A}^T \hat{\Sigma}^{-1} \mathbf{A})^{-1} (\mathbf{A}^T \hat{\Sigma}^{-1} \mathbf{y}) \tag{10.6}$$

and the cycle repeated until convergence. Computational methods are discussed in, amongst other sources, Mardia and Marshall (1984) and Ripley (1988).

Consider the case of area data and fitting a first-order simultaneous autoregression. The autoregressive parameter is ρ and the connectivity matrix represented by $\mathbf{W}$ and allowing for non-constant (heteroscedastic) variance then (see (9.14)):

$$\Sigma = (\mathbf{I} - \rho\mathbf{W})^{-1} \Lambda (\mathbf{I} - \rho\mathbf{W})^{-1} \tag{10.7}$$

where Λ is a diagonal matrix with elements $\sigma^2[d_1, d_2, \ldots, d_n]$ where $\sigma^2 d_i$ is the error variance for the ith case and the $\{d_i\}$ are pre-specified. For example population averages by area, computed from equally variable observations, will have smaller variances in areas with large populations than in areas with small populations. For the first-order conditional autoregressive model (10.7) becomes (see 9.11):

$$(\mathbf{I} - \tau\mathbf{W})^{-1} \Lambda \tag{10.8}$$

where τ is now the autoregressive parameter.

One approach to fitting is to base the choice of order of trend surface on an exploratory analysis of the data (see chapter 7) and then having decided the order follow the procedure described above. A more computationally demanding procedure is to use inference theory to decide on the order of trend surface. Fit a series of full models (10.4) starting with a zero order model then a full first-order model, then a full second-order model and so on. Tests of significance on the model parameters can be carried out using the results in Mardia and Marshall (1984). These generalize the results in Ord (1975) and are

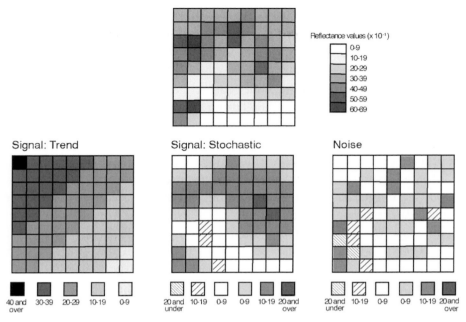

Figure 10.2 A map of reflectance values additively decomposed into trend and stochastic signals and stochastic noise

summarized in Cliff and Ord (pp. 241–2) and Haining (1991, pp. 134–5). Akaike's information criterion can also be called on to help decide when to stop. Haining (1991, pp. 253–60) adopts a procedure along these lines in modelling pollution levels.

Once the model has been fitted the different components of the data consisting of signal and error can be mapped. The deterministic component of the signal (the trend) is obtained by mapping $A\tilde{\theta}$. The random noise component is mapped by:

$$\tilde{e} = \tilde{L}^{-1}(y - A\tilde{\theta}) \tag{10.9}$$

where $\tilde{L}\tilde{L}^T = \Sigma$. A map of the stochastic component of the signal is obtained by mapping $(y - A\tilde{\theta} - \tilde{e})$. This amounts to a decomposition of the data into three different scales of variation. These components are trend which is the large- or macro-scale regional component of variation, noise which is the local- or micro-scale of variation associated with the individual spatial units and stochastic signal which is the intermediate- or meso-scale variation.

Figure 10.2 shows a map of reflectance values that has been decomposed into trend and stochastic signal (spatially structured random variation) and stochastic noise (spatially unstructured random variation). The trend is first order and the stochastic signal is a first-order simultaneous autoregression.

10.1.2 Models for discrete-valued variables

In section 9.1.2(c) the autologistic, autobinomial and auto-Poisson models were defined. The more familiar non-spatial forms of these models obtained by setting all the interaction parameters $\{b(i, j)\}$ to 0 in (9.18), (9.19) and (9.20) respectively are examples of the general linear model and parameters can be estimated by maximum likelihood using an iterative weighted least squares procedure (see, e.g., Dobson, 1999). The 'intra-site' component of these models represented by $\alpha(i)$ for the purpose of descriptive modelling is specified by the appropriate order of trend surface model (section 9.1.1).

The full spatial models cannot be properly fitted by the standard methods developed for generalized linear modelling, since these assume that observations on the response (Y) are independent. Two estimation methods were originally suggested – a pseudo-likelihood and a coding method (Besag, 1974, 1975, 1978; Cross and Jain, 1983). Pseudo-likelihood estimation maximizes the sum of the logs of the conditional probabilities with respect to the unknown parameters using all the data – as if the function were the true likelihood. The coding method uses a subset where the observations are conditionally independent. Gumpertz et al. (1997, p. 136) refer to recent developments in Markov Chain Monte Carlo methods for estimating autologistic models (see section 11.2.2).

For illustration take the autologistic model (9.18) and suppose we want to fit a model where $\alpha(i) = \theta_{0,0} + \theta_{1,0}\, s_1(i) + \theta_{0,1}\, s_2(i)$ and $\Sigma_{j \in N(i)}\, b(i, j)\, y(j) = \beta \Sigma_{j \in N(i)} w(i, j)\, y(j)$. The neighbours of any case i, $N(i)$, are the set of adjacent (first-order) sites and $w(i, j)$ is 1 if j is a neighbour of i otherwise it is 0. Figure 10.3 shows the sites and each site's neighbours for two configurations, one regular and one irregular, together with a first-order coding. The pseudo-likelihood estimation maximizes with respect to the unknown parameters, $(\theta_{0,0}, \theta_{1,0}, \theta_{0,1}, \beta)$, the conditional log likelihood:

$$\Sigma_i \ln\{\exp[y(i)(\alpha(i) + \beta \Sigma_{j \in N(i)}\, w(i, j)\, y(j))]/[1 + \exp[(\alpha(i) $$
$$+ \beta \Sigma_{j \in N(i)}\, w(i, j)\, y(j))]]\} \tag{10.10}$$

The coding estimator also maximizes an expression like (10.10) except that the summation includes only those sites in any particular coding $h(\Sigma_{c(h)})$. For a regular pattern of sites, two codings using disjoint subsets of the data will be possible and so the two sets of estimates have to be reconciled – such as by taking the average. In the case of an irregular pattern of sites, depending on the neighbourhood structure, several codings may be possible but with considerable loss of information for each estimation.

Pseudo-likelihood estimation will tend to provide the better point estimates of the parameters with higher levels of precision but coding estimation will

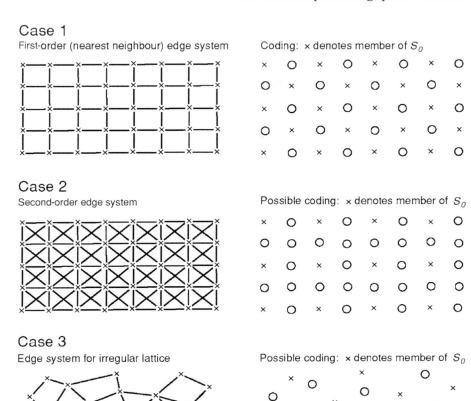

Case 1
First-order (nearest neighbour) edge system

Coding: × denotes member of S_0

Case 2
Second-order edge system

Possible coding: × denotes member of S_0

Case 3
Edge system for irregular lattice

Possible coding: × denotes member of S_0

Figure 10.3 Coding schemes: (a) First order on a grid; (b) Second order on grid; (c) Coding on an irregular lattice

allow inference on the parameters. The goodness of fit of the models can be assessed from the cross-classification matrix of correctly and incorrectly predicted 0s and 1s and by comparing the observed counts in the matrix with the expected values under the null hypothesis. The minimized Akaike Information Criterion can be used to decide between competing models and Gumpertz et al. (1997) also recommend implementing cross-validation to check the predictive ability of the model.

Parameter estimates and cross-classification tables for either of these two estimation methods can be obtained from standard packages where for the illustrative example the data input for case i (if it is included) will be: $\{y(i), s_1(i), s_2(i), \Sigma_{j \in N(i)} w(i, j) y(j)\}$. Note however that in the case of pseudo-likelihood

Table 10.1 *Fitting orders of logistic and autologistic model to TB control adoption data*

| | Model type and order | | | | |
	Log(0)	Autolog(0)	Log(1)	Autolog(1)	Log(2)
Constant	−0.37	−1.69	−1.97	−2.37	−1.90
(east–west)			0.24	0.32	0.81
(south–north)			1.96	0.38	2.09
(east–west)^2					−0.58
(south–north)^2					0.02
(south–north)×(east–west)					−0.56
β		0.46		0.53	
log−likelihood	−150.78	−95.86	−137.93	−92.88	−135.69

estimation the printed standard errors are not valid and likelihood ratio statistics are not chi-squared distributed for large samples (Gumpertz et al., 1997, p. 136). Gumpertz et al. (pp. 143–4) compute parametric bootstrap standard errors for their pseudo-likelihood estimates.

Haining (1990, pp. 288–91) provides an example of fitting an autologistic model with trend by pseudo-likelihood estimation to the distribution of Swedish farms which had/had not adopted TB controls by 1935 (see figure 10.4(a)). Table 10.1 shows the results of fitting an ordinary logistic regression model (Log) using a succession of trend surface models from order 0 to 2. Corresponding autologistic models (Autolog) which include the interaction term β are fit for trend orders 0 and 1. A site has the value 1 ($y(i) = 1$) if farm i is an adopter, otherwise it has the value 0. In fitting the autologistic model the weighting $w(i, j)$ between any two farms (i and j) was specified as $\exp(-1.2d(i, j))$ where $d(i, j)$ is the distance between the two farms.

There is strong evidence in the maps and the fit of the model for a nearest neighbour component to the pattern of adoption with some evidence of a general increase in the number of adopters towards the north and west of the map. The parameters are significant but according to Akaike's criterion there is no benefit in fitting Autolog(1) over Autolog(0). In figure 10.4(b) we choose to show the map of correctly and incorrectly classified farms from the Autolog(1) model. If the probability of a farm being an adopter is, according to the model, greater than 0.5 then it is classified as an adopter, otherwise as a non-adopter. The white and black circles denote farms correctly classified by the model, whilst the grey shades denote different forms of misclassification. Table 10.2 shows the misclassification matrix. Expected values under the null hypothesis of a random classification are in brackets. 71% of the farms are

(a)

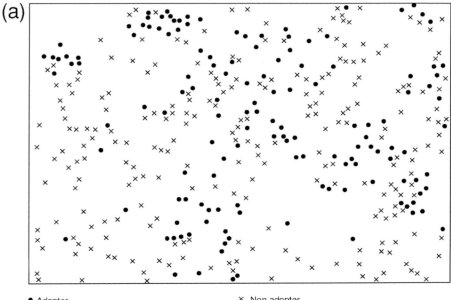

● Adopter   × Non-adopter

(b)

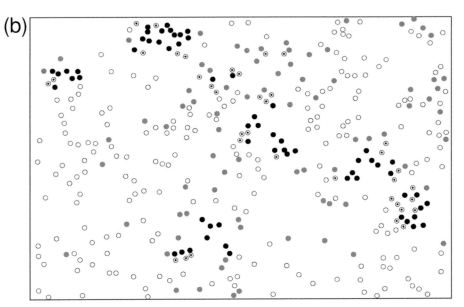

● Adopter correctly classified by the model.   ● Adopter classified by the model as a non-adopter.
○ Non-adopter correctly classified by the model.   ⊙ Non-adopter classified by the model as an adopter.

Figure 10.4 (a) Map of adopters and non-adopters of TB controls in an area of
Sweden, (b) Map showing the classification of farms according to an autologistic
model with linear trend

Table 10.2 *Misclassification matrix for the autologistic model with linear trend*

		Did not adopt	Adopted
		0	1
Model prediction	0	208 (180.49)	73 (100.51)
	1	38 (65.51)	64 (36.49)

correctly classified by the model; 62.7% of the adopting farms are correctly classified by the Autolog (1) model.

10.2   Some general problems in modelling spatial variation

As in other areas of spatial data analysis distinctive problems can arise in descriptive modelling that stem from the spatial nature of the problem. Trend surface modelling may be affected by *frame* and *boundary effects*. The fit of a trend surface model, particularly a high-order surface, may be distorted to lie parallel to the long axis of the study region particularly if the distribution of datapoints reflects the orientation of the study region. The second-order trend surface model fit to cancer mortality data and reported in Haining (1990, p. 373) shows evidence of such a frame effect (see figure 10.5).

Boundary effects are problematic in the case of fitting semi-variogram and covariance functions since datapoints in the study area but close to the boundary may lose their nearest neighbours. A commonly adopted procedure is to create a buffer zone by extending the spatial coverage beyond the area of interest. This is appropriate providing the process in the buffer zone is the same as the process in the study area. It might not be appropriate to adopt this approach say in the case of an air pollution study for an urban area in which the buffer zone was defined as the adjoining rural area.

Spatial sampling may result in *uneven coverage* of the area with some areas intensively sampled and others with only a sparse coverage. The tendency will be to fit a descriptive model that is weighted towards the area that is most intensively sampled. The effect of this will be lessened if generalized least squares fitting is adopted in which the contribution of sample points that lie in clusters of spatially autocorrelated values are downweighted in the estimation. A few isolated points in an otherwise undersampled part of the map will tend to have a strong influence on the model in that part of the map. This is the spatial parallel to the leverage effect in normal linear regression. It also means that error in a data value in a sparsely sampled area of the map will have a more serious effect than the equivalent sized error in a sample point in a well-sampled

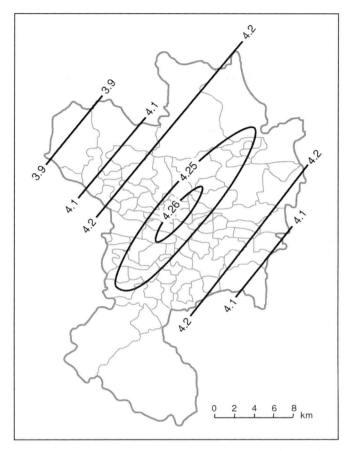

Figure 10.5 Possible frame effect in fitting a second-order trend surface to cancer mortality data. Glasgow

area. This suggests the need to take particular care in checking isolated data values for error. Unwin and Wrigley (1987a, b) investigate the seriousness of the problem of clustered datapoints.

Uneven spatial sampling has implications for some aspects of kriging. The estimator for attribute variability $(\text{Var}(Y(o)))$ that appears in the prediction error (see (4.38) and (4.41)) could be a poor estimator of the global variance if there is substantial oversampling in certain areas of the map relative to other areas.

10.3   Hierarchical Bayesian models

Bayesian inference makes inference about parameters from a model that provides the joint distribution for parameters (φ) and observations $(\mathbf{y})$ denoted $p(\varphi, \mathbf{y})$. In this discussion φ will be the vector of relative risks and $\mathbf{y}$

the vector of observed numbers of cases. This joint distribution is the product of the sampling distribution of the observations given the parameters ($p(\mathbf{y} \mid \varphi)$) and the prior distribution of the parameters $p(\varphi)$. The posterior distribution of the parameters given the observations ($p(\varphi \mid \mathbf{y})$) is obtained by conditioning on the known data values using Bayes rule:

$$p(\varphi \mid \mathbf{y}) = p(\varphi, \mathbf{y}) / p(\mathbf{y}) = p(\mathbf{y} \mid \varphi) \, p(\varphi) / p(\mathbf{y}) \qquad (10.11)$$

Since $p(\mathbf{y})$ does not depend on φ and the values of $\mathbf{y}$ are known $p(\mathbf{y})$ can be considered a constant so that:

$$p(\varphi \mid \mathbf{y}) \propto p(\mathbf{y} \mid \varphi) \, p(\varphi) \qquad (10.11a)$$

The aim of Bayesian inference is to develop a model for $p(\varphi, \mathbf{y})$ and then summarize the properties of $p(\varphi \mid \mathbf{y})$ (Gelman et al., 1995, p. 8).

Mapping problems where the data represent a sample and the quantity of interest is a parameter that is not directly measurable (such as the risk of being a victim of a particular crime or educational attainment by area) can be tackled using this methodology. Craglia, Haining and Wiles (2000) give an example of estimating the risk of burglary by small area. Bayesian methods are used to obtain more robust estimates for small geographical areas using data from an area-wide census. A national or regional survey may have been undertaken without geographical stratification so that any estimates computed for geographical subareas have unacceptably large sampling errors. The local-area estimate, based on a small number of observations, is an unbiased estimate of the local-area quantity of interest but has low precision. The national or regional estimate, based on many observations, has high precision but when spatial heterogeneity is expected, is a biased estimate of any local-area quantity of interest. Bayesian methods may be used to estimate small-area quantitites of interest by 'borrowing strength' from the national or regional-level survey (see Ghosh and Rao, 1994; Longford, 1999).

There is common ground between this methodology and that encountered in earlier chapters on map interpolation and missing data estimation (chapter 4). As noted by Clayton and Bernardinelli (1992, p. 206) if no data were available for estimating relative risk in a particular area, but data were available in all other areas in the same region, we would be willing to estimate the missing rate using information from the other areas. This amounts to using prior information to estimate the missing value. Bayesian modelling integrates these two types of information – such information as is available for the particular area and the 'prior' information available from the other areas.

We consider the case of relative risk estimation in more detail and one particular model for relative risk estimation. The observed mortality counts of a non-contagious disease in an area i are assumed to be independently Poisson distributed with parameter $E(i)r(i)$, where $E(i)$ is the expected number of cases based on overall age- and sex-specific rates and $r(i)$ are the unknown area-specific relative risks of mortality from the disease. It is the $\{r(i)\}$ that are the parameters of interest. In terms of (10.11), this distribution is $p(\mathbf{y} \mid \varphi)$. The $r(i)$ are assumed to be random variables, independently gamma distributed with shape parameter v and scale parameter α. This distribution is $p(\varphi)$ in (10.11). So the $r(i)$ in any particular area, i, is assumed to be a sample from this gamma distribution which has a mean (v/α) and variance (v/α^2).

Under these assumptions the posterior distribution of $r(i)$, conditional on $O(i)$, denoted $p(\varphi \mid \mathbf{y})$ in (10.11), is also gamma distributed but with scale parameter $(E(i) + \alpha)$ and shape parameter $(O(i) + v)$. It follows that the posterior mean of $r(i)$ is:

$$[O(i) + v]/[E(i) + \alpha] = (O(i)/E(i))[E(i)/(E(i) + \alpha)]$$
$$+ (v/\alpha)[1.0 - [E(i)/(E(i) + \alpha)]] \tag{10.12a}$$
$$\equiv \{(O(i)/E(i))\,w(i)\} + \{(v/\alpha)(1.0 - w(i))\} \tag{10.12b}$$

This is often called the empirical Bayes estimate of relative risk when estimates of the unknown parameters, v and α, are substituted (Clayton and Kaldor, 1987). This estimator for $r(i)$ is a weighted function of the SMR for area i and the mean of the (prior) gamma distribution (v/α). It is a compromise between these two quantities. Before discussing $w(i)$ in (10.12b) note the following in (10.12a). The smaller α for given $E(i)$ the closer any Bayes adjusted rate will be to its SMR $(O(i)/E(i))$ as can be seen directly from (10.12a) since $[E(i)/E(i) + \alpha] \to 1$. This is reasonable. The smaller the value of α for fixed v the larger the variation in $\{r(i)\}$ across the region so there is less benefit in borrowing information from other areas than when α is large. Since $E(i)$ is a function of the size of the population at risk, areas with large populations will have adjusted rates that remain close to their SMR which again follows directly from (10.12a). It is areas with small populations that will have adjusted rates that are shifted most towards the overall mean (v/α). A large $E(i)$ implies a standardized mortality ratio with a higher precision than when $E(i)$ is small (see section 6.2.1), so this pattern of differential adjustment is intuitively sensible.

The form of the estimator (10.12b) can be compared with the map interpolators and smoothers discussed in sections 4.4 and 7.1.3. In kriging, models are selected and parameters estimated in order to minimize a cross-validation measure (see section 10.1.1(b)). The spatial smoothers described by Kafadar (1996) aim to generate maps with better variance properties than the maps constructed from (non-spatial) statistics that treat each area in isolation. In 'Bayesian smoothing' a likelihood is specified for the smoothing parameter(s) (Bernardinelli and Clayton, 1992, p. 207). The parameter $w(i)$ in the version of the empirical Bayes estimator (10.12b) is the smoothing parameter. It can be expressed as a function of the prior estimate of the geographical variability in the rates (v/α^2) since:

$$w(i) = [E(i)/(E(i) + \alpha)] = [v/\alpha^2]/\{[v/\alpha^2] + [v/\alpha E(i)]\} \tag{10.13}$$

Note again that when $E(i)$ is large – that is the population at risk in area i is large – the last term on the right-hand side in (10.13) will tend to be close to 0 and so the right-hand side close to 1.

The marginal distribution of the $\{O(i)\}$ is independent negative binomial random variables with parameters $E(i)v$ and α. It follows from the properties of the mean and variance of a negative binomial random variable that:

$$E[O(i)] = E(i)v/\alpha \quad \text{and} \quad \text{Var}[O(i)] = [1 + (1/\alpha)] E[O(i)] \tag{10.14}$$

In a Poisson model with intensity parameter λ, the mean ($E[.]$) and variance ($\text{Var}[.]$) are identical (and equal to λ). So in this model the observed counts show more dispersion than would be expected from the simple Poisson model and $(1/\alpha)$ captures this overdispersion or extra-Poisson variation. Manton et al. (1981) provide another derivation of this model based on the assumption of a population that is heterogeneous in terms of lifestyle, genetics and environmental exposures.

For the empirical Bayes approach the two parameters v and α are estimated from the $\{O(i)\}$ by maximum likelihood using the methods of generalized linear modelling (see, e.g., Clayton and Kaldor, 1987, p. 673; Dobson, 1999) and their estimates substituted into (10.12). However this situation is the exception rather than the norm. With many models the posterior mean cannot be written down in closed form and easily calculated. In these situations, rather than estimating the posterior mean, the posterior mode may be obtained and this is equivalent to using the method of maximum penalized likelihood. In (10.11a) assuming independence so that:

$$\Pi_i p(\varphi \mid y_i) \propto \Pi_i p(y_i \mid \varphi) \, p(\varphi) \tag{10.15}$$

estimation proceeds by maximizing with respect to the unknown parameters (φ):

$$[\Sigma_i \log L(y_i \mid \varphi) + \log p(\varphi)] \tag{10.16}$$

where $p(y_i \mid \varphi)$ in (10.15) is now regarded as a function of φ for fixed $\{y_i\}$, and is called the likelihood function, $L(y_i \mid \varphi)$. The second term in (10.16) acts as a penalty function handicapping solutions that conflict with the prior model (Clayton and Bernardinelli, 1992).

The use of a gamma prior for the relative risks does not allow the introduction of spatial correlation into the relative risks. In addition it is more difficult to introduce covariates into the model at a later stage when the objective might be to try and explain the variation in relative risk. For these reasons a log normal prior for the relative risks may be preferred. Another feature of the log normal prior is that it shrinks extreme values towards the overall mean less severely than the gamma prior. This prior, like the gamma, has the effect of producing an empirical Bayes estimate that is a compromise between the local and the global estimate of relative risk. There is further discussion of this model in Clayton and Kaldor (1987, pp. 674–5) together with a likelihood method of estimation for the posterior mean based on a normal approximation to the posterior distribution of the relative risks. For a discussion of other Gaussian models applied to regional maps of relative risk see Cressie (1992).

The log normal prior for relative risk can be adapted to include spatially structured variation in the relative risks – see the intrinsic Gaussian autoregression and the convolution Gaussian models in section 9.1.4. The intrinsic autoregression smooths area-specific relative risks based on small populations towards the neighbourhood not the global rate. The smoothing process of the convolution model depends on the relative importance of the spatially structured and unstructured components. Mollie (1996) provides an illustration in the case of gall-bladder and bile-duct cancer mortality in France.

Other forms of local spatial 'borrowing' might be considered depending on the specific purpose of the analysis. Lawson and Clark (2002) use a two component spatial mixture model of the form:

$$\log(r(i)) = e(i) + p_i \, v(i) + (1 - p_i) \, t(i) \tag{10.17}$$

The term $e(i)$ is independent normal and $v(i)$ is an intrinsic normal autoregression describing smooth spatial variation. The first two terms on the right-hand side (ignoring the term p_i) correspond to the convolution prior. The additional term $t(i)$ is a prior that measures the total absolute difference between neighbours. That is, the distribution is of the form:

$$\zeta^{-1} \exp[\zeta^{-1} \Sigma_{N(i,j)} \, |t(i) - t(j)|] \tag{10.18}$$

where $N(i, j)$ denotes that i and j are neighbours and ζ is a parameter. The $\{t(i)\}$ component measures discrete jumps in relative risk. The relative importance of the two spatial components depends on the value of p_i which lies in the interval $[0, 1]$. Lawson and Clark (2002) suggest the use of this model for detecting hot spots rather than estimators such as (10.12) or the various Poisson log normal models described above. The posterior mean estimators from these previous models are seen as suitable for providing an estimate of overall map variation but they tend to oversmooth and hence are unlikely to be appropriate if the analyst is mainly interested in checking for extreme values (hot spots or cold spots). Wright et al. (2002) develop another approach to hot spot detection based on defining a loss function on the posterior expectations and then attaching most weight to the highest ranks (highest rates).

Denison and Holmes (2001) use Bayesian partition models to obtain disease rate maps (Ferreira et al., 2002). They are also mainly concerned with the detection of hot spots or detecting raised incidence around a putative disease source. Their method leads to the division of the geographic space into regions or tiles (Dirichlet cells or Voronoi polygons). The purpose of Bayesian inference is to estimate the posterior distribution of the centroids of the Dirichlet cells given the data on counts and the locations of these counts. (Each subarea is in turn represented by a centroid such as the population-weighted centroid.) From these centroids the Dirichlet cells can then constructed (see chapter 2). Different models can be used to represent variation in each region so the final risk map at any point borrows strength from other areas in the same region or tile. These distributions are however assumed to be independent which means that in terms of the model (although not the resulting maps which are obtained by averaging over the different posterior models which yield different distributions of centroids) there are sharp discontinuities at the region boundaries. The data in each region are assumed to arise from the same probability distribution and the same expected risk. The use of partition models for disease mapping has drawn criticism (see, e.g., Lawson and Clark, 2002, p. 360). A similar approach is adopted by Knorr-Held and Rasser (2000) which involves clustering subareas into regions on the basis of similar risk levels.

Spatial smoothing based on strict spatial adjacency may not be a good basis on which to borrow information because such neighbours are not necessarily similar (e.g. across urban/rural boundaries or where there are physical barriers). Lawson and Clark's model might be appropriate to apply in circumstances where such discontinuities are present in the geography. Consideration might also be given to borrowing strength from other areas which though not adjacent (e.g. two urban areas) are similar in terms of the factors that influence disease rates. This approach would be implemented by how the connectivity matrix ($\mathbf{W}$) is defined.

Models such as the Poisson–gamma model are mathematically convenient because closed-form expressions can be obtained for the posterior distribution and the posterior mean estimated by maximum likelihood. However other models, including some of those discussed above, may be more realistic but give rise to complex posterior distributions that cannot be written down in closed form. In order to estimate the parameters of these models other approaches are needed. There is a further consideration. In empirical Bayes modelling the parameters for the prior distribution (called hyperparameters), for example v and α in the case of the Poisson–gamma model, are treated as fixed quantities. These parameters are estimated and in the case of the Poisson–gamma their values substituted into the expression for the posterior mean (10.12). This approach does not allow for any uncertainty in the estimation of these quantities. A consequence is that this will tend to produce underestimates of the probability intervals associated with the estimated relative risk. This then leads to further oversmoothing to the map of rates – see Devine and Louis (1994) and Devine et al. (1996) who develop a constrained empirical Bayes estimator.

In a fully Bayesian approach, distibutions are specified for the hyperparameters as well. Bernardinelli and Montomoli (1992) provide a comparison of relative risk rates from empirical and fully Bayesian analyses and emphasize the advantages of adopting a fully Bayesian approach. There is further reported evidence in Bernardinelli et al. (1995) who also comment on the 'relative insensitivity of these methods to the choice of priors' (p. 2421). As additional complexity is added to any model in the form of prior distributions that do not allow integration to closed-form expressions or the introduction of distributions for the hyperparameters, then methods other than likelihood-based estimation are needed. Even where a likelihood function can be written down, when the number of areas is large there is the problem of handling large matrix manipulations (including matrix inversions in the case of spatial models) and problems in evaluating confidence intervals on the parameters. The approach that is used for inference in fully Bayesian analysis and some forms of empirical Bayes analysis is Markov Chain Monte Carlo (MCMC) simulation.

MCMC is a class of simulation algorithms for obtaining samples from the required posterior distributions of model parameters (e.g. $p(\varphi \mid \mathbf{y})$ in the case of (10.11)). Large samples (size N) are drawn from this posterior distribution and then properties such as the posterior mean, or mode or quantiles obtained by Monte Carlo integration on the marginal distributions of the parameters of interest. The posterior mean of a parameter is obtained by taking the average of the N samples. The term 'Markov chain' in the description of the algorithm is because the samples are dependent and come from a Markov Chain which has the required posterior distribution as its stationary distribution. The

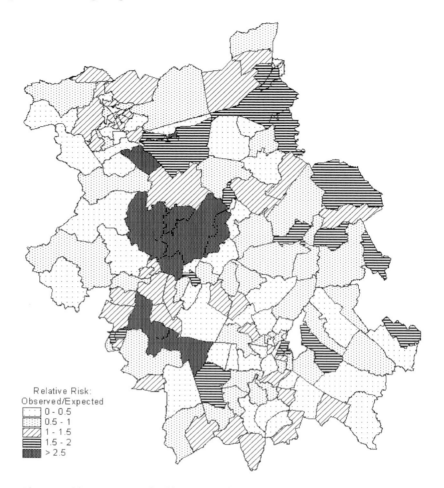

Figure 10.6(a) Lung cancer incidence maps: Cambridgeshire 1998. Standardized incidence ratio map (observed/expected)

general Markov Chain sampling algorithm is the Metropolis–Hastings algorithm. A special case that can be used for some problems is the Gibbs sampler. A full introduction to this methodology can be found in Gilks et al. (1996). MCMC using the special case of the Gibbs sampler is used in the WinBUGS software (see appendix I). An introduction to the Gibbs sampler can be found in Casella and George (1992). A comparison of the different algorithms can be found in Gelman et al. (1995, chapter 11).

We now illustrate some aspects of this methodology. Figure 10.6 shows maps of standardized incidence ratios and Bayes adjusted ratios by ward for lung cancer in Cambridgeshire, 1998. Figure 10.6(a) shows the usual

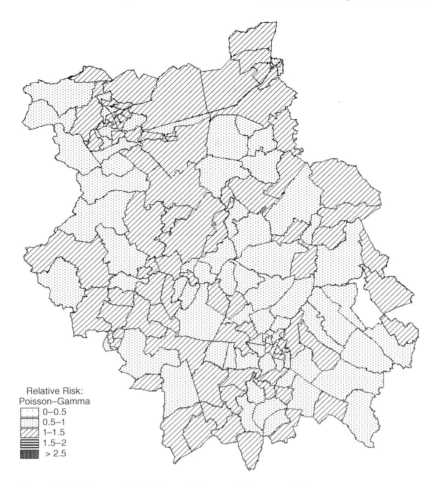

Figure 10.6(b) Bayes-adjusted map: Poisson–Gamma model

standardized incidence ratio map which is the ratio of the observed to expected counts. Expected counts are based on indirect standardization. The county's population is taken as the standard population. This is the maximum likelihood estimate of relative risk under the assumption that the counts are independently Poisson distributed, each with intensity parameter that is the product of the expected count for the given area and the area-specific relative risk. The map suggests considerable heterogeneity of these ratios but this may in part be an artefact of population size variation and the fact that if a ward has no cases in the interval of time then the relative risk for that ward is 0.

Figure 10.6(b) shows the results of using the Poisson–gamma model (see section 9.1.4) whilst 10.6(c) shows the map of Bayes-adjusted ratios obtained

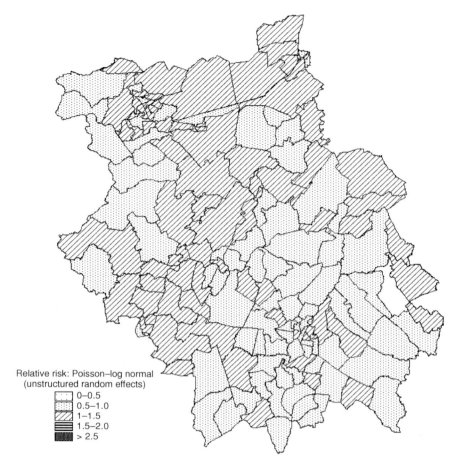

Relative risk: Poisson–log normal
(unstructured random effects)
   0–0.5
   0.5–1.0
   1–1.5
   1.5–2.0
   > 2.5

Figure 10.6(c) Bayes-adjusted map: Poisson–log normal with spatially
unstructured extra-Poisson variation

by treating the relative risks as log normal with spatially unstructured extra-
Poisson variation. Both of these maps show much less spatial heterogeneity
than is apparent in figure 10.6(a). The wards with relative risks above 1.5 have
been adjusted to less than 1.5.

Figure 10.6(d) uses a convolution prior with spatially structured and unstructured extra-Poisson variation. The connectivity matrix is based on simple adjacency. There appears to be a general tendency for risk to fall from the
north and west of the county towards the south and east. The significance of
ratios (above or below 1.0) can be evaluated by computing the probability interval (above or below 1.0) associated with the posterior density. If 5% or less of
the posterior distribution for any ward lies below 1.0 (therefore 95% above 1.0)

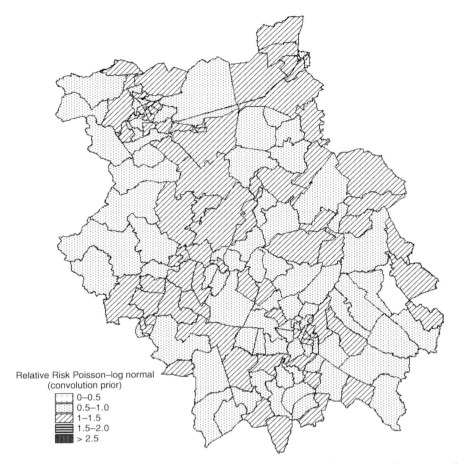

Relative Risk Poisson–log normal
(convolution prior)
- 0–0.5
- 0.5–1.0
- 1–1.5
- 1.5–2.0
- > 2.5

Figure 10.6(d) Bayes-adjusted map: Poisson–log normal with spatially structured
and unstructured extra-Poisson variation

then this would be taken as indicative that the ward had a significantly high
ratio for the county. No wards have significantly high or low ratios.

The data are reported in appendix II. The WinBUGS code for the Poisson log
normal models is similar to that used in the example in section 11.2.5 but without the covariates. There will be further discussion of this methodology and the
use of WinBUGS for these types of analysis in chapter 11.

Note that in all applications, an essential part of parameter estimation is to
closely monitor convergence of the parameters during the sampling process in
WinBUGS.

11

Statistical modelling of spatial variation: explanatory modelling

This chapter examines models for explaining spatial variation in a response variable in terms of predictor variables (factors or covariates). Since the models refer to spatial data at a single point in time they are not process models, nor are they causal models in the usual sense of that word. To avoid confusion with the models of chapter 10 they are called 'explanatory' models. In fact these models only describe the 'here and now' of some response variable in terms of a set of other variables where the direction of the relationship is from the set of exogeneous predictor variables to the endogeneous response variable.

Section 11.1 describes methodologies for modelling spatial data distinguishing between what are termed 'classical' and 'econometric' approaches and discussing a third approach that may be more widely applicable for spatial data analysis. Section 11.2 reviews some applications that cover different modelling situations. Examples include fitting normal linear regression models and regression models with spatially lagged terms in the predictors. This is followed by applications of modelling count data using maximum likelihood and Bayesian methods. An example of multi-level modelling applied to spatial data is described. The emphasis throughout is on the special issues that modelling spatial data raises.

11.1 Methodologies for spatial data modelling

11.1.1 The 'classical' approach

The classical approach to regression modelling is characterized as follows. Regression theory is applied to the results of a series of experiments. The levels of different treatments $(X_1, \ldots, X_k)$ assumed to influence the level of the response variable (Y) are controlled. The levels of the treatments are varied according to an experimental design and the response measured. Model errors

350

(*e*) are due to random measurement error associated with measuring treatment levels or response outcomes. Errors can also be due to the effects of other factors which influence the level of the response but their effects are random (with at most a few unknown parameters) and are uncorrelated with the treatment variables that are included in the model.

In the classical use of regression therefore the specification of the normal linear model:

$$Y(i) = \mu(i) + e(i) \quad i = 1, \ldots, n \tag{11.1}$$
$$\mu(i) = \beta_0 + \beta_1 X_1(i) + \cdots + \beta_k X_k(i)$$

is assumed to be correct. Ordinary least squares is used to estimate the parameters of the mean $(\beta_0, \beta_1, \ldots, \beta_k)$ and the errors (σ^2). Hypothesis testing is used to establish whether certain treatments have a statistically significant effect on the experimental outcome. The analysts job is to guard against failure to satisfy statistical assumptions such as correlation amongst the errors, multicollinearity and non-constant error variance. Tests are available to detect the presence of such problems and if they are present then remedial steps are implemented. If a parameter sign is not what the analyst is expecting this is often taken as indicative of either some experimental error or that regression assumptions have been violated. A good model is one that does not violate statistical assumptions, where there are no serious data effects and a high goodness of fit (R^2) measure is obtained. The theory of regression modelling, its assumptions, diagnostics and what remedial steps can be taken are described and illustrated in for example Chatterjee and Price (1991).

It is argued that application of the classical approach to modelling in non-experimental science is based on this conceptualization of the data. Leamer (1978, p. 4) argues that this use of statistical theory is not, a priori, unreasonable but in following this route the analyst is implicitly accepting as a statement of faith what he calls the *axiom of correct specification*. The axiom states that the set of explanatory variables that determine the response are unique, complete, small in number and observable; other determinants have a probability distribution with at most a few unknown parameters; all unknown parameters are constant. It follows that the analyst dealing with non-experimental data needs to decide whether these are reasonable assumptions.

In spatial data analysis this approach to modelling is exemplified by the analyst who fits a regression model (11.1) on the basis of what he believes are strong a priori grounds in terms of which predictors to include. Then, as part of a programme of diagnostic checking he includes a test for spatial autocorrelation in the residuals using a Moran test (Cliff and Ord, 1981, pp. 197–228). This tests the null hypothesis of a random distribution of the errors against a general (non-specific) alternative that the errors are spatially autocorrelated. The form of the test is similar to that described in chapter 7 (section 7.2) except that critical values for the test need to be adjusted to reflect that the residuals will be correlated. This is because even if the population errors are independent the errors sum to 0 by construction (Cliff and Ord, 1981, p. 200; Tiefelsdorf and Boots, 1995; Hepple, 1998).

If spatial autocorrelation is detected then a spatial error model may be fitted (see 9.30) using maximum likelihood. This is done to 'soak-up' the residual spatial autocorrelation thereby patching the model so that valid inferences can be drawn on the predictors. Fitting a regression model by ordinary least squares when the errors are spatially autocorrelated means that the least squares estimators of the regression parameters are inefficient and the estimator of σ^2 is downward biased. The consequence of the latter is that the observed R^2 goodness of fit measure is inflated and tests of hypothesis have higher type 1 error levels than suggested by ordinary least squares theory. The result of the spatial autocorrelation test (if the null hypothesis of independence is rejected) is warning the analyst of the following. (a) The model is not doing as good a job in explaining the variation in the response variable as is implied by the R^2 measure. (b) There is information in the residuals about the behaviour of the response variable that is not being used. (c) Predictors may have been retained in the model at the α level of significance when in fact they may not be significant at that level. However in adopting this approach the analyst is maintaining faith with the original set of predictors – invoking the axiom of correct specification.

A variant of this patching process that allows the introduction of some new predictors is to use an appropriate Lagrange Multiplier (LM) test (Anselin, 1988; Anselin and Rey, 1991). The LM tests against a specific rather than a general alternative hypothesis. These tests can be used to provide an indication as to whether spatially lagged structures might be introduced into the model on one or more of the covariates or the response variable. Note that no new variables are introduced only spatial functions of variables already in the model. Following the application of the LM test, models such as (9.31) or (9.32) or some combination are fitted, depending on the diagnostic evidence, using maximum likelihood. If spatial interaction ought to be present in the specification (that is the response variable, spatially lagged, should appear as a predictor) then failure to include such a term means that estimates for the other covariates will tend to be biased upwards. This is because some of the effect due to spatial interaction will be allocated to the covariates – particularly those that have a spatial structure that is similar to the spatial structure in the response.

These approaches have in common that they start with a (simple) well-defined initial model and move towards a more general specification based on diagnostic evidence. The approach has been described as simple → general (Gilbert, 1986).

The use of the simple → general methodology has been described as an 'act of faith' because the axiom of correct specification is untestable. There may be other problems. If the model is misspecified then the diagnostics used to identify problems may be misleading and that in revising the model the analyst may be drawn down a wrong track. Notwithstanding these criticisms this is a widely adopted approach to modelling.

11.1.2   The econometric approach

In any non-experimental science, it is often argued, the absence of a controlled experiment means it is not clear what model should be fit to data, since, in the absence of an experimental set-up, it is not clear what set of predictors might be responsible for the observed outcomes. The problem is further complicated when there are competing explanations or theories for the outcomes and the problem is then to decide which one is best supported by the current set of data. Gilbert (1986, p. 284) gives an illustration of such a difference of view. The fundamental problems are: (i) model specification – what model should be fit to the data? and (ii) model assessment or model validation – is the final model that is reported scientifically acceptable? At issue is not whether the model is right or wrong (it is almost certainly wrong) but whether it is useful or misleading with respect to the purpose for which it was constructed.

Leamer (1978) argues that, faced with the difference between the natural science model for conducting science and the situation encountered in the social scences, econometricians have adopted a number of different positions and have engaged in a number of different approaches to model specification. Leamer classifies specification searches found in the econometrics literature into six types. Each reflects responses to the failure to meet one or more of the elements of the axiom of correct specification. The different types of searches are:

1. *Hypothesis testing search*: if the set of predictors is not uniquely defined then several possible regression models can be justified and the analyst uses hypothesis testing to see which ones are best supported by the data.
2. *Interpretive searching*: if the set of predictors is not small, one regression model is chosen and the analyst tries to ensure the model fits the data perhaps by imposing constraints and/or undertakes 3 below.

3. *Simplification search*: the analyst reduces the complexity of the model whilst retaining adequacy of fit.
4. *Proxy searches*: if some predictors are not observable or can be measured in different ways different predictor definitions are compared to see which provides the best fit.
5. *Data selection searches*: if the unmeasured or unobserved predictors that are likely to influence the response variable have complex effects or if the unknown parameters cannot be assumed to be constant then models are fit on different subsets of the data and results compared.
6. *Post-data model construction*: if the predictor set is known not to be complete a purely inductive search is carried out, usually involving looking for additional variables, to try and account for as much variation in the response as possible.

One methodology that has been proposed to address the specification/ validation debate in economics is that developed by Hendry and Sargan. It is referred to by Hepple (1996) as the 'LSE–Oxford' econometric methodology. This approach contrasts with the classical approach by moving from the general to the simple: general → simple (Gilbert, 1986). The initial specification encompasses competing explanations of the data and this overparametrized specification is then simplified by using diagnostics and carrying out tests on the parameters until a model is arrived at that is an acceptable representation of the data.

Gilbert (1986) identifies six criteria by which to judge whether a model is acceptable or 'congruent with the evidence provided by the data'. These six criteria are:

1. The model is data admissable, that is it is logically possible for the model to have generated the data. For example, a normal linear model would not be admissable for a response that was restricted to the interval 0 to 1.
2. The model is consistent with at least one theory.
3. It must be possible to condition on the predictors, that is they must be exogenous and hence their levels not a function of the response variable they are being used to explain.
4. Parameters must be constant so that, if the data set is split, parameter values, within the limits of sampling variation, must remain the same.
5. The difference between the observed and fitted values of the response must be random.
6. The model must be able to encompass rival models and explain why competing explanations are not as good.

These general criteria can be applied to spatial modelling although with some issues of special note. Competing theories (criterion 2) may specify alternative spatial lag structures that do not nest within one another. The definition of a first-order spatial lag is not unambiguously defined in the way a first-order time series lag is defined and the problems mount when higher-order spatial lags are assumed. 'The modelling strategy for spatial (analysis) must therefore give more attention to non-nested models' (Hepple, 1996, p. 243). Splitting data means dividing the data into discrete *areas* (criterion 4) so that if spatial heterogeneity is present it must be possible to account for that heterogeneity within the encompassing model. If the differences between observed and fitted values, the residuals, (criterion 5) are not spatially random then the implication is that it is not enough to simply try and patch up the model in the sense of 11.1.1. Rather it is necessary to respecify the model, taking into account competing representations of the horizontal or spatial interaction effects between observations. Finally some types of spatial models may be problematic, such as the spatial regression where the response variable in spatially lagged form appears as a covariate (9.32), because of the presence of reciprocal dependence encountered in cross-sectional spatial data analysis (criterion 3).

Models should be designed, according to the design criteria discussed by Gilbert (1986, p. 295), to be: fit-for-purpose, robust, parsimonious and with residuals that are uncorrelated. 'Fit-for-purpose' means that the model must enable the analyst to answer the question that was the reason why the model was developed in the first place. Model robustness means that there is no serious multicollinearity (correlation) amongst the predictors that would lead to model shifts if the design matrix ($\mathbf{X}$) were to change and also that parameters are interpretable in theory. Parsimony means that where there is a choice always choose the simpler model. The multicollinearity problem raises a particular problem in the case of spatial modelling since a covariate X and its spatially lagged form $\mathbf{W}X$ will be collinear if X is spatially autocorrelated. If residuals are not white noise then this implies there may be information in the residuals about the process not captured by the current specification.

We now discuss two examples of the general $\rightarrow$ simple methodology of model specification in order to illustrate the methodology.

(a)   A general spatial specification

A general normal linear spatial model based on the models of (9.29) to (9.32) can be written in matrix notation as follows:

$$\mathbf{Y} = \alpha\mathbf{1} + \rho\mathbf{W}\mathbf{Y} + \mathbf{X}\boldsymbol{\beta} + \mathbf{W}\mathbf{X}\boldsymbol{\delta} + (\mathbf{I} - \phi\mathbf{W})^{-1}\mathbf{e} \qquad (11.2)$$

Y is an $n \times 1$ column vector of responses. **WY** is the matrix product of the $n \times n$ pre-specified connectivity matrix **W** describing the spatial neighbours ($\{N(i)\}$) and quantifying the strength of the neighbourliness of each observation. So the ith element of **WY** is:

$$(\mathbf{WY})_i = \Sigma_{h \in N(i)} \, w(i, h) \, Y(h) \tag{11.3}$$

X is the $n \times k$ matrix for the k predictor variables. So **WX** is an $n \times k$ matrix where entry (i, j) is:

$$(\mathbf{WX})_{i,j} = \Sigma_{h \in N(i)} \, w(i, h) \, X_j(h) \tag{11.4}$$

The last term in (11.2) is the error term. The $n \times 1$ vector of errors (**e**) are independent normal with mean 0 and variance σ^2. The matrix operation on **e**, $(\mathbf{I} - \phi\mathbf{W})^{-1}$, generates a simultaneous spatial autoregression on the errors (see section 9.1.2) so that the errors are not independent. Finally the parameters to be estimated are the scalars α, ρ and ϕ and the $k \times 1$ vectors $\boldsymbol{\beta}$ and $\boldsymbol{\delta}$.

Model (11.2) is a general specification. A more general form can be generated by allowing the connectivity matrix **W** to have a different structure in different parts of the model and to encompass different assumptions about the form of **W**. However within the structure given by (11.2) a number of competing spatial models can be defined by imposing restrictions on the parameters. For example suppose we fit:

$$\mathbf{Y} = \gamma_0 \mathbf{1} + \gamma_1 \mathbf{WY} + \mathbf{X}\gamma_2 + \mathbf{WX}\gamma_3 + \mathbf{e} \tag{11.5}$$

where the models parameters are now given by $\gamma_1\gamma_2$ and γ_3. If the condition $-\gamma_3 = \gamma_1\gamma_2$ is satisfied this implies that a normal linear regression model with spatially autocorrelated errors is a convenient simplification of the general model defined by (11.5). If $\gamma_1 = 0$ then a normal linear regression model with spatially lagged effects on the covariates (but not the response) is an acceptable simplification.

(b)   Two models of spatial pricing

Haining (1983a, b) and Sheppard et al. (1991) developed different models to explain spatial patterns of petrol pricing in interdependent spatial markets at the intra-urban scale. The Haining model proposed a spatially autoregressive model (9.32) to capture local competition effects between retailers. It also included in the specification, site effects ($\mathbf{X}_{\mathrm{site}}$) such as whether the site sold other automobile services and whether the site was a major or minor brand retailer. Location effects ($\mathbf{X}_{\mathrm{location}}$) such as whether a retailer was on

a
main road or not were also included. The Haining hypothesis is specified
as:

$$H_{\text{Haining}}: \mathbf{Y} = \alpha\mathbf{1} + \rho\mathbf{WY} + \mathbf{X}_{\text{site}}\boldsymbol{\beta}_s + \mathbf{X}_{\text{location}}\boldsymbol{\beta}_1 + \mathbf{e} \quad \mathbf{e} \sim IN(0, \sigma^2)$$
(11.6)

where the response vector $\mathbf{Y}$ is the retail price of petrol at each of n sites and the
$\mathbf{X}$ matrix has been partitioned into covariates that refer to site characteristics
and those that refer to location characteristics. This model has been recently
tested in Ning and Haining (2002).

The model derived from Sheppard et al. (1992) was tested in Plummer et al.
(1998). The Sheppard hypothesis is specified as:

$$H_{\text{Sheppard}}: \mathbf{Y} = \alpha\mathbf{1} + \mathbf{Z}_{\text{access}}\boldsymbol{\gamma}_a + \mathbf{Z}_{\text{site}}\boldsymbol{\gamma}_s + \mathbf{Z}_{\text{location}}\boldsymbol{\gamma}_1 + \mathbf{e} \quad \mathbf{e} \sim IN(0, \sigma^2)$$
(11.7)

The site- and location-specific covariates in the Sheppard model overlap with
those in the Haining model. The difference lies in the replacement of the
autoregressive term by a measure of accessibility ($\mathbf{Z}_{\text{access}}$). This measures how
accessible a site is in terms of flows of consumers within the urban area. The
Sheppard model therefore focuses on consumer behaviour whereas the earlier
Haining model focuses on retailer neighbourhood competition.

The Haining hypothesis is tested with respect to Sheppard's by identifying those $\mathbf{Z}$ variables referring to site and location that are not collinear with
the $\mathbf{X}$ variables ($\mathbf{Z}^*$) including the access variable and specifying a nesting
hypothesis:

$$H_{\text{nesting}}: \mathbf{Y} = \alpha\mathbf{1} + \rho\mathbf{WY} + \mathbf{X}_{\text{site}}\boldsymbol{\beta}_s + \mathbf{X}_{\text{location}}\boldsymbol{\beta}_1 + \mathbf{Z}^*_{\text{site}}\boldsymbol{\phi}_s + \mathbf{Z}^*_{\text{location}}\boldsymbol{\phi}_1$$
$$+ \mathbf{Z}_{\text{access}}\boldsymbol{\phi}_a + \mathbf{e} \quad \mathbf{e} \sim IN(0, \sigma^2)$$
(11.8)

and the Haining hypothesis implies $\boldsymbol{\phi}_s = \boldsymbol{\phi}_1 = \boldsymbol{\phi}_a = \mathbf{0}$ The Sheppard model
can be tested with respect to the Haining model by formulating the model
in terms of $\mathbf{Z}$ and $\mathbf{X}^*$. The Haining model requires that the spatial structure
of inter-site competition is specified through the connectivity matrix $\mathbf{W}$.
In comparing the Shepppard and Haining hypotheses or indeed different
versions of the same models it should be noted that neither of the theories
specifies in detail the site- and location-specific covariates that are important
nor the spatial structure of inter-site competition or how to measure accessibility. The former is the less problematic since data can be collected on many
covariates that are relevant to the two hypotheses. The latter is more problematic because different spatial structures do not necessarily nest and there are

many ways of specifying the structure of **W** just as there are many ways to measure accessibility.

11.1.3   A 'data-driven' methodology

The econometric approach appears to raise some difficulties when applied to spatial modelling. Furthermore, in many areas where spatial analysis is used there may not be strong competing explanations for phenomena which it is appropriate to formally oppose according to the 'LSE–Oxford' methodology. This is also in part because in spatial modelling the analyst is typically looking for associations rather than trying to develop explanations. Prior knowledge or prior theorizing is often factored into spatial modelling in much more informal ways as part of a programme of 'data-driven' modelling more akin to the classical route although not with the same level of certainty about the underlying experiment. There is also a greater willingness to re-specify to include spatial effects or to add new covariates. The analyst who finds residual spatial autocorrelation would consider introducing other predictors into the model. The spatial pattern of the residuals would be examined for clues as to what additional predictors might be added – and a spatial error model only used as a last resort. Note that heteroscedasticity and spatial autocorrelation can be difficult to disentangle on purely statistical grounds as part of a programme of diagnostic checking and model re-specification (see 7.2.1). If heteroscedasticity is present in the residuals this may justify fitting models such as (9.30), (9.34) or (9.36).

Figure 11.1 summarizes data-driven modelling. Exploratory data analysis provides an important input into model specification. Model diagnostics are used to help re-specify the model if that is thought to be necessary (Anselin, 1988). The italics indicate where in practice substantive issues may also influence the conduct of modelling. Theoretical considerations or prior research in the field may be used in the initial specification of the model and in any subsequent re-specification required as a result of carrying out diagnostic checks. Theoretical considerations are also used to evaluate the final model – perhaps helping to choose between fitted models that in purely statistical terms are equally acceptable.

11.2   Some applications of linear modelling of spatial data

The range of examples desribed here provides illustrations of practice. In each case the method of analysis is described, a brief summary of findings and then some additional issues raised. In most cases the reader can refer back to an original paper for full details.

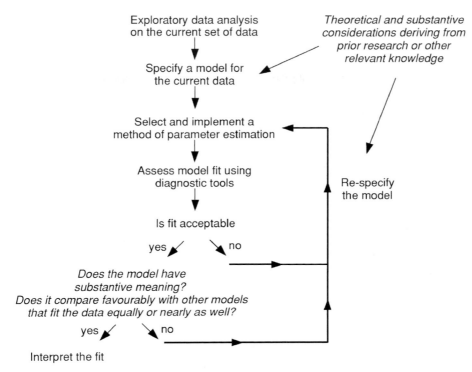

Figure 11.1 Data-driven statistical modelling

11.2.1 Testing for regional income convergence

Rey and Montouri (1999) report findings from a state-level analysis of regional income convergence in the United States over the period 1929–94 and two subperiods within it. Their analysis provides an interesting application and comparison of normal linear spatial regression models.

Let $\mathbf{y}(t)$ denote the vector of real per capita income for all states at time t. In standard (non-spatial) tests for 'β-convergence' in per capita income levels the following model is specified and fit to data:

$$\ln[\mathbf{y}(t+k)/\mathbf{y}(t)] = \alpha + \beta \ln[\mathbf{y}(t)] + \mathbf{e}(t) \tag{11.9}$$

where $\ln[.]$ denotes log to the base e, $\mathbf{e}(t)$ is the vector of independent errors and α and β are parameters to be estimated. The parameter β is the parameter of interest because, if negative and significant, this provides evidence that inter-state convergence has occurred with growth rates in per capita income over the period of k years being negatively correlated with incomes at the start of the period.

Table 11.1 *Spatial dependence models: selected diagnostics*

Model	LM_{error} p-value	$LM_{spatial\ lag}$ p-value	Moran test p-value	p-value for spatial coef.	β
(11.9)	0.020	0.144	0.000		−0.427
Spatial errors (11.10)		0.665		ξ: 0.000	−0.399
Spatial lag on $\mathbf{y}[(t+k)/\mathbf{y}(t)]$	0.146			ρ: 0.002	−0.338
Spatial lag on $\mathbf{y}(t)$			0.000	τ: 0.155	−0.370

Rey and Montouri fit (11.9) by ordinary least squares and for the period 1929–45, $\beta = -0.427$ ($p = 0.000$). The model has an R^2 of 0.823. The Moran test for residual spatial autocorrelation shows autocorrelated residuals but does not provide evidence on the form of misspecification. Two Lagrange Multiplier (LM) tests are performed to provide evidence to guide a re-specification of (11.9). One LM test is against an alternative where $e(t)$ in (11.9) is replaced by:

$$\mathbf{u}(t) = \xi \mathbf{W}\mathbf{u}(t) + \mathbf{e}(t) \tag{11.10}$$

a first-order spatial autoregressive error model with parameter ξ (see 9.30). The other is against an alternative where the term $\rho \mathbf{W} \ln[\mathbf{y}(t + k)/\mathbf{y}(t)]$ is introduced into (11.9), called a spatial lag model and where the spatial lag is on the response variable. They also consider a third re-specification where the term $\tau \mathbf{W} \ln[\mathbf{y}(t)]$ is introduced into (11.9), that is a spatial lag effect on the predictor variable.

The first row of table 11.1 summarizes the evidence of these tests. The evidence suggests that the spatial error model re-specification (11.10) should be preferable to the inclusion of $\rho \mathbf{W} \ln[\mathbf{y}(t + k)/\mathbf{y}(t)]$. These tests do not indicate whether either of these will be superior to introducing a spatial lag on the predictor variable. A Breusch–Pagan test for heteroscedastic (non-constant) variance of the errors is not significant even though the data refer to areas of varying population size. This may be due to the large population sizes of the US states.

The spatial error version and the spatial lag on $\mathbf{y}[(t + k)/\mathbf{y}(t)]$ models are both fitted to the data by maximum likelihood and LM tests carried out for the alternative model. The model with a spatial lag on $[\mathbf{y}(t)]$ is also fitted and for this ordinary least squares can be used. Table 11.1 rows 2 to 4 summarize some of the important results.

The evidence of the LM tests favour the spatial error model over the spatial lag model, a finding that is also confirmed by evaluating Akaike's Information Criterion in the form given by:

$$-2L(\mathbf{y} \mid \alpha, \beta \text{ and } \xi \text{ or } \rho \text{ or } \tau) + 2K$$

where $L(\mathbf{y} \mid .)$ is the maximized log likelihood and K the number of unknown parameters. The third model is unsatisfactory. The spatial coefficient (τ) is not significant and the residuals are spatially autocorrelated.

The evidence of their model fitting is that the original version of the model to test for β-convergence is misspecified. The result of re-specifying the model is that the estimate of β shifts quite significantly (from -0.427 to -0.399) thus altering the estimate of the convergence rate from (0.035 to 0.032) obtained by computing $\ln[\beta + 1]/-k$. Their decision to fit spatial models is also supported by the results of exploratory data analysis and theoretical arguments about why convergence effects might have a spatial structure. They point to the effects of technology spillovers between adjacent states ('substantive spatial dependence') and the effects of analysing a convergence process in terms of an artifical spatial framework (the states) which does not correspond to market processes ('nuisance dependence').

The choice of connectivity matrix $(\mathbf{W})$ is problematic in the case of economic processes where inter-sectoral and inter-firm linkages for example need not reflect spatial contiguity. Predictions on how any spatial shock in one part of the system will diffuse spatially will be heavily dependent on the assumed connectivity matrix. Also it seems likely that $\mathbf{y}(t)$ will be correlated with other unmeasured covariates that affect per capita income growth rates. Because these covariates remain unspecified in the model they will be present in the error term. This introduces a downward bias into the estimate of β. This is one of the problems of coping with the effects of spatial autocorrelation in residuals through a spatial error model rather than trying to specify the covariates that might account for the autocorrelation. The price of overcoming the problems created by unmeasured spatially autocorrelated covariates may be an overly conservative estimate of the convergence parameter. The inclusion of covariates in a fully specified 'conditional convergence' model provides another way to handle these types of processes.

A number of examples of this type of modelling, with references, can be found in amongst other sources Anselin (1988), Haining (1990) and Griffith and Layne (1999). There is a further discussion of diagnostics for spatial regression models in Haining (1990a and 1994a).

11.2.2   Models for binary responses

(a)   A logistic model with spatial lags on the covariates

Craglia et al. (2001) develop a logistic model to predict police-defined, high-intensity crime areas (HIAs) in English cities. HIAs are areas with supposedly large amounts of often violent crime where many of the residents are either unwilling or afraid to assist the police in their enquiries, thus adding to

Table 11.2 *Best fitting final model for the HIA crime data*

Variable	Regression coefficient		Correctly classified (%)	Incorrectly classified (%)
INDEX	0.2351	% of EDs in	52.9	47.1
TERRACE	0.0102	HIAs classified		
TURNOVER	0.0229	% of EDs not in	79.9	20.1
Constant	−2.3465	HIAs classified		

the difficulty and cost of policing such areas. The context for the work was a possible adjustment to the police funding formula, by police force area (PFA) in England and Wales.

The response variable $y(i)$ was binary the value depending on whether the ith ED was in a police defined HIA (1) or not (0). The intial model specification examined the link with a number of census variables that captured different aspects of social and economic disadvantage, population instability and ethnic mix. Variation in these area-level attributes have been found to be associated with varying levels of violent crime. Table 11.2 identifies the covariates and their coefficient estimates that appeared in the final model based on police data provided by three PFAs (Greater Manchester, Merseyside and Northumbria). The variable *INDEX* was the DETR index of deprivation which is based on five variables: unemployment, overcrowded households, households lacking amenities, children in low-earning households and households with no cars. *TERRACE* measures the proportion of dwellings in an ED that are terraced and *TURNOVER* is the proportion of residents with different addresses one year before the 1991 Census. The model identifies HIAs as deprived areas where households live at high density and where there is relatively high levels of social instability as measured by the inflow and outflow of residents.

The probability that the ith ED (enumeration district) is in an HIA according to the final model is given by:

$$\exp(\mathbf{X}(i)\beta)/[1.0 + \exp(\mathbf{X}(i)\beta)]$$
$$\mathbf{X}(i)\beta = -0.23465 + 0.2351(INDEX(i)) + 0.0102(TERRACE(i))$$
$$+ 0.0229(TURNOVER(i)) \tag{11.11}$$

Figure 11.2 shows the geography of the correctly and incorrectly classified EDs for the Greater Manchester PFA after adjusting the probability for predicting an ED as being in an HIA so that the police count and the model count were the same. In general the model performed well in that it did not generate many 'false positives' (clusters of EDs predicted to be in an HIA that had not been designated an HIA by the police). Also it did well in predicting the locations of police defined HIAs. Errors of classification often arose on the borders of the

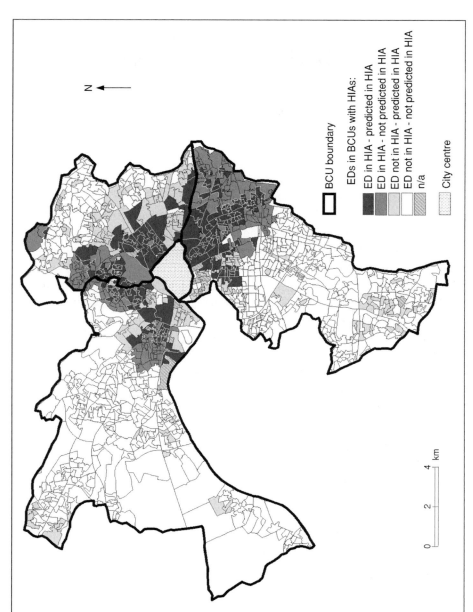

Legend:

BCU boundary

EDs in BCUs with HIAs:

ED in HIA - predicted in HIA
ED in HIA - not predicted in HIA
ED not in HIA - predicted in HIA
ED not in HIA - not predicted in HIA
n/a

City centre

N

0   2   4   km

Figure 11.2 Cross classification of EDs by model prediction and police classification: Greater Manchester PFA (from Craglia et al., 2001, p. 1931)

Table 11.3 *Best fitting final model for HIA data after including spatially averaged variables*

Variable	Regression coefficient		Correctly classified (%)	Incorrectly classified (%)
INDEX	0.1801	% of EDs in	53.68	46.32
TERRACE	0.0091	HIAs classified		
TURNOVER	0.0201	% of EDs not in	79.9	20.1
W*INDEX	0.1447	HIAs classified		
W*TERRACE	0.0056			
Constant	−2.9470			

HIAs – that is the spatial extent of HIAs was less well captured. The join-count test, based on counting the number of times across the spatial distribution differently misclassified EDs were adjacent and under the assumption of non-free sampling, showed there was strong positive spatial correlation in the difference map between the police-defined and model-defined maps.

There are no substantive reasons to anticipate a spread process underlying the formation and spatial extent of an HIA that would lend justification to fitting an autologistic model (see sections 9.1.2(c) and 9.2.2). There is some evidence however that particularly difficult policing areas may be those areas that are embedded within the geographical cores of more extensive tracts of social and economic disadvantage. For this reason the model was re-fitted with additional variables that were the spatial average values of the corresponding variables (**W***TERRACE; **W***INDEX, **W***TURNOVER). Spatial averages were computed over just the nearest neighbour EDs which shared a common boundary. The results of this model fitting are displayed in table 11.3.

Two of the spatially averaged predictors are significant but the improvement in the misclassification matrix is marginal. Higher-order spatial averaging might have captured better an embeddedness effect if there is one. The paper also explores the question of whether part of the reason for the spatial structure in ED misclassification might lie in the translation of police defined HIAs from street plans (with no reference to ED boundaries) to ED geography. The translation process required decisions to be taken as to whether to include or exclude in an HIA any ED where the boundary of the police defined HIA cut across the ED.

Dubin (1997) models innovation adoption using a logistic regression model. Spatial effects are also handled through the explanatory variables and the paper reviews other approaches to this problem.

(b)   Autologistic models with covariates
Augustin et al. (1996) fit an autologistic model with covariates to data on the presence/absence of red deer by 1km grid squares in the Grampian

region of Scotland. Interaction effects between deer suggest the need for an autologistic model. The variate $y(i)$ is 1 if red deer are present in grid square i and
0 otherwise. $\theta(i)$ is the probability that deer are present in grid square i. Their
model is given by:

$$\log(\theta(i)/(1 - \theta(i)) = \beta_0 + \beta_1(alt^2(i)) + \beta_2(s_1(i)) + \beta_3(s_2(i)) + \beta_4(mires(i))$$
$$+ \beta_5(pine(i)) + \beta_6(\Sigma_j w(i, j) y(j) / \Sigma_j w(i, j))$$

where s_1 and s_2 denote the terms for a linear trend surface and *alt* (altitude),
mires and *pine* are environmental covariates. The last term is the spatial interaction effect, and the spatial extent of the neighbours used to select the appropriate connectivity matrix ($\{w(i, j)\}$ was determined by looking at the reduction
in the deviance. The analysis was complicated by the presence of unsurveyed
squares and presence/absence was predicted at the unsurveyed sites before fitting the model.

Gumpertz et al. (1997) use the autologistic model to analyse the spatial pattern of disease on bell peppers on a regular grid. The model covariates measure
soil water content and soil pathogen density and the purpose is to estimate the
effects of these soil variables after controlling for the way the disease spreads
between adjacent plants. Their example contrasts a pure logistic model, an autologistic model with covariates and an autologistic model with no covariates.
If spatial dependence is not included in the model specification then the effects
of the soil variables are overestimated.

As noted earlier (10.1.2) this model can be fit by standard generalized linear modelling software if the analyst is only interested in efficient parameter
estimation but not if inference is to be undertaken. Gumpertz et al. (1997) use
bootstrapping methods to estimate standard errors and another approach is
to fit the model on a coding scheme (see section 10.1.2). Huffer and Wu (1998)
use MCMC to estimate parameters and obtain confidence intervals. Another
approach is to re-specify the problem as a (Bayesian) hierarchical model with
covariates and with unstructured and spatially structured random effects (see
sections 9.2.3, 11.2.4 and 11.2.5). There is further discussion of approaches in
Heagerty and Lele (1998).

11.2.3 Multi-level modelling

Sampson et al. (1997) use multi-level modelling to analyse the influence of 'collective efficacy' on perceived levels of violence and violent victimization amongst Chicago residents sampled across 343 'neighbourhood clusters'.
The term collective efficacy refers to the 'linkage of mutual trust and willingness to intervene for the common good' (p. 919), a concept that is closely akin
to that of 'social capital' but defined here in terms of 'inhibiting the occurrence

of personal violence' (p. 919). The neighbourhood clusters were based on geographically contiguous census tracts and were constructed to be internally homogeneous in terms of a number of census variables, including socio-economic status and ethnic composition.

The analytical model that underlies their findings is a Gaussian multi-level structure that with some simplification can be represented as follows:

$$Y(i, j, k) = \pi(j, k) + e(i, j, k) \tag{11.12a}$$

$$\pi(j, k) = \eta(k) + \Sigma_q \, \beta_q \, Z_q(j, k) + r(j, k) \tag{11.12b}$$

$$\eta(k) = \gamma_0 + \Sigma_p \, \gamma_p \, \Sigma_p \, X_p(k) + u(k) \tag{11.12c}$$

$Y(i, j, k)$ denotes the ith response of person j in neighbourhood cluster k regarding their perception of violence. The measure of perception of violence was based on their knowledge of different forms of violence in the neighbourhood. The parameter $\pi(j, k)$ represents the score for person j in neighbourhood k. The term $e(i, j, k)$ is the measurement error term and is assumed to be independent and of constant variance (σ_e^2). The parameter $\pi(j, k)$ is a function of $q = 11$ individual-level characteristics $(Z_1, \ldots, Z_{11})$ that include age, gender, marital status, ethnicity, socio-economic status and period of residence and an area-level parameter $\eta(k)$. $r(j, k)$ is an independent normal error term with mean 0 and constant variance (σ_r^2) and measures random effects in the individual-level score. The parameter $\eta(k)$ is a function of $p = 4$ area-level charateristics $(X_1, \ldots, X_4)$. These are: a measure of the concentration of economic disadvantage in the neighbourhood, residential stability, immigrant concentration and a measure of collective efficacy. The term $u(k)$ is an independent normal error with mean 0 and constant variance (σ_u^2).

Sampson et al. (1997) also develop a multi-level model for violent victimization in which the response $Y(j, k)$ is binary: 0 (individual j in area k has not been a victim of violence) or 1 (has been a victim of violence). So $Y(j, k)$ is Bernoulli with parameter $p(j, k)$ which is the probability that individual j in area k is victimized. Now using the same notation as earlier:

$$\text{logit}[p(j, k)] = \log(p(j, k)/(1 - p(j, k)))$$
$$= \eta(k) + \Sigma_q \, \beta_q \, Z_q(j, k) + r(j, k)$$
$$\eta(k) = \gamma_0 + \Sigma_p \, \gamma_p \, X_p(k) + u(k)$$

They also model the number of homicides in each neighbourhood cluster as a Poisson log linear model with overdispersion where $Y(k)$ is the number of homicides in neighbourhood k and $E[Y(k)] = N(k)\lambda(k)$ where $N(k)$ is the population size and $\lambda(k)$ is the homicide rate. Then $\eta(k) = \log[\lambda(k)]$ and the right-hand side of (11.12c) can be used to model variation in $\log[\lambda(k)]$.

The model does not include any specification of possible spatial autocorrelation amongst the individuals in the sample. Individuals in the same neighbourhood are unlikely to be exposed to identical experiences of crime because even neighbourhoods (unless they are very small) will not be homogeneous in terms of relative risk. The risk of becoming a victim of crime can, geographically, be highly localized. Whilst geographical proximity within the neighbourhood may be only one aspect of the autocorrelation that may exist amongst the population of individuals it is worth considering. To ignore this aspect of the problem could raise inference problems as described in section 11.1 at the second level of the model (11.12b) because the amount of information in the sample is futher reduced from that identified by the conventional statistical theory for the multi-level model. With vary large sample sizes this might not be a problem but would need attention if sample sizes were small.

The model also treats each neighbourhood cluster in isolation. There is no specification of the spatial relationships between the 343 neighbourhoods. There are two reasons to consider this problematic. First individuals may live in different neighbourhoods but be geographically close because they live either side of a neighbourhood boundary and so experience broadly the same events if not at their place of residence then through their patterns of movement within the local area. Second the geographical incidence of violence does not follow neighbourhood boundaries – in the same way that economic processes do not observe state boundaries in the example in section 11.2.1. The spatial framework in terms of which the third level of the model is constructed in reality consists of spatially overlapping entities. Spatial units are not discrete and well defined. For these reasons at the third level of the model (11.12c) a refinement of the specification would allow for the location of the neighbourhoods relative to each other, perhaps through the use of a spatial model on the $\{u(k)\}$. Other specifications that recognize the locations of the neighbourhoods relative to each other should also be considered.

It has been suggested that multi-level modelling allows the analyst to avoid the problems raised by autocorrelation in spatial modelling but for reasons suggested through this example this cannot be taken for granted.

11.2.4 Bayesian modelling of burglaries in Sheffield

Data are available on the total number of burglaries by ward in 1995 (see appendix III). $O(i)$ denotes the count for ward i ($i = 1, \ldots, 29$). In this analysis we shall ignore certain complications. Our data refer to the total number of burglaries not the number of households burgled. A household once burgled is at greater risk of being burgled again (repeat victimization) than another household that has not been burgled for a period of perhaps as long as six months. If one household is burgled in a given period of time, nearby households also have a higher risk of being burgled in the same period.

Notwithstanding these qualifications, $\{O(i)\}$ are assumed to be independent binomial random variables with parameter $NU(i)$, the number of households in ward i, and $P(i)$ the probability that any household in i is burgled. Given the previous comments and the heterogeneity of household types, the assumption of a constant $P(i)$ within any ward i is a strong assumption. Figure 11.3 shows a multiplicative decomposition of the ward-level burglary counts into two maps: the number of households in each ward and the burglary rate by ward $(O(i)/NU(i))$ which is the maximum likelihood estimate of $P(i)$. This decomposes the count into that due to the size of the population at risk and the risk itself and shows how these vary across Sheffield at the ward scale.

The parameter $P(i)$ is modelled through a logit transformation $(\text{logit}(P(i)))$. Logit$(P(i))$ was initially expressed as a function of spatially structured $(v(i))$ and spatially unstructured $(e(i))$ random effects:

$$\text{logit}(P(i)) = \log[P(i)/(1.0 - P(i))] = \mu + v(i) + e(i) \tag{11.13}$$

where μ is the intercept term (mean). $v(i)$ is an intrinsic Gaussian autoregression (see 9.25 and 9.26). The neighbours of ward i are those that share a common boundary with i. $e(i)$ is an independent normal variable.

Appendix III.3(a) shows the WinBUGS code where beta0 $= \mu$. A uniform prior on the whole real line $(d$ flat$())$ has been imposed on beta0 and a gamma prior on the precision of the $\{e(i)\}$ and $\{v(i)\}$. The posterior distribution of each $P(i)$ divided by the average rate for the whole of Sheffield was generated $(SR[i])$ and the proportion of each of these distributions that lie above 1.0 or below 1.0, computed. If 95% of the posterior distribution lies below (above) 1.0 then the burglary rate is deemed to be significantly less than (greater than) the Sheffield average. If neither of these conditions is true then the rate for the ward is not significantly different from the Sheffield average. Figure 11.4 identifies those wards with burglary rates that are significantly greater or less than the average rate, based on their posterior distributions from the binomial-logistic model. The burglary rates are based on large counts (burglary is not a rare event) and the same map is obtained by computing the usual confidence intervals around the estimate of $P(i)$ given by the number of burglaries in ward i divided by the number of households in ward i. Contrast this with the effect of the equivalent Bayesian adjustment of standardized incidence ratios for the Cambridgeshire lung cancer data (see 10.3).

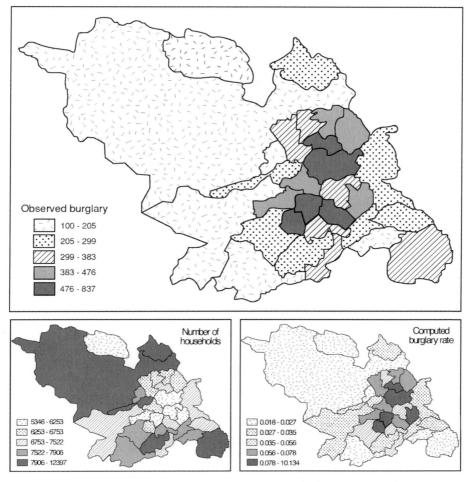

Figure 11.3 Decomposition of the count of burglaries by ward in Sheffield: (a) Observed count (b) Number of households (c) Burglary rate

The estimate of the variance of $v(i)$, κ^2, is 0.4544 whilst that of $e(i)$, σ^2, is 0.1737. The ratio, 2.616, indicates that, in the as yet unexplained part of the variation in $\{P(i)\}$, spatially structured random effects dominate spatially unstructured random effects. This suggests that there remains unexplained spatial structure in the distribution of the $\{P(i)\}$. Figure 11.5 shows the maps of $\exp(v(i))$ and $\exp(e(i))$ that show the variation in the odds decomposition around 1.0 for the spatially structured and spatially unstructured random effects. Both maps display variation.

Two covariates representing deprivation ($X_1(i)$) and population turnover ($X_2(i)$), two variables that as they increase are likely to be associated with

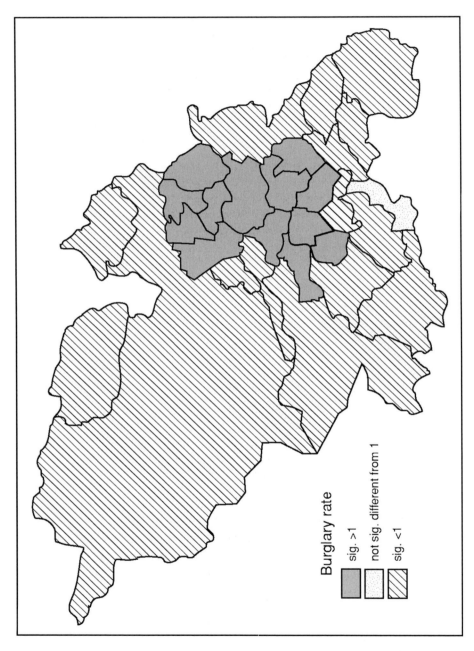

Burglary rate

sig. >1

not sig. different from 1

sig. <1

Figure 11.4 Classification of wards into those with statistically significant rates of burglary either above or below the Sheffield average

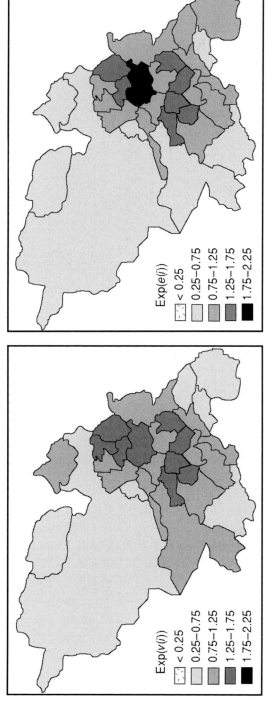

Figure 11.5 Maps of (a) exp($v(i)$) and (b) exp($e(i)$) for the model without covariates

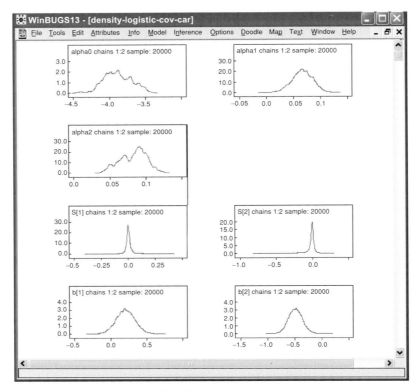

Figure 11.6 Selected posterior distributions: the three regression parameters (alpha0 = β_0; alpha1 = β_1; alpha2 = β_1); the spatially structured ($S(i) = v(i)$) and unstructured ($b(i) = e(i)$) random effects

increasing rates of burglary, were included in the model so that (11.3) becomes:

$$\text{logit}(P(i)) = \log[P(i)/(1.0 - P(i)] = \mu(i) + v(i) + e(i) \qquad (11.14)$$

$$\mu(i) = \beta_0 + \beta_1 X_1(i) + \beta_2 X_2(i) \qquad (11.15)$$

The WinBUGS code is given in appendix III.4(a).

The posterior distributions of the regression parameters β_0, β_1 and β_2 are shown in figure 11.6 (alpha0, alpha1 and alpha2). The parameter estimates, the posterior means, are: $\beta_0 = -3.884$; $\beta_1 = 0.066$; $\beta_2 = 0.082$ (see appendix AIII.4(b)). The corresponding odds ratios are $\exp(\beta_1) = 1.068$; $\exp(\beta_2) = 1.086$. A one unit increase in deprivation and population turnover produces a nearly 16% (1.068 × 1.086) increase in the odds ratio or the relative risk of a house being burgled. The covariates are statistically significant and the signs are consistent with expectations. As a result of introducing the two covariates, the ratio of the variance of $v(i)$ to $e(i)$ is reduced to 0.025 indicating that the introduction of the covariates means that now spatially unstructured heterogeneity

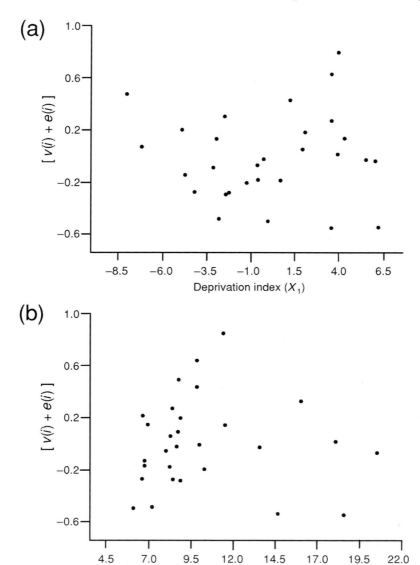

Figure 11.7 Plot of $[v(i) + e(i)]$ against (a) deprivation ($X_1(i)$) and (b) population turnover ($X_2(i)$)

dominates the error map. Figures 11.7(a) and (b) show the plots of $[v(i) + e(i)]$ against $X_1(i)$ and $X_2(i)$. The errors show no evidence of any association with the values of either of the two covariates.

Figure 11.8 shows a multiplicative decomposition of the odds map $\{(P(i)/(1-P(i))\}$ shown in (a), where the estimate of each $P(i)$ is the mean of

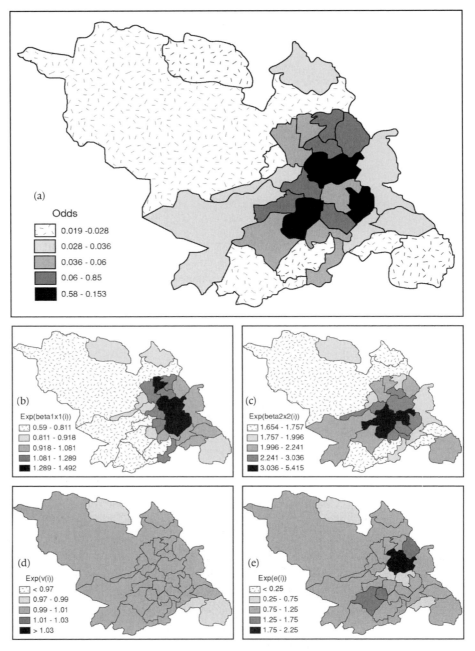

Figure 11.8 Multiplicative decomposition of the odds map for burglary in Sheffield by ward (a). The decomposition is into the explained components due to deprivation (b) and population turnover (c). There are two unexplained components: spatially structured (d) and spatially unstructured (e). The decomposition is in terms of posterior means

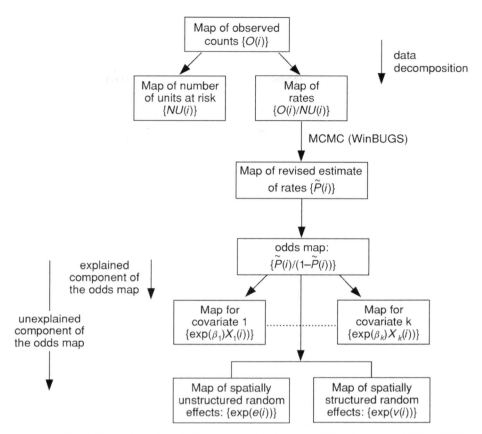

Figure 11.9 Data and posterior means odds decompositions in figures 11.3 and 11.8

the corresponding posterior distribution obtained from the binomial-logistic model. The decomposition (which is approximate since it is based on posterior means) is into the explained ($\exp(\beta_1 X_1)$; $\exp(\beta_2 X_2)$) and unexplained (spatially structured: $\exp(v)$; spatially random: $\exp(e)$) components. Maps (b) and (c) show the area-by-area increased risk associated with each area's level of deprivation and population turnover. The odds for the spatially structured random effect (d) is uniformly close to 1 whilst the spatially random component (e) shows areas where the model is unable to predict some of the higher and lower risks of burglary. The numerical values are given in table AIII.5. Figure 11.9 shows the two levels of map decomposition associated with the data (figure 11.3) and the Bayesian modelling (figure 11.8).

The spatial component of the model of this section (represented by $v(i)$), unlike the models discussed in sections 11.2.1 and 11.2.2, does not contain a 'spatial parameter' that allows the strength of the similarity between neighbouring values to be estimated. This simplifies the fitting procedure but at the price of artificially 'fixing' the mean of the spatially structured element in the random effects as a local spatial average (but see the reference to WinBUGS in appendix 1). The decomposition of the random effects term into $[v(i) + e(i)]$ is similar in concept to but not the same as the decompostion undertaken in, say, the regression model with spatial errors (see (9.30) and discussed in relation to the example in section 10.1.1(*c*) and illustrated in figure 10.2).

11.2.5　Bayesian modelling of children excluded from school

Appendix IV reports the number of children excluded from school by ward in Sheffield. The count $\{O(i)\}$ refers to a rare event and each is assumed an independent Poisson random variable with mean $E(i)R(i)$ where $E(i)$ is the expected number excluded from school and $R(i)$ is the relative risk of exclusion for area i. The expected count for ward i is just the rate of exclusions (total number of exclusions in the city divided by the number of children in the age category in the city) multiplied by the number of children in the age category in ward i.

The parameter $\log[R(i)]$ is modelled as a function of a ward index of deprivation X_1 and spatially structured $(v(i))$ and spatially unstructured $(e(i))$ random effects. The variable % of single parent families was also included in an initial specification but found not to be significant. The WinBUGS code is given in table AIV.2 in appendix IV. The results of the analysis are reported in table AIV.3 where it can be seen that the deprivation variable is strongly significant and the parameter yields an odds ratio of 1.22 or a 22% increase in risk for every unit increase in the Townsend index of deprivation. The variance ratio of the spatially structured random effect to the spatially unstructured random effect is $(622/1422) = 0.437$. Spatially unstructured random effects dominate the unexplained variation. Table AIV.4 provides an area-by-area decomposition of the posterior mean of the relative risk into three components. The maps for this decomposition of the relative risk are shown in figure 11.9. These maps show for each ward the contribution of each of the three elements of the model to the relative risk of children being excluded from school.

The diagnostics for these spatial models are relatively crude. This appears to be an area that is currently receiving attention (see Spiegelhalter et al., 2001). Required are diagnostics that will help distinguish between competing models (overall goodness of fit measures – the Bayesian information criterion and the deviance information criterion) and diagnostics that will help to assess how well a model fits the data (local goodness of fit measures – standardized posterior predictive residuals). These statistics are discussed in a spatial context in Lawson and Clark (2002, pp. 365–9).

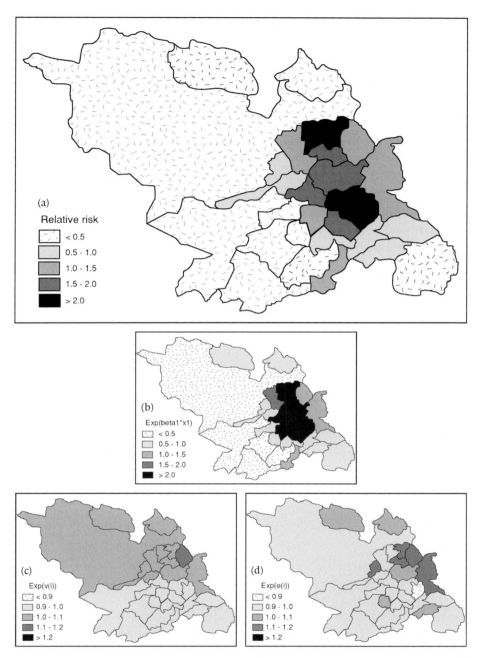

Figure 11.10 Multiplicative decomposition of the relative risk map for children excluded from school in Sheffield by ward (a). The decomposition is into the explained component due to deprivation (b) and the unexplained component. There are two unexplained components: spatially structured (c). and spatially unstructured (d). The decomposition is in terms of posterior means

11.3   Concluding comments

In the models described in this chapter and the previous one, in addition to the important issue of data quality (including the quality of the spatial referencing) two issues that have substantive significance and affect model interpretation recur in the conduct of spatial analysis. The first is the choice of spatial framework (the spatial units) in terms of which data values are recorded. The second is the definition of the spatial relationships between the individual units. The first is a problem of long-standing concern in the area of spatial analysis, the second has assumed greater importance with the growing use of the sorts of spatial models discussed in this book. The analyst needs, so far as it is possible, to construct spatial frameworks that are meaningful in terms of the processes under investigation. In many areas of the social and environmental sciences the analyst also needs to define spatial relationships between areas in ways that are meaningful in terms of the underlying processes that are responsible for the spatial variation in the data.

Appendix I Software

The following is a short listing of software that is available for implementing some of the methods described in this book. A good source of information on specialist spatial analysis software is provided by the website of the Centre for Spatially Integrated Social Science: www.csiss.org: There is a spatial tools menu which includes a search engine.

Spatial statistical capability has been added to some geographic information systems. ESRI's ArcGIS (version 8.1) provides *Geostatistical Analyst*. This undertakes kriging, including modelling the semi-variogram, exploratory spatial data analysis (including identifying global and local outliers, trends and spatial autocorrelation) and interpolation (inverse distance weighting, local and global polynomial surface fitting and kriging). The module *Spatial Analyst* enables raster and vector data to be integrated. The website for information is: www.esri.com/software

Semi-variogram and kriging software are available in GENSTAT and SAS and in MLP from the Numerical Algorithms Group at Oxford.

S+SpatialStats is an add-on to S-PLUS from Mathsoft (www.mathsoft.com). It has geostatistical capability (variogram modelling, ordinary and universal kriging). Spatial regression modelling (with conditional autoregressive, simultanous autoregressive or moving average errors) is included. There are autocorrelation tests and point process methods.

The package Stat! from Biomedware (www.biomedware.com) has a number of tests including clustering tests and Moran's spatial autocorrelation test as well as Oden's modification. The Biomedware site provides access to Luc Anselin's Space-Stat, a package that has many exploratory spatial data analysis tools and also has advanced spatial regression modelling (spatial econometric) tools including modelling with spatially lagged response variables as predictors. Other models include the regression model with spatial errors, the trend surface model and the spatial expansion method model. It includes an excellent tutorial workbook. A number of spatial clustering and cluster detection tests (including the Besag-Newell test and Ripley's K test) are available at www.terraseer.com/csr/clusterseer_methods.html

This site is currently linked to the Biomedware site. It also has software with boundary detection and analysis methods. James LeSage has developed a spatial econometrics library that can be viewed at www.spatial-econometrics.com (see also www.spatial-statistics.com by Kelly Pace).

Generalized linear modelling of logistic, binomial and Poisson models (but not the 'auto' versions except on coding schemes or by pseudo-likelihood) can be implemented in packages that include BMDP, GENSTAT, GLIM, MINITAB, SAS, SPSS, STATA and SYSTAT. Multi-level modelling can be performed in ML-wiN (http://multilevel.ioe.ac.uk), HLM (http://www.ssicentral.com) and VARCL (http://www.assess.com).

There are increasing amounts of software available from the web that are currently free of charge or relatively inexpensive. WinBUGS for Bayesian modelling using Gibbs sampling is available from the MRC Unit of Biostatistics at Cambridge University (www.mrc-bsu.cam.ac.uk). GeoBUGS is being developed to model spatial data. In the current (beta) version (1.4) GeoBUGS is subsumed within WinBUGS. It contains an extended range of spatial priors including a CAR model with a spatial parameter.

SPLANCS is available from the Department of Mathematics website (www.maths.lancs.ac.uk/~rowlings/Splancs) at the University of Lancaster and includes kernel density estimation and K-function clustering tests.

MANET for exploring data with missing values is available from the website www1.math.uni.augsburg.de/Manet/

Kulldorff's scan test is available from the US National Cancer Institute, Division of Cancer prevention: dcp.nci.nih.gov/BB/SaTScan.html

The crime pattern analysis package, CrimeStat, is available from the US National Institute of Justice Crime Mapping Research Centre. The web address for this package is: www.ojp.usdoj.gov/cmrc/tools/welcome.html. Version 2.0 is available from www.icpsr.umich.edu/NACJD/crimestat.html

Geographically weighted regression analysis software is available from the geography department website at the University of Newcastle on Tyne. Openshaw's geographical analysis (cluster detection) machine (GAM) is available along with other software via the Geography department website at Leeds University. SAGE (Spatial Analysis in a GIS environment) developed at Sheffield University can be accessed via the GIS research centre of the Departments of Geography and Town and Regional Planning (SCGISA).

The spatial visualization package CommonGIS developed from DESCARTES can be downloaded from www.commongis.de. This is understood to be the visualization tool for the 2001 UK Census, downloadable from the MIMAS website (www.mimas.ac.uk). See also www.mimas.ac.uk/argus/ICA/J.Dykes/ by Jason Dykes.

A useful source for data sets, some of which are spatial, can be found at: http://lib.stat.cmu.edu/datasets/. See also the UK data archive at www.data-archive.ac.uk.

Appendix II Cambridgeshire lung cancer data (observed and expected counts and ward neighbours)

Ward no.	Observed incidence $O(i)$	Expected incidence $E(i)$	Adjacent wards
1	7	3.600	4, 6, 10, 12, 121, 128
2	4	4.500	3, 7, 14, 114
3	2	3.900	2, 8, 9, 14, 102, 109, 114
4	3	4.000	1, 5, 11, 12, 107, 110, 126, 128
5	5	4.800	4, 10, 11, 12, 13
6	5	5.100	1, 7, 8, 10, 14, 121, 128
7	3	3.200	2, 6, 14, 114, 121
8	3	2.200	3, 6, 9, 10, 13, 14
9	1	3.100	3, 8, 13, 97, 102
10	1	3.000	1, 5, 6, 8, 12, 13, 14
11	5	5.200	4, 5, 13, 110
12	5	3.100	1, 4, 5, 10
13	2	4.100	5, 8, 9, 10, 11, 97, 110, 113
14	0	3.700	2, 3, 6, 7, 8, 10
15	1	1.800	19, 30, 128, 130, 131
16	2	3.000	19, 27, 30, 32
17	3	1.500	32
18	1	1.800	20, 22, 26, 29, 31, 34, 40
19	1	1.300	15, 16, 30, 32, 94, 130
20	3	2.300	18, 21, 22, 26, 27
21	4	2.200	20, 22, 27, 28
22	2	3.500	18, 20, 21, 28, 31
23	4	1.800	25, 27
24	2	2.100	28, 29, 31, 61, 103, 133
25	2	1.000	23, 27
26	8	4.000	18, 20
27	3	4.700	16, 20, 21, 23, 25, 28, 30, 131
28	1	1.000	21, 22, 24, 27, 31, 103, 131
29	1	1.800	18, 24, 31, 34, 36, 61
30	0	1.200	15, 16, 19, 27, 131
31	2	1.200	18, 22, 24, 28, 29

(Continued)

(Continued)

Ward no.	Observed incidence $O(i)$	Expected incidence $E(i)$	Adjacent wards
32	0	1.300	16, 17, 19
33	1	1.300	34, 35, 43, 51, 53, 80, 90
34	2	1.800	18, 29, 33, 35, 36, 40, 53
35	2	1.200	33, 34, 36, 37, 90
36	0	0.800	29, 34, 35, 37, 61, 84
37	3	1.300	35, 36, 84, 90
38	4	2.000	41, 42, 43, 45, 47, 51, 57
39	3	2.900	44, 46, 55, 57
40	1	0.800	18, 34, 45, 53
41	4	5.100	38, 42, 43, 45, 53
42	1	3.100	38, 41, 43
43	1	3.700	33, 38, 41, 42, 51, 53
44	1	1.000	39
45	1	0.500	38, 40, 41, 53
46	3	2.500	39, 57, 138
47	1	2.000	38, 48, 49, 50, 51, 52, 138, 152
48	2	1.000	47, 49, 51, 52
49	3	1.700	47, 48, 50, 51
50	3	1.000	47, 49, 51
51	2	1.300	33, 38, 43, 47, 48, 49, 50, 52, 67, 80, 150, 152
52	3	1.500	47, 48, 51, 152
53	0	1.000	33, 34, 40, 41, 43, 45
54	1	2.900	56, 57
55	1	2.600	39, 56, 57
56	2	1.500	54, 55, 57
57	7	4.600	38, 39, 46, 54, 55, 56
58	2	2.400	59, 64, 69, 75, 85, 87, 88
59	2	1.500	58, 78, 85, 87
60	3	1.000	80, 89, 90
61	0	1.600	24, 29, 36, 77, 81, 84, 90, 123, 133
62	4	2.100	63, 66, 78, 79, 85
63	2	2.700	62, 66, 70
64	1	1.000	58, 76, 83, 85, 88
65	0	1.000	86, 91, 140, 150, 157
66	4	3.400	62, 63, 70, 79
67	2	0.800	51, 80, 86, 91, 150
68	0	1.300	71, 72, 77, 82, 105
69	4	2.800	58, 71, 73, 74, 75, 87, 124
70	1	1.200	63, 66, 79, 87, 93, 99, 108
71	1	0.800	68, 69, 72, 73, 105, 124
72	1	1.700	68, 71, 73, 82
73	5	1.100	69, 71, 72, 74, 81, 82, 89, 90
74	6	4.000	69, 73, 75, 88, 89
75	4	3.800	58, 69, 74, 88
76	0	1.000	64, 85

(Continued)

Ward no.	Observed incidence O(*i*)	Expected incidence E(*i*)	Adjacent wards
77	1	1.300	61, 68, 81, 82, 105, 123, 127
78	2	1.600	59, 62, 79, 85, 87
79	5	2.800	62, 66, 70, 78, 87
80	5	4.000	33, 51, 60, 67, 86, 89, 90
81	1	2.900	61, 73, 77, 82, 90
82	6	3.800	68, 72, 73, 77, 81
83	1	2.400	64, 86, 88, 89
84	4	1.400	36, 37, 61, 90
85	3	1.300	58, 59, 62, 64, 76, 78
86	1	1.800	65, 67, 80, 83, 89, 91
87	3	1.100	58, 59, 69, 70, 78, 79, 99, 124
88	0	1.900	58, 64, 74, 75, 83, 89
89	2	0.800	60, 73, 74, 80, 83, 86, 88, 90
90	6	2.100	33, 35, 37, 60, 61, 73, 80, 81, 84, 89
91	0	2.900	65, 67, 86, 150
92	0	1.200	94, 104, 107, 115, 116, 125, 126, 132
93	0	0.800	70, 99, 108, 120, 122, 129
94	0	1.500	19, 92, 100, 107, 116, 130
95	1	2.400	102, 103, 105, 109, 114, 118
96	1	1.000	106, 112, 113, 119, 120, 122
97	0	1.000	9, 13, 101, 102, 113
98	2	1.700	119, 120, 129
99	2	0.800	70, 87, 93, 105, 111, 122, 124
100	0	0.800	94, 116
101	0	1.300	97, 102, 111, 113, 122
102	1	1.000	3, 9, 95, 97, 101, 105, 109, 111
103	1	3.000	24, 28, 95, 114, 118, 131, 133
104	0	0.800	92, 106, 115, 132
105	0	1.000	68, 71, 77, 95, 99, 102, 111, 118, 124, 127
106	0	1.000	96, 104, 112, 115, 119, 132
107	2	2.600	4, 92, 94, 126, 128, 130
108	2	1.800	70, 93
109	3	2.200	3, 95, 102, 114
110	4	2.900	4, 11, 13, 113, 117, 125, 126
111	0	1.600	99, 101, 102, 105, 122
112	1	1.200	96, 106, 113, 117, 132
113	1	1.000	13, 96, 97, 101, 110, 112, 117, 122
114	1	4.300	2, 3, 7, 95, 103, 109, 121, 131
115	1	0.800	92, 104, 106, 119, 132
116	3	2.300	92, 94, 100
117	1	1.000	110, 112, 113, 125, 132
118	1	0.800	95, 103, 105, 123, 127, 133
119	3	2.300	96, 98, 106, 115, 120

(Continued)

(Continued)

Ward no.	Observed incidence $O(i)$	Expected incidence $E(i)$	Adjacent wards
120	1	1.000	93, 96, 98, 119, 122, 129
121	0	1.400	1, 6, 7, 114, 128, 131
122	2	1.300	93, 96, 99, 101, 111, 113, 120, 129
123	1	1.300	61, 77, 118, 127, 133
124	1	0.800	69, 71, 87, 99, 105
125	5	3.600	92, 110, 117, 126, 132
126	0	1.200	4, 92, 107, 110, 125
127	1	0.800	77, 105, 118, 123
128	3	1.400	1, 4, 6, 15, 107, 121, 130, 131
129	1	1.000	93, 98, 120, 122
130	2	1.000	15, 19, 94, 107, 128
131	2	2.400	15, 27, 28, 30, 103, 114, 121, 128
132	0	1.300	92, 104, 106, 112, 115, 117, 125
133	1	1.600	24, 61, 103, 118, 123
134	1	1.500	140, 143, 157
135	5	3.600	137, 139, 142, 147, 149, 156
136	5	4.300	137, 138, 142, 147, 148
137	1	3.700	135, 136, 138, 139, 147, 152
138	3	3.000	46, 47, 136, 137, 141, 148, 152
139	8	3.900	135, 137, 145, 150, 152, 156
140	2	1.500	65, 134, 141, 143, 144, 145, 146, 150, 151, 154, 156, 157
141	1	0.800	
142	4	3.700	138, 140, 143, 148, 154, 155
143	2	1.500	135, 136, 147, 148, 149, 153
144	3	3.500	134, 140, 141
145	3	3.300	140, 149, 151, 153, 154, 155
146	4	3.900	139, 140, 150, 156
147	6	4.100	140, 145, 150
148	3	3.000	135, 136, 137, 142
149	5	3.400	136, 138, 141, 142, 153, 155
150	1	2.000	135, 142, 144, 151, 153, 156
151	1	2.800	51, 65, 67, 91, 139, 140, 145, 146, 152
152	5	4.000	140, 144, 149, 156
153	1	2.800	47, 51, 52, 137, 138, 139, 150
154	1	2.300	142, 144, 148, 149, 155
155	6	4.300	140, 141, 144,
156	2	4.300	141, 144, 148, 153, 154
157	0	1.000	135, 139, 140, 145, 149, 151, 65, 134, 140

Appendix III Sheffield burglary data

Table AIII.1 *Burglary data for Sheffield: ward number, ward name, census ID, number of households, number of burglaries in 1995, Townsend deprivation index (large positive implies deprived; large negative implies affluent); population turnover (% residents having moved within one year of the 1991 census)*

Ward no.	Ward Name	Ward ID	No. of households $NU(i)$	No. of burglaries $O(i)$	Townsend deprivation $X_1(i)$	Turnover $X_2(i)$
1	Beauchief	05CGFA	8575	267	−4.93	6.71
2	Birley	05CGFB	7856	156	−0.04	6.14
3	Brightside	05CGFC	6565	476	1.18	9.84
4	Broomhill	05CGFD	5601	436	−3.18	20.60
5	Burngreave	05CGFE	6253	837	3.90	11.42
6	Castle	05CGFF	5998	336	6.06	14.61
7	Chapel Green	05CGFG	8902	299	−2.96	6.87
8	Darnall	05CGFH	7396	252	0.64	8.22
9	Dore	05CGFJ	7684	172	−4.78	6.77
10	Ecclesall	05CGFK	7561	299	−7.99	8.76
11	Firth Park	05CGFL	6995	708	3.49	9.89
12	Hallam	05CGFM	7328	205	−7.17	8.71
13	Handsworth	05CGFN	7522	214	−0.65	6.80
14	Heeley	05CGFP	7532	331	−0.28	10.07
15	Hillsborough	05CGFQ	7559	196	−2.36	8.43
16	Intake	05CGFR	7906	275	−0.71	8.04
17	Manor	05CGFS	5346	443	6.01	13.54
18	Mosborough	05CGFT	12397	329	−2.50	8.87
19	Nether Edge	05CGFU	6438	533	−2.52	16.08
20	Nether Shire	05CGFW	6684	429	3.54	8.37
21	Netherthorpe	05CGFX	6428	413	3.49	18.55
22	Norton	05CGFY	6995	326	1.86	8.32
23	Owlerton	05CGFZ	6753	383	2.06	8.87
24	Park	05CGGA	7436	561	4.28	11.54
25	Sharrow	05CGGB	5957	642	3.85	18.17
26	Southey Green	05CGGC	6407	361	5.58	8.70
27	South Wortley	05CGGD	9429	188	−4.27	6.67
28	Stocksbridge	05CGGE	5486	100	−2.79	7.23
29	Walkley	05CGGF	7909	275	−1.30	10.31

Table AIII.2 *Wards and their adjacent neighbours*

Ward number	Adjacent wards
1	10, 19, 14, 22, 9
2	13, 16, 18
3	27, 20, 8, 11, 5
4	21, 12, 25, 10, 19
5	3, 23, 8, 11, 21, 6
6	8, 5, 21, 17, 25, 24
7	27
8	3, 5, 6, 17, 13, 16
9	12, 10, 1, 22
10	12, 4, 19, 1, 9
11	20, 3, 26, 23, 5
12	27, 29, 21, 4, 10, 9
13	8, 16, 18, 2
14	25, 24, 16, 19, 1, 22
15	27, 23, 29
16	8, 17, 24, 13, 14, 2, 22
17	8, 6, 24, 16
18	13, 2
19	4, 25, 10, 14, 1
20	27, 3, 26, 11
21	23, 5, 29, 6, 12, 4, 25
22	16, 14, 1, 9
23	27, 26, 11, 15, 5, 29, 21
24	6, 17, 25, 16, 14
25	21, 6, 4, 24, 19, 14
26	27, 20, 23, 11
27	28, 7, 20, 3, 26, 23, 15, 29, 12
28	27
29	27, 23, 15, 21, 12

Table AIII.3(a) *WinBUGS code for binomial-logistic model with spatially structured* $(v[i])$ *and spatially unstructured* $(e[i])$ *random effects (convolution Gaussian prior)*

```
Model {
     for (i in 1:N) {
             O[i] ~ dbin(P[i], NU[i])                                    # binomial for observed counts
             logit (P[i])< − beta0 + v[i] + e[i]                         # logistic regression
             e[i] ~ dnorm(0,prec.e)                                      # normal prior
             SR[i]< − P[i] / avrate                                      # avrate: average burglary rate
             PP[i]< − step(SR[i] − 1)                                    # posterior probability rate >1
             CPP[i]< − step(−(SR[i] × 1))                                # posterior probability rate<1
     }
         v[1:N] ~ car.normal (adj[], weights[], num[], prec.v)           # CAR prior
         for(k in 1:sumNumNeigh) {weights[k] < − 1}                      # equal weight
           beta0 ~ dflat()                                               # flat prior
           prec.v ~ dgamma(0.5, 0.0005)                                  # prior on spatial precision
           v.v < − 1/prec.v                                              # spatial variance (conditional)
           sigma < − sqrt(1 / prec.v)
           sd < − sd(v[ ])
           prec.e ~ dgamma(0.001, 0.001)                                 # prior on unstructured precision
           v.e < − 1 / prec.e                                            # unstructured variance
           sigma.e < − 1 / sqrt(prec.e)
     }
```

Table AIII.3(b) *Sample results from WinBUGS*

A 3000 update burn in followed by a further 5000 updates with 2 chains gave the parameter estimates

Node statistics

node	mean	sd	MC error	2.5%	median	97.5%	start	sample
beta0	−3.032	0.07102	0.005129	−3.191	−3.036	−2.881	3001	10000
v.e	0.1737	0.1588	0.01204	0.001125	0.1336	0.5191	3001	10000
v.v	0.4544	0.4161	0.03134	1.983E-4	0.4141	1.348	3001	10000

Summary statistics

node	mean	sd	sample
P[1]	0.031	0.002	10000
P[2]	0.020	0.002	10000
.			
P[29]	0.035	0.002	10000
v[1]	−0.190	0.220	10000
v[2]	−0.479	0.385	10000
.			
v[29]	−0.140	0.197	10000
e[1]	−0.212	0.236	10000
e[2]	−0.380	0.390	10000
.			
e[29]	−0.152	0.241	10000

Table AIII.4(a) *WinBUGS code for binomial-logistic model with two covariates, spatially structured (v[i]) and spatially unstructured (e[i]) random effects*

```
Model {
    for (i in 1 : N) {
            O[i] ~ dbin(P[i], NU[i])                                    # binomial for observed counts
            logit (P[i])< − beta0 + beta1 * x1[i] + beta2 * x2[i] + v[i] + e[i]
                                                                        # logistic regression
            e[i] ~ dnorm(0,prec.e)                                      # normal prior
            SR[i]< − P[i] / avrate                                      # avrate:average burglary rate
            PP[i]< − step(SR[i] − 1)                                    # posterior probability rate >1
            CPP[i]< − step(−(SR[i] − 1))                                # posterior probability rate <1
    }
            v[1:N] ~ car.normal(adj[ ], weights[ ], num[ ], prec.v)    # CAR prior
            for (k in 1:sumNumNeigh) {weights[k]<-1}                   # equal weight
            beta1 ~ dnorm(0.0, 1.0E-5)                                 # normal prior
            beta2 ~ dnorm(0.0, 1.0E-5)                                 # normal prior
            beta0 ~ dnorm(0.0, 1.0E−5)                                 # normal prior
            prec.v ~ dgamma(0.5, 0.0005)                               # prior on spatial precision
            v.v < − 1 / prec.v                                         # spatial variance (conditional)
            sigma < − sqrt(1 / prec.v)
            sd < − sd(v[ ])
            prec.e ~ dgamma(0.001, 0.001)                              # prior on unstructured precision
            v.e < − 1 / prec.e                                         # unstructured variance
            sigma.e < − 1 / sqrt(prec.e)
    }
```

Table AIII.4(b) *Sample results from WinBUGS*

A 5000 update burn in followed by a further 10 000 updates with 2 chains gave the parameter estimates

Node statistics

node	mean	sd	MC error	2.5%	median	97.5%	start	sample
beta0	−3.884	0.1964	0.01343	−4.244	−3.894	−3.506	5001	20000
beta1	0.06642	0.01911	0.001056	0.02771	0.06677	0.1021	5001	20000
beta2	0.08264	0.01846	0.001266	0.04607	0.0547	0.1148	5001	20000
v.e	0.1358	0.04264	8.87E-4	0.07422	0.1287	0.2396	5001	20000
v.v	0.003411	0.01209	7.111E-4	1.415E-4	7.619E-4	0.02749	5001	20000

Summary statistics

node	mean	sd	sample
P[1]	0.031	0.002	20000
P[2]	0.020	0.002	20000
.			
P[29]	0.035	0.002	20000
v[1]	0.004	0.031	20000
v[2]	−0.012	0.055	20000
.			
v[29]	−0.001	0.028	20000
e[1]	0.208	0.133	20000
e[2]	−0.486	0.128	20000
.			
e[29]	−0.201	0.099	20000

Table AIII.5 *Area by area decomposition of the posterior mean of the odds:* $P(i)/(1-P(i))$

Ward no.	$P(i)=\dfrac{O(i)}{NU(i)}$	$\dfrac{P(i)}{1-P(i)^{+}}$ from WinBUGS	$\exp(\text{beta1}^{*}\,X_1(i))^{+}$	$\exp(\text{beta2}^{*}\,X_2(i))^{+}$	$\exp(v(i))^{+}$	$\exp(e(i))^{+}$
1	0.031	0.032	0.722	1.733	1.004	1.231
2	0.020	0.020	0.997	1.654	0.988	0.615
3	0.073	0.078	1.081	2.241	1.006	1.537
4	0.078	0.085	0.811	5.415	1.003	0.922
5	0.134	0.153	1.294	2.552	1.005	2.223
6	0.056	0.060	1.492	3.315	0.998	0.584
7	0.034	0.034	0.823	1.757	1.003	1.155
8	0.034	0.035	1.043	1.962	0.998	0.839
9	0.022	0.024	0.729	1.743	1.002	0.875
10	0.040	0.041	0.590	2.051	1.006	1.616
11	0.101	0.111	1.259	2.249	1.007	1.887
12	0.028	0.029	0.623	2.042	1.002	1.088
13	0.028	0.030	0.958	1.747	0.991	0.857
14	0.044	0.046	0.982	2.284	1.001	0.989
15	0.026	0.027	0.856	1.996	0.998	0.763
16	0.035	0.036	0.954	1.933	0.997	0.947
17	0.083	0.091	1.487	3.036	0.998	0.964
18	0.027	0.028	0.848	2.070	0.989	0.764
19	0.083	0.089	0.847	3.738	1.005	1.359
20	0.064	0.068	1.263	1.987	1.005	1.305
21	0.064	0.070	1.259	4.579	1.000	0.576
22	0.047	0.049	1.131	1.978	1.001	1.052
23	0.057	0.059	1.146	2.069	1.003	1.214
24	0.075	0.081	1.326	2.576	1.000	1.148
25	0.108	0.121	1.289	4.438	1.001	1.010
26	0.056	0.059	1.445	2.040	1.003	0.973
27	0.02	0.02	0.754	1.728	1.000	0.764
28	0.018	0.019	0.832	1.810	0.987	0.626
29	0.035	0.036	0.918	2.330	0.999	0.818

[+] Calculation uses the posterior mean

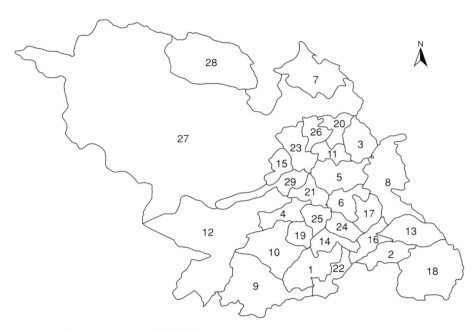

Figure AIII.1 Map of Sheffield wards

Appendix IV Children excluded from school: Sheffield

Table AIV.1 *Numbers of children excluded from school by ward: Sheffield. Ward number, observed number excluded; expected numbers based on average for the city times the number of children in the ward; deprivation index*

Ward no.	Observed number ($O(i)$)	Expected number ($E(i)$)	Deprivation index $X_1(i)$
1	1	5.130	−4.930
2	5	5.410	−0.040
3	13	5.510	1.180
4	0	2.910	−3.180
5	11	5.030	3.900
6	8	4.650	6.060
7	3	6.090	−2.960
8	10	6.070	0.640
9	1	4.880	−4.780
10	0	6.240	−7.990
11	15	5.910	3.490
12	0	4.820	−7.170
13	3	4.500	−0.650
14	3	4.950	−0.280
15	7	4.940	−2.360
16	1	5.120	−0.710
17	5	3.810	6.010
18	2	9.930	−2.500
19	3	3.970	−2.520
20	14	5.020	3.540
21	4	2.550	3.490
22	4	3.900	1.860
23	4	4.600	2.060
24	6	4.490	4.280
25	3	4.140	3.850
26	13	5.300	5.580
27	1	6.560	−4.270
28	2	3.790	−2.790
29	2	3.740	−1.300

Table AIV.2 *WinBUGS code for Poisson log-normal model with a single covariate and spatially structured (v[i]) and spatially unstructured (e[i]) random effects*

```
model
{
        v[1 : N] ~ car.normal(adj[ ], weights[ ], num[ ], prec.v)
        v.mean <−mean(v[ ])
    for (i in 1 : N) {
                O[i] ~ dpois(mu[i])
                log(mu[i]) <−log(E[i]) + beta0 + beta1* x1[i] +v[i]+e[i] #x1:tdi
                e[i] ~ dnorm(0, prec.e)
                R[i] <−exp(beta0+beta1*x1[i]+v[i]+e[i])
                }
        for (k in 1:sumNumNeigh) {weights[k] <−1 }
        # other priors
            beta0 ~ dflat()
            beta1 ~ dnorm(0.0, 1.0E-5)
            prec.v ~ dgamma (0.5,0.0005)          # prior on spatially structured precision
            v.v <- 1/prec.v                       # spatially structured variance
            sigma.v <−1 / sqrt(prec.v)            # spatially structured standard deviation
            prec.e ~ dgamma(0.5, 0.0005)          # prior on unstructured precision
            v.e <−1/prec.e                        # unstructured variance
            sigma.e <−sqrt(1 / prec.e)            # unstructured standard deviation
            }
```

Table AIV.3 *Sample results from WinBUGS*

A 5000 update burn in followed by a further 5000 updates with two chains gave the parameter estimates

Node statistics

node	Mean	sd	MC error	2.5%	median	97.5%	start	sample
beta0	−0.2509	0.1213	0.003788	−0.5103	−0.243	−0.03141	5001	10000
beta1	0.1997	0.03112	8.243E-4	0.1434	0.198	0.267	5001	10000
prec.e	622.1	1159.0	61.68	3.522	135.7	4027.0	5001	10000
prec.v	1422.0	1811.0	102.2	2.105	765.3	6424.0	5001	10000

Summary statistics

node	mean	sd	sample
R[1]	0.289	0.089	10000
R[2]	0.795	0.186	10000
.			
R[29]	0.625	0.153	10000
v[1]	−0.041	0.142	10000
v[2]	−0.023	0.135	10000
.			
v[29]	0.020	0.108	10000
e[1]	−0.017	0.201	10000
e[2]	0.025	0.176	10000
.			
e[29]	−0.010	0.192	10000

Table AIV.4 *Area by area decomposition of the posterior mean of the relative risk of children excluded from school in Sheffield: Poisson log normal convolution prior*

Ward No.	R_i^+ from WINBUGS	$\exp(\text{beta1}^* X_1(i))^+$	$\exp(v(i))^+$	$\exp(e(i))^+$
1	0.289	0.374	0.960	0.983
2	0.795	0.992	0.977	1.025
3	1.352	1.266	1.105	1.195
4	0.394	0.530	0.957	0.951
5	1.879	2.179	1.043	1.044
6	2.366	3.354	0.971	0.910
7	0.471	0.554	1.034	1.016
8	1.052	1.136	1.033	1.124
9	0.299	0.385	0.961	0.985
10	0.157	0.203	0.955	0.960
11	1.926	2.008	1.084	1.111
12	0.189	0.239	0.980	0.964
13	0.686	0.878	0.978	0.996
14	0.715	0.946	0.962	0.981
15	0.644	0.624	1.087	1.164
16	0.620	0.868	0.961	0.920
17	2.257	3.321	0.951	0.889
18	0.430	0.607	0.937	0.926
19	0.494	0.605	0.969	1.043
20	2.002	2.028	1.096	1.126
21	1.588	2.008	0.995	0.999
22	1.105	1.450	0.966	0.988
23	1.183	1.509	1.028	0.958
24	1.694	2.351	0.955	0.946
25	1.484	2.157	0.949	0.901
26	2.505	3.048	1.044	0.995
27	0.346	0.426	1.039	0.958
28	0.489	0.573	1.035	1.010
29	0.625	0.771	1.020	0.990

+ Calculation uses the posterior mean

References

Acheson, D. (1998). *Report of the Independent Enquiry into Inequalities in Health*. London: HMSO.

Aitkin, M. and Longford, N. (1986). Statistical modelling issues in school effectiveness studies (with discussion). *Journal Royal Statistical Society*, **A**, **149**, 1–43.

Akaike, H. (1973). Information theory and an extension of the maximum likelihood principle. *Second International Symposium on Information Theory*, eds. Petrov, B.N. and Csáki, F., pp. 267–81. Budapest: Akadémiai Kiadó.

Alvanides, S. and Openshaw, S. (2000). Zone design for planning and policy analysis. *Geographical Information and Planning: European Perspectives*, eds. Geertman, S., Openshaw, S. and Stillwell, J., pp. 299–315. Berlin: Springer-Verlag.

Anas, A. and Eum, S.J. (1984). Hedonic analysis of a housing market in disequilibrium. *Journal of Urban Economics*, **15**, 87–106.

Andrienko, G.L. and Andrienko, N.V. (1999). Interactive maps for visual data exploration. *International Journal of Geographical Information Science*, **13**, 355–74.

Andrienko, G. and Andrienko, N. (2001). Exploring spatial data with dominant attribute map and parallel co-ordinates. *Computers, Environment and Urban Systems*, **25**, 5–15.

Anselin, L. (1988). *Spatial Econometrics: Methods and Models*. Dordrecht: Kluwer Academic.

Anselin, L. (1995). Local indicators of spatial association – LISA. *Geographical Analysis*, **27**, 93–115.

Anselin, L. (1996). The Moran scatterplot as an ESDA tool to assess local instability in spatial association. *Spatial Analytical Perspectives on GIS*, eds. Fischer, M., Scholten, H.J. and Unwin, D., pp. 111–25. London: Taylor & Francis.

Anselin, L. (1998). Exploratory spatial data analysis in a geocomputational environment. *Geocomputation: A Primer*, eds. Longley, P.A., Brooks, S.M., McDonnell, R. and Macmillan, W., pp. 77–84. New York: Wiley.

Anselin, L. and Bao, S. (1997). Exploratory spatial data analysis linking Space Stat and Arc View. *Recent Developments in Spatial Analysis*, eds. Fischer, M. and Getis, A., pp. 35–59. Berlin: Springer-Verlag.

Anselin, L., Dodson, R. and Hudak, S. (1993). Linking GIS and spatial data analysis in practice. *Geographical Systems*, **1**, 3–23.

Anselin, L. and Rey, S.J. (1991). Properties of tests for spatial dependence in linear regression models. *Geographical Analysis*, **23**, 112–31.

Anselin, L. and Smirnov, O. (1996). Efficient algorithms for constructing proper higher order spatial lag operators. *Journal of Regional Science*, **36**, 67–89.

Arbia, G. (1989). *Spatial Data Configuration in Statistical Analysis of Regional Economic and Related Problems*. Dordrecht: Kluwer.

Arbia, G., Benedetti, R. and Espa, G. (1996). Effects of the MAUP on image classification. *Geographical Systems*, **3**, 123–41.

Arbia, G., Griffith, D.A. and Haining, R.P. (1998). Error propagation modelling in raster GIS: overlay operations. *International Journal of Geographical Information Science*, **12**, 145–67.

Arbia, G., Griffith, D.A. and Haining, R.P. (1999). Error propagation modelling in raster GIS: adding and ratioing operations. *Cartography and Geographic Information Science*, **26**, 297–315.

Arbia, G., Griffith, D. and Haining, R.P. (2003). Spatial error propagation when computing linear combinations of spectral bands: the case of vegetation indices. *Environmental and Ecological Statistics*, **10** (to appear).

Arbia, G. and Lafratta, G. (1997). Evaluating and updating the sample design in repeated environmental surveys: monitoring air quality in Padua. *Journal of Agricultural, Biological and Environmental Statistics*, **2**, 451–66.

Armstrong, B. (2001). Comments on the papers by Guthrie, Sheppard and Best et al., Chambers, Steel and Darby et al. *Journal of the Royal Statistical Society*, A, **164**, 205–7.

Armstrong, H.W. and Taylor, J. (2000). *Regional Economics and Policy*. New York: Harvester Wheatsheaf.

Asimov, D. (1985). The Grand Tour: a tool for viewing multi-dimensional data. *SIAM Journal on Scientific and Statistical Computing*, **6**, 128–43.

Assunção and Reis, E.A. (1999). A new proposal to adjust Moran's I for population density. *Statistics in Medicine*, **18**, 2147–62.

Augustin, N.H., Mugglestone, M.A. and Buckland, S.T. (1996). An autologistic model for the spatial distribution of wildlife. *Journal of Applied Ecology*, **33**, 339–47.

Bailey, N.T.J. (1967). The simulation of stochastic epidemics in two dimensions. *Proceedings, Fifth Berkeley Symposium on Mathematics and Statistics*, **4**, 237–57. Berkeley and LosAngeles, CA: University of California.

Bailey, N.T.J. (1975). *The Mathematical Theory of Infectious Diseases and Its Applications*. London: Charles Griffin & Co.

Bailey, T.C. and Gatrell, A.C. (1995). *Interactive Spatial Data Analysis*. Harlow: Longman.

Bao, S. and Henry, M. (1996). Heterogeneity issues in local measurements of spatial association. *Geographical Systems*, **3**, 1–14.

Barcelo, J.A. and Pallares, M. (1998). Beyond GIS: the archaeology of social spaces. *Archeologia e Calcolatori*, **9**, 47–80.

Barro, R.J. and Sala-i-Martin, X. (1995). *Economic Growth*. New York: McGraw-Hill.

Bartlett, M.S. (1935). Some aspects of the time correlation problem in regard to tests of significance. *Journal Royal Statistical Society*, **98**, 536–43.

Bartlett, M.S. (1957). Measles periodicity and community size. *Journal of the Royal Statistical Society*, A, **120**, 48–70.

Bartlett, M.S. (1960). The critical community size for measles in the United States. *Journal of the Royal Statistical Society*, A, **123**, 37–44.

Batty, M. (1998). Urban evolution on the desktop: simulation with the use of extended cellular automata. *Environment and Planning*, **A**, 30, 1943–67.

Baumol, W.J. (1994). Multivariate growth patterns: contagion and common forces as possible sources of convergence. *Convergence of Productivity – Cross National Studies and Historical Evidence*, eds. Baumol, W.J., Nelson, R.R. and Wolff, E.N., pp. 62–85. New York: Oxford University Press.

Bavaud, F. (1998). Models for spatial weights: a systematic look. *Geographical Analysis*, **30**, 153–71.

Becker, N.G. (1989). *Analysis of Infectious Disease Data: Monographs on Statistics and Applied Probability*. London: Chapman & Hall.

Becker, R.A., Cleveland, W.S. and Shyu, M-J. (1996). The visual design and control of trellis display. *Journal of Computational and Graphical Statistics*, **5**, 123–55.

Becker, R.A., Cleveland, W.S. and Wilks, A.R. (1987). Dynamic graphics for data analysis. *Statistical Science*, **2**, 355–95.

Belsley, D.A., Kuh, E. and Welsch, R.E. (1980). *Regression Diagnostics*. New York: Wiley.

Benenson, I. (1998). Multi-agent simulations of residential dynamics in the city. *Computing, Environment and Urban Systems*, **22**, 25–42.

Bernardinelli, L., Clayton, D. and Montomoli, C. (1995). Bayesian estimates of disease maps: how important are priors? *Statistics in Medicine*, **14**, 2411–31.

Bernardinelli, L. and Montomoli, C. (1992). Empirical Bayes versus fully Bayesian analysis of geographical variation in disease risk. *Statistics in Medicine*, **11**, 983–1007.

Bernstein, R., Lotspiech, J.B., Meyers, H.J., Kolsky, H.G. and Lees, R.D. (1984). Analysis and processing of LANDSAT-4 sensor data using advanced image processing techniques and technologies. *IEEE Transactions on Geoscience and Remote Sensing*, GE-22, 192–221.

Berry, B.J.L. (1966). Essays on commodity flows and the spatial structure of the Indian economy. Research Paper No. 111, Department of Geography, University of Chicago, 334 pp.

Berry, B.J.L. and Marble, D.F. (1968). *Spatial Analysis*. Englewood Cliffs, NJ: Prentice Hall.

Bertin, J. (1983). *Semiology of Graphics: Diagrams, Networks, Maps*. Madison, University of Wisconsin Press (First published 1967).

Besag, J.E. (1974). Spatial interaction and the statistical analysis of lattice systems. *Journal of the Royal Statistical Society*, **B**, 36, 192–225.

Besag, J.E. (1975). Statistical analysis of non-lattice data. *The Statistician*, **24**, 179–95.

Besag, J.E. (1977). Errors in variables estimation for Gaussian lattice schemes. *Journal of the Royal Statistical Society*, **B**, 39, 73–8.

Besag, J.E. (1978). Some methods of statistical analysis for spatial data. *Bulletin of the International Statistical Institute*, **47**, 77–92.

Besag, J.E. (1986). On the statistical analysis of dirty pictures. *Journal of the Royal Statistical Society*, **B**, 48, 259–302.

Besag, J. and Clifford, P. (1989). Generalized Monte Carlo significance tests. *Biometrika*, **76**, 633–42.

Besag, J. and Kooperberg, C. (1995). On conditional and intrinsic autoregressions. *Biometrics*, **82**, 733–46.

Besag, J. and Newell, J. (1991). The detection of clusters in rare diseases. *Journal of the Royal Statistical Society*, **A**, 154, 143–55.

Besag. J.E., York, J. and Mollié, A. (1991). Bayesian image restoration with two applications in spatial statistics. *Annals of the Institute of Statistical Mathematics*, **43**, 1–21.

Bierkens, M.F.P, Finke, P.A. and de Willigen, P. (2000). *Upscaling and Downscaling Methods for Environmental Research*. London: Kluwer Academic Publishers.

Bithell, J.F. (1990). An application of density estimation to geographical epidemiology. *Statistics in Medicine*, **9**, 691–701.

Bithell, J.F. (1995). The choice of test for detecting raised disease risk near a point source. *Statistics in Medicine*, **14**, 2309–22.

Bivand, R. (1980). A Monte Carlo study of correlation coefficient estimation with spatially autocorrelated observations. *Quaestiones Geographicae*, **6**, 5–10.

Block, R. (1979). Community, environment and violent crime. *Criminology*, **17**, 46–57.

Bloom, L.M. and Kentwell, D.J. (1998). A geostatistical analysis of cropped and uncropped soil from the Jimperding Brook catchment of Western Australia. In *Geo ENV II – Geostatistics for Environmental Applications*, eds. Gomez-Hernández, J., Soares, A. and Froidevaux, R. pp. 369–379. London: Kluwer Academic.

Bolthausen, E. (1982). On the central limit theorem for stationary mixing random fields. *Annals of Probability*, **10**, 1047–50.

Boots, B. (1982). Comments of the use of eigenfunctions to measure structural properties of geographic networks. *Environment and Planning*, **A, 14**, 1063–72.

Boots, B. (1984). Evaluating principal eigenvalues as measures of network structure. *Geographical Analysis*, **16**, 270–5.

Borgeson, W.T., Baston, R.M. and Keiffer, H.H. (1985). Geometric accuracy of LANDSAT-4 and LANDSAT-5 Thematic Mapper images. *Photogrammetric Engineering and Remote Sensing*, **51**, 1893–98.

Bottoms, A.E., Mamby, R.I. and Walker, M.A. (1987). A localised crime survey in contrasting areas of a city. *British Journal of Criminology*, **27**, 125–54.

Bottoms, A.E. and Wiles, P. (1997). Environmental criminology. *The Oxford Handbook of Criminology*, second edition, eds. Maguire, M., Morgan, R. and Reiner, R., pp. 305–59. Oxford: Oxford University Press.

Bowers, K.J. and Hirschfield, A. (1999). Exploring links between crime and disadvantage in N.W. England: an analysis using Geographical Information Systems. *International Journal of Geographical Information Science*, **13**, 159–84.

Bowie, W.R., King, A.S., Werker, D.H., Isaac-Renton, J.L., Bell, A., Eng, S.B. and Marion, S.A. (1997). The outbreak of toxoplasmosis associated with municipal drinking water. *The Lancet*, **350**, 173–77.

Bracken, I. and Martin, D. (1989). The generation of spatial population distribution from census centroid data. *Environment and Planning*, **A, 21**, 537–43.

Brantingham, P.J. and Brantingham, P.L. (1991). *Environmental Criminology*, second edition. Prospect Heights, IL: Waveland Press.

Bras, R.L. and Rodriguez-Iturbe, I. (1976). Network design for the estimation of the areal mean of rainfall events. *Water Resources Research*, **12**, 1185–95.

Brassel, K., Bucher, F., Stephan, E-M. and Vckovski, A. (1995). *Completeness in Elements of Spatial Data Quality*. eds. Guptill, S.G. and Morrison, J.L., pp. 81–108. Oxford: Elsevier Science.

Brassel, K.E. and Weibel, R. (1988). A review and framework of automated map generalization. *International Journal of Geographical Information Systems*, **2**, 229–44.

Breusch, T. and Pagan, A. (1979). A simple test for heteroskedasticity and random coefficient variation. *Econometrica*, **47**, 1287–94.

Brindley, P., Wise, S.M., Maheswaran, R. and Haining, R.P. (2002). Small area based population exposure estimates for modelled nitrogen oxide pollution data. Submitted to *Computers Environment and Urban Systems*.

Bronfenbrenner, U. (1979). *The Ecology of Human Development*. Cambridge, MA: Harvard University Press.

Brook, D. (1964). On the distinction between the conditional probability and the joint probability approaches in the specification of nearest neighbour systems. *Biometrika*, **51**, 481–3.

Brooks-Gunn, J., Duncan, G.J., Klebanov, P.K. and Sealand, N. (1993). Do neighbourhoods influence child and adolescent development? *American Journal of Sociology*, **99**, 353–95.

Brunsdon, C. (2000). The Comap: Investigating geographical pattern via conditional spatial distributions. Proceedings of the GIS Research UK Conference, University of York, April 2000, pp. 97–101.

Brunsdon, C., Fotheringham, A.S. and Charlton, M.E. (1998). An investigation of methods for visualizing highly multivariate datasets. *Case Studies of Visualization in the Social Sciences*, eds. Unwin, D. and Fisher, P., pp. 55–79. Advisory group on Computer Graphics, Technical Report Series No. 43 (ISSN 1356–9066).

Brus, D.J. and deGruijter, J.J. (1997). Random sampling or geostatistical modelling? Choosing between design-based and model-based sampling strategies for soil (with Discussion). *Geoderma*, **80**, 1–44.

Brusegard, D. and Menger, G. (1989). Real data and real problems: dealing with large spatial databases. *Accuracy of Spatial Databases*, pp. 177–85. London: Taylor & Francis.

Buck, S.F. (1960). A method of estimation of missing values in multivariate data suitable for use with an electronic computer. *Journal of the Royal Statistical Society*, **B, 22**, 302–6.

Buja, A., Cook, D. and Swayne, D. F. (1996). Interactive high-dimensional data visualization. *Journal of Computational and Graphical Statistics*, **5**, 78–99.

Bunge, W. (1962). Theoretical geography. *Lund Studies in Geography*. Lund: Gleerup.

Burgess, T.M. and Webster, M.R. (1980). Optimal interpolation and isarithmic mapping of soil properties I. The semi-variogram and punctual kriging. *Journal of Soil Science*, **31**, 315–31.

Burgess, T.M., Webster, M.R. and McBratney, A.B. (1981). Optimal interpolation and isarithmic mapping of soil properties, IV Sampling. *Journal of Soil Science*, **32**, 643–59.

Burrough, P.A., Bregt, A.K., de Heus, M.J. and Kloosterman, E.G. (1985). Complementary use of thermal imagery and spectral analysis of soil properties and wheat yields to reveal cyclic patterns in the Flevopolders. *Journal of Soil Science*, **36**, 141–52.

Burrough, P.A. and Frank, A.U. (1995). Concepts and paradigms in spatial information: are current GIS truly generic? *International Journal of Geographical Information Systems*, **9**, 101–16.

Bursik, R.J. and Grasmick, H.G. (1993). *Neighbourhoods and Crime*. New York: Lexington.

Buttenfield, B.P. and Beard, M.K. (1994). Graphical and geographical components of data quality. *Visualization in Geographic Information Systems*, eds. Hearnshaw, H.M. and Unwin, D.J., pp. 150–7. New York: Wiley.

Buttenfield, B.P. and Mark, D.M. (1994). Expert systems in cartographic design. *Geographic Information Systems: The Microcomputer and Modern Cartography*, ed. Taylor, D.R.F., pp. 129–50. Oxford: Pergamon.

Câmara, A.S. and Raper, J. (1999). *Spatial Multi-Media and Virtual Reality*. London: Taylor and Francis.

Carr, D.B., Wallin, J.F. and Carr, D.A. (2000). Two new templates for epidemiology applications: linked micro map plots and conditioned choropleth maps. *Statistics in Medicine*, **19**, 2521–38.

Carter, J.R. (1992). The effect of data precision on the calculation of slope and aspect using gridded DEMS. *Cartographica*, **29**, 22–34.

Casella, G. and George, E.I. (1992). Explaining the Gibbs sampler. *The American Statistician*, **46**, 167–74.

Ceccato, V., Haining, R.P. and Signoretta, P.E. (2002). Exploring offence statistics in Stockholm City using spatial analysis tools. *Annals of the Association of American Geographers*, **92**, 29–51.

Cerioli, A. (1997). Modified tests of independence in 2×2 tables with spatial data. *Biometrics*, **53**, 619–28.

Chatterjee, S. and Price, B. (1991). *Regression Analysis by Example*, second edition. New York: Wiley.

Chilès, J-P. and Delfiner, P. (1999). *Geostatistics: Modeling Spatial Uncertainty*. New York: Wiley.

Chorley, R.J. (1972). *Spatial Analysis in Geomorphology*. London: Methuen.

Chou, Y.H.H. (1991). Map resolution and spatial autocorrelation. *Geographical Analysis*, **23**, 228–46.

Chow G. (1960). Tests of equality between sets of coefficients in two linear regressions. *Econometrica*, **28**, 591–605.

Choynowski, M. (1959). Maps based on probabilities. *Journal of the American Statistical Association*, **54**, 385–8.

Chrisman, N.R. (1989). Modelling error in overlaid categorical maps. *Accuracy of Spatial Databases*, eds. Goodchild, M. and Gopal, S., pp. 21–34. London: Taylor & Francis.

Christakos, G. (1984). On the problem of permissable covariance and variogram models. *Water Resources Research*, **20**, 251–65.

Cislaghi, C., Biggeri, A., Braga, M., Lagazio, C. and Marchi, M. (1995). Exploratory tools for disease mapping in geographical epidemiology. *Statistics in Medicine*, **14**, 2363–81.

Clarke, D.L. (ed.) (1977). *Spatial Archaeology*. London: Academic Press.

Clayton, D. and Bernardinelli, L. (1992). Bayesian methods for mapping disease risk. *Geographical and Environmental Epidemiology: Methods for Small Area Studies*, eds. Elliott, P., Cuzick, J., English, D. and Stern, R., pp. 205–20. Oxford: Oxford University Press.

Clayton, D.G., Bernardinelli, L. and Montomoli, C. (1993). Spatial correlation in ecological analysis. *International Journal of Epidemiology*, **22**, 1193–202.

Clayton, D. and Kaldor, J. (1987). Empirical Bayes estimates of age – standardised relative risks for use in disease mapping. *Biometrics*, **43**, 671–81.

Cleveland, W.S. and Devlin, S.J. (1988). Locally weighted regression: an approach to regression analysis by local fitting. *Journal of the American Statistical Association*, **83**, 596–610.

Cleveland, W.S. (1993). Rejoinder: a model for studying display methods of statistical graphics. *Journal of Computational and Graphical Statistics*, **2**, 361–4.

Cleveland, W.S. (1993a). *Visualizing Data*. AT&T Bell Laboratories, Murray Hill, New Jersey, USA.

Cleveland, W.S. (1994). *The Elements of Graphing Data*, second edition, AT&T Bell Laboratories. Murray Hill, New Jersey, USA.

Cliff, A.D., Haggett, P. and Ord, J.K. (1985). *Spatial Aspects of Influenza Epidemics*. London: Pion.

Cliff, A.D., Haggett, P., Ord, J.K., Bassett, K. and Davies, R.B. (1975). *Elements of Spatial Structure: A Quantitative Approach*. Cambridge: Cambridge University Press.

Cliff, A.D., Haggett, P. and Smallman-Raynor, M. (1993). *Measles: An Historical Geography*. Oxford: Blackwell.

Cliff, A.D. and Kelly, F.P. (1977). Regional taxonomy using trend surface coefficients and invariants. *Environment and Planning*, **A, 9**, 945–55.

Cliff, A.D. and Ord, J.K. (1981). *Spatial Processes*. London: Pion.

Clifford, P. and Richardson, S. (1985). Testing the association between two spatial processes. *Statistics and Decisions*, Suppl. No. 2, 155–60.

Clifford, P., Richardson, S. and Hémon, D. (1989). Assessing the significance of the correlation between two spatial processes. *Biometrics*, **45**, 123–34.

Coale, A.J. and Stephan, F.F. (1962). The case of the Indians and teenage widows. *Journal, American Statistical Association*, **57**, 338–47.

Cockings, S., Fisher, P.F. and Langford, M. (1997). Parametrization and visualization of the errors in areal interpolation. *Geographical Analysis*, **29**, 314–28.

Cohen, J. and Tita, G. (1999). Editor's introduction. *Journal of Quantitative Criminology*, **15**, 373–8.

Collins, S. (1995). Modelling spatial variations in air quality using GIS. *GIS and Health*, eds. Gatrell, A. and Löytönen, M., London: Taylor & Francis.

Collins, S.E., Haining, R.P., Bowns, I.R., Crofts, D.J., Williams, T.S., Rigby, A.S. and Hall, D.M.B. (1998). Errors in postcode to enumeration district mapping and their effect on small area analyses of health data. *Journal of Public Health Medicine*, **20**, 325–30.

Congalton, R. (1991). A review of assessing the accuracy of classifications of remotely sensed data. *Remote Sensing and the Environment*, **37**, 35–46.

Congdon, P. (2001). *Bayesian Statistical Modelling*. Chichester: Wiley.

Congdon, P. (2002). A model for mental health needs and resourcing in small geographic areas: a multivariate spatial perspective. *Geographical Analysis*, **34**, 168–86.

Conley, T.G. (1999). GMM estimation with cross-sectional dependence. *Journal of Econometrics*, **92**, 1–45.

Cook, D., Buja, A., Cabrera, J. and Hurley, C. (1995). Grand tour and projection pursuit. *Journal of Computational and Graphical Statistics*, **4**, 155–72.

Cook, D.G. and Pocock, S.J. (1983). Multiple regression in geographical mortality studies with allowance for spatially correlated errors. *Biometrics*, **39**, 361–71.

Coombes, M. and Openshaw, S. (2001). Contrasting approaches to identifying 'localities' for research and public administration. *Life and Motion of Socio-economic Units*, eds. Frank, A., Raper, J. and Cheylan, J.P., pp. 301–15. London: Taylor & Francis.

Coppock, J.T. (1955). The relationship between farm and parish boundaries. *Geographical Studies*, **1**, 12–25.

Couclelis, H. (1985). Cellular worlds: a framework for modelling micro-macro dynamics. *Environment and Planning*, **B**, **17**, 585–96.

Court, A. (1970). Map comparisons. *Economic Geography*, **46**, 435–8.

Craglia, M., Haining, R.P. and Signoretta, P.E. (2001). Modeling high intensity crime areas in English cities. *Urban Studies*, **38**, 1921–41.

Craglia, M., Haining, R.P. and Signoretta, P.E. (2002). Identifying areas of multiple social need: a case study in the preparation of Children Service plans. *Environment and Planning*, C (to appear)

Craglia, M., Haining, R.P. and Wiles, P. (2000). A comparative evaluation of approaches to urban crime pattern analysis. *Urban Studies*, **37**, 711–29.

Craig, P., Haslett, J., Unwin, A. and Wills, G. (1989). Moving statistics – an extension of brushing for spatial data. *Computing Science and Statistics. Proceedings of the 21st Symposium on the Interface*, pp. 170–4. Berlin: Springer Verlag.

Craig, R.G. (1979). Autocorrelation in LANDSAT data. In *Proceedings of the 13th International Symposium on Remote Sensing of the Environment*, pp. 1517–24. Ann Arbor, Michigan.

Craig, R.G. and Labovitz, M.L. (1980). Sources of variation in LANDSAT autocorrelation. In *Proceedings of the 14th International Symposium on Remote Sensing of the Environment*, pp. 1755–67. San Jose, Costa Rica.

Cressie, N. (1984). Towards resistant geostatistics. In *Geostatistics for Natural Resources Characterization*, eds. Verly, G., David, M., Journel, A.G. and Marechal, A., pp. 21–44. Dordrecht: Reidel.

Cressie, N. (1985). Fitting variogram models by weighted least squares. *Journal of the International Association of Mathematical Geology*, **17**, 563–86.

Cressie, N. (1991). *Statistics for Spatial Data*. New York: Wiley.

Cressie, N. (1992). Smoothing regional maps using empirical Bayes predictors. *Geographical Analysis*, **24**, 75–95.

Cressie, N. (1996). Change of support and the modifiable areal unit problem. *Geographical Systems*, **3**, 159–80.

Cressie, N. and Chan, N.H. (1989). Spatial modeling of regional variables. *Journal of the American Statistical Association*, **84**, 393–401.

Cressie, N. and Hawkins, D.M. (1980). Robust estimation of the variogram, I. *Journal of the International Association of Mathematical Geology*, **12**, 115–25.

Cressie, N. and Read, T.R.C. (1989). Spatial data analysis of regional counts. *Biometrical Journal*, **31**, 699–719.

Cromley, R.G. (1996). A comparison of optimal classification strategies for choropleth displays of spatially aggregated data. *International Journal of Geographical Information Systems*, **10**, 405–24.

Cross, G.R. and Jain, A.K. (1983). Markov random field texture models. *IEEE Transactions on Pattern Analysis and Machine Intelligence*, **PAM 1–5**, 1, 25–39.

Cuff, D. and Mattson, M.T. (1982). *Thematic Maps: Their Design and Production*. New York: Methuen.

Curran, P. (1980). Multispectral remote sensing of vegetation amount. *Progress in Physical Geography*, **4**, 315–41.

Cuzick, J. and Edwards, R. (1990). Spatial clustering for inhomogeneous populations. *Journal of the Royal Statistical Society*, **B**, **52**, 73–104.

Cuzick, J. and Elliott, P. (1992). Small area studies: purpose and methods. *Geographical and Environmental Epidemiology: Methods for Small Area Studies*, eds. Elliott, P., Cuzich, J., English, D. and Stern, R., pp. 14–21. Oxford: Oxford University Press.

Decker, D. (2001). *GIS Data Sources*. New York: Wiley.

deLepper, M.J.C., Scholten, H.J. and Stern, R.M. (eds.) (1995). *The Added Value of Geographical Information Systems in Public and Environmental Health*. Dordrecht: Kluwer Academic.

Dempster, A.P. and Rubin, D.B. (1983). Overview in incomplete data in sample surveys. *Vol II: Theory and Annotated Bibliography*, eds. Madow, W.G., Olkin, I. and Rubin, D.B., pp. 3–10, New York: Academic Press.

Denison, D.G.T. and Holmes, C.C. (2001). Bayesian partitioning for estimating disease risk. *Biometrics*, **57**, 143–9.

Dent, B.D. (1985). *Principles of Thematic Map Design*. London: Addison-Wesley.

DETR (2000). Indices of deprivation 2000: regeneration research summary. See www.regeneration.detr.gov.uk

Devine, O.J. and Louis, T.A. (1994). A constrained empirical Bayes estimator for incidence rates in areas with small populations. *Statistics in Medicine*, **13**, 1119–33.

Devine, O.J., Louis, T.A. and Halloran, M.E. (1996). Identifying areas with elevated disease incidence rates using empirical Bayes estimators. *Geographical Analysis*, **28**, 187–99.

Diggle, P.J. and Chetwynd, A.D. (1991). Second-order analysis of spatial clustering for inhomogeneous populations. *Biometrics*, **47**, 1155–63.

Diggle, P., Morris, S. and Morton-Jones, T. (1999). Case control isotonic regression for investigation of elevation in risk around a point source. *Statistics in Medicine*, **18**, 1605–13.

Dobson, A.J. (1999). *An Introduction to Generalized Linear Models*. Boca Raton: Chapman & Hall.

Dockery, D.W., Pope, C.A., Xu, X., Spengler, J.D., Ware, J.H., Fay, M.E., Ferris, B.G. and Speizer, F.E. (1993). An association between air pollution and mortality in six US cities. *New England Journal of Medicine*, **329**, 1753–9.

Doreian, P. and Hummon, N.P. (1976). *Modelling Social Processes*. New York: Elsevier.

Dorling, D. (1992). Stretching space and splicing time: from cartographic animation to interactive visualization. *Cartography and Geographic Information Systems*, **19**, 215–27.

Dorling, D. (1994). Cartograms for visualizing human geography. *Visualization in Geographic Information Systems*, eds. Hearnshaw, H.M. and Unwin, D.J., pp. 85–102. New York: Wiley.

Dorling, D. (1995). Visualizing the 1991 Census. *Census Users Handbook*, ed. Openshaw, S., pp. 167–211. Cambridge: GeoInformation International.

Dow, M.M., Burton, M.L. and White, D.R. (1982). Network autocorrelation: a simulation study of a foundational problem in regression and survey research. *Social Networks*, **4**, 169–200.

Dowd, P.A. (1984). The variogram and kriging: robust and resistant estimators. *Geostatistics for Natural Resources Characterization*, eds. Verly, G., David, M., Journel, A.G. and Marechal, A., pp. 91–106. Dordrecht: Reidel.

Drummond, J. (1995). Positional Accuracy. In *Elements of Spatial Data Quality*, eds. Guptill, S.C. and Morrison, J.L., pp. 31–58. Oxford: Elsevier Science.

Dubin, R. (1997). A note on the estimation of spatial logit models. *Geographical Systems*, **4**, 181–93.

Dunn, R. (1987). Variable width framed rectangle charts for statistical mapping. *American Statistician*, **41**, 153–6.

Dunn, R. (1989). Approaches to two-variable color mapping. *American Statistician*, **43**, 245–52.

Dunn, R. and Harrison, A.R. (1993). Two dimensional systematic sampling of land use. *Applied Statistics*, **42**, 585–601.

Dunn, R., Harrison, A.R. and White, J.C. (1990). Positional accuracy and measurement error in digital databases of land use: an empirical study. *International Journal of Geographical Information Systems*, **4**, 385–98.

Dykes, J.A. (1997). Exploring spatial data representation with dynamic graphics. *Computers and Geosciences*, **23**, 345–70.

Dykes, J. (1998). Cartographic visualization: exploratory spatial data analysis with local indicators of spatial association using Tcl/Tk and cdv. *The Statistician*, **47**, 3, 485–97.

Earnshaw, R.A. and Wiseman, N. (eds) (1992). *An Introductory Guide to Scientific Visualization*. Berlin: Springer-Verlag.

Elliott, P., Cuzick, J., English, D. and Stern, R. (1992). *Geographical and Environmental Epidemiology: Methods for Small Area Studies*. Oxford: Oxford University Press.

Elliott, P., Wakefield, J., Best, N. and Briggs, D. (2000). *Spatial Epidemiology: Methods and Applications*. Oxford: Oxford University Press.

Emerson, J.D. and Hoaglin, D.C. (1983). Analysis of two way tables by medians. *Understanding Robust and Exploratory Data Analysis*, eds. Hoaglin, D.C., Mosteller, F.M. and Tukey, J.W., pp. 166–210. New York: Wiley.

English, D. (1992). Geographical epidemiology and ecological studies. *Geographical and Environmental Epidemiology: Methods for Small Area Studies*, eds. Elliott, P., Cuzich, J., English, D. and Stern, R. pp. 3–13. Oxford: Oxford University Press.

ESRC (2001). Health Variations Programme. www.lancs.ac.uk/users/apsocsci/hvp.htm

Everett, B. (1979). *Cluster Analysis*. London: Heinemann.

Faminow, M.D. and Benson, B.L. (1990). Integration of spatial markets. *American Journal of Agricultural Economics* , **70**, 49–62.

Farrington, D.P. and Dowds, E.A. (1984). Disentangling criminal behaviour and police reaction. *Reactions to Crime*, eds. Farrington, D.P. and Gunn, J. Chichester: Wiley.

Fedra, K. (1993). GIS and environmental modelling. *Environmental Modelling with GIS*, eds. Goodchild, M.F., Parks, B.O. and Steyaert, L.T., pp. 35–50. New York: Oxford University Press.

Ferreira, J.T.A.S., Denison, D.G.T. and Holmes, C.C. (2002). Partition modeling. Technical report at http://stats.ma.ic.ac.uk/~dgtd

Fieldhouse, E.A. and Tye, R. (1996). Deprived people or deprived places? Exploring the ecological fallacy in studies of deprivation with the samples of anonymised records. *Environment and Planning*, A, **28**, 237–59.

Fik, T.J. (1988). Spatial competitional price reporting in related food markets. *Economic Geography*, **64**, 29–44.

Fik, T.J. (1991). Price patterns in competitively clustered markets. *Environment and Planning*, A, **23**, 1545–60.

Firebaugh, G. (1978). A rule for inferring individual relationships from aggregate data. *American Sociological Review*, **43**, 557–72.

Fischer, M. Scholten, H.J. and Unwin, D. (1996). *Spatial Analytical Perspectives on GIS*. London: Taylor & Francis.

Fisher, P. and Langford, M. (1995). Modelling the errors in areal interpolation between zonal systems by Monte Carlo simulation. *Environment and Planning*, **A, 27**, 211–24.

Fisher, R. (1935). *The Design of Experiments*. Edinburgh: Oliver & Boyd.

Flowerdew, R. and Green, M. (1989). Statistical methods for inference between incompatible zonal systems. *Accuracy of Spatial Databases*, eds. Goodchild, M. and Gopal, S., pp. 239–47. London: Taylor & Francis .

Flowerdew, R., Green, M. and Kehris, E. (1991). Using areal interpolation methods in geographic information systems. *Papers in Regional Science*, **70**, 303–15.

Follmer, H. (1974). Random economies with many interacting agents. *Journal of Mathematical Economics*, **1**, 51–62.

Ford, G.E. and Zanelli, C.I. (1985). Analysis and quantification of errors in the geometric correction of images. *Photogrammetric Engineering and Remote Sensing*, **51**, 1725–34.

Forster, B.C. (1980). Urban residential ground cover using LANDSAT digital data. *Photogrammetric Engineering and Remote Sensing*, **46**, 547–58.

Fotheringham, A.S., Brunsdon, C. and Charlton, M. (2000). *Quantitative Geography: Perspectives on Spatial Data Analysis*. London: Sage.

Fotheringham. A.S. and Charlton, M. (1994). GIS and exploratory spatial data analysis: an overview of some research issues. *Geographical Systems*, **1**, 315–27.

Fotheringham, A.S. and Rogerson, P. (1994). *Spatial Analysis and GIS*. London: Taylor & Francis.

Fotheringham, A.S. and Wegener, M. (2000). *Spatial Models and GIS*. London: Taylor & Francis.

Freedman, D.A., Klein, S.P., Sachs, J., Smyth, C.A. and Everett, C.G. (1991). Ecological regression and voting rights (with discussion). *Evaluation Review*, **15**, 673–711.

Freedman, D.A., Ostland, M. and Roberts, M.R. (1998). A solution to the ecological inference problem. *Journal of the American Statistical Association*, **93**, 1518–22.

Freedman, D.A., Ostland, M. and Roberts, M.R. (1999). The future of ecological inference research: a comment on Freedman et al. – a response to King's comment. *Journal of the American Statistical Association*, **94**, 355–7.

Freund, J. (1992). *Mathematical Statistics*, fifth edition, Englewood Cliffs, NJ: Prentice-Hall.

Friedman, G.D. (1994). *Primer of Epidemiology*. New York: McGraw Hill.

Friendly, M. (1995). Conceptual and visual models for categorical data. *The American Statistician*, **49**, 153–60.

Frogbrook, Z.L. and Oliver, M.A. (2000). The effects of sampling on the accuracy of predictions of soil properties in precision agriculture. *Proceedings of the 4th International Symposium on Spatial Accuracy Assessment in Natural Resources and Environmental Sciences*, pp. 225–32. Delft: Delft University Press.

Fuller, W.A. (1975). Regression analysis for sample survey. *Sankhya*, **37**(c), 117–32.

Gahegan, M. (1999). Four barriers to the development of effective exploratory visualization tools for the geosciences. *International Journal of Geographical Information Sciences*, **13**, 289–309.

Gallup, J.L., Sacks, J.D. and Mellinger, A.D. (1999). Geography and economic development. *International Regional Science Review*, **22**, 179–232.

Galtung, J. (1967). *Theory and Methods of Social Research*. New York: Columbia University Press.

Gardiner, M.J. (1989). Review of reported increases in childhood cancer rates in the vicinity of nuclear installations in the UK. *Journal of the Royal Statistical Society*, **A, 152**, 307–25.

Gatrell, A.C. (1983). *Distance and Space: A Geographical Perspective*. Oxford: Clarendon Press.

Gatrell, A.C. (1998). Structures of geographical and social space and their consequences for human health. *Geografiska Annaler*, **79**, 141–54.

Gatrell, A.C., Bailey, T.C., Diggle, P.J. and Rowlingson, B.S. (1996). Spatial point pattern analysis and its application in geographical epidemiology. *Transactions, Institute of British Geographers*, **21**, 256–74.

Gatrell, A.C. and Dunn, C.E. (1995). Geographical information systems and spatial epidemiology: modelling the possible association between cancer of the larynx and incineration in North West England. *The Added Value of Geographical Information Systems in Public and Environmental Health*, eds. deLepper, M.J.C., Scholten, H.J. and Stern, R.M., pp. 215–35. Dordrecht: Kluwer Academic.

Gatrell, A.C. and Löytönen, M. (eds.) (1998). *GIS and Health*. London: Taylor & Francis.

Geary, R.C. (1954). The contiguity ratio and statistical mapping. *The Incorporated Statistician*, **5**, 115–45.

Gehlke, C.E. and Biehl, K. (1934). Certain effects of grouping upon the size of the correlation coefficient in census tract material. *Journal of the American Statistical Association*, **29**, 169–70.

Gelman, A., Price, P.N. and Lin, C-Y. (2000). A method of quantifying artefacts in mapping methods illustrated by application to headbanging. *Statistics in Medicine*, **19**, 2309–20.

Gelman, A., Carlin, J.B., Stern, H.S. and Rubin, D.B. (1995). *Bayesian Data Analysis*. London: Chapman & Hall.

Gelman, A., Park, D.K., Ansolabehere, S., Price, P.N. and Minnite, L. (2001). Models, assumptions and model checking in ecological regressions. *Journal of the Royal Statistical Society*, **A, 164**, 101–18.

Gelman, A. and Price, P.N. (1999). All maps of parameter estimates are misleading. *Statistics in Medicine*, **18**, 3221–34.

Geman, S. and Geman, D. (1984). Stochastic relaxation, Gibbs distributions and the Bayesian restoration of images. *IEEE, Transactions in Pattern Analysis and Machine Intelligence*, **6**, 721–41.

Getis, A. and Ord, J.K. (1992). The analysis of spatial association by use of distance statistics. *Geographical Analysis*, **24**, 189–206 (with correction, 1993, **25**, p. 276).

Getis, A. and Ord, J.K. (1995). Local spatial autocorrelation statistics: distributional issues and an application. *Geographical Analysis*, **27**, 286–306.

Getis, A. and Ord, J.K. (1996). Local spatial statistics: an overview. *Spatial Analysis: Modelling in a GIS environment*, eds. Longley, P. and Batty, M., pp. 261–77. Cambridge: Geoinformation International.

Ghosh, M. and Rao, J.N.K. (1994). Small area estimation: an appraisal. *Statistical Science*, **9**, 55–93.

Gilbert, C.L. (1986). Professor Hendry's econometric methodology. *Oxford Bulletin of Economics and Statistics*, **48**, 283–307.

Gilks, W.R., Richardson, S. and Spiegelhalter, D.J. (1996). *Markov Chain Monte Carlo in Practice: Interdisciplinary Statistics*. London: Chapman and Hall.

Godambe, V.P. and Thompson, M.E. (1971). Bayes, fiducial and frequency aspects of statistical inference in regression analysis in survey sampling. *Journal of the Royal Statistical Society*, **B, 33**, 361–90.

Goldsmith, V., McGuire, P.G., Mollenkopf, J.H. and Ross, T.A. (eds). (2000). *Analyzing Crime Patterns: Frontiers of Practice*. Thousand Oaks: Sage.

Goldstein, H. (1994). Multi-level cross-classified models. *Sociological Methods and Research*, **22**, 364–75.

Good, I.J. (1983). The philosophy of exploratory data analysis. *Philosophy of Science*, **50**, 283–95.

Goodchild, M.F. (1989). Modelling error in objects and fields. *Accuracy of Spatial Databases*, eds. Goodchild, M. and Gopal, S. pp. 107–13. London: Taylor & Francis.

Goodchild, M.F. (1995). Attribute accuracy. *Elements of Spatial Data Quality*, eds. Guptill, S.C. and Morrison, J.L. pp. 59–79. Oxford: Elsevier Science.

Goodchild, M.F., Anselin, L. and Deichmann, U. (1993). A framework for the areal interpolation of socio-economic data. *Environment and Planning*, **A, 25**, 383–97.

Goodchild, M.F. and Gopal, S. (1989). *Accuracy of Spatial Databases*. New York: Taylor & Francis.

Goodchild, M.F., Guoqing, S. and Shiren, Y. (1992). Development and test of an error model for categorical data. *International Journal of Geographical Information Systems*, **6**, 87–104.

Goodchild, M.F. and Lam, N.S-M. (1980). Areal interpolation: a variant of the traditional spatial problem. *Geo-Processing*, **1**, 297–312.

Goodchild, M.F., Parks, B.O. and Steyaert, L.T. (1993). *Environmental Modelling with GIS*. New York: Oxford University Press.

Goodman, L. (1953). Ecological regressions and the behaviour of individuals. *American Sociological Review*, **18**, 663–6.

Goovaerts, P. (1997). *Geostatistics for Natural Resources Evaluation*. Oxford: Oxford University Press.

Gordon, A. and Womersley, J. (1997). The use of mapping in public health and planning health services. *Journal of Public Health Medicine*, **19**, 139–47.

Greater Glasgow Health Board (1981). Census 1981, Maps for Community Medicine Areas. Greater Glasgow Information Services Unit. 51 pp.

Green, M. and Flowerdew, R. (1996). New evidence on the modifiable areal unit problem. *Spatial Analysis: Modelling in a GIS Environment*, eds. Longley, P. and Batty, M., pp. 41–54. Cambridge: GeoInformation International.

Griffith, D.A. (1978). A spatially adjusted ANOVA model. *Geographical Analysis*, **10**, 296–301.

Griffith, D.A. (1982). Dynamic characteristics of spatial economic systems. *Economic Geography*, **58**, 177–96.

Griffith, D.A. (1989). Distance calculations and errors in geographic databases. *Accuracy of Spatial Databases*, eds. Goodchild, M. and Gopal, S., pp. 81–90. New York: Taylor & Francis.

Griffith, D.A. (1996). Spatial autocorrelation and eigenfunctions of the geographic weights matrix accompanying geo-referenced data. *Canadian Geographer*, **40**, 351–67.

Griffith, D.A., Bennett, R.J. and Haining, R.P. (1989). Statistical analysis of spatial data in the presence of missing observations: a methodological guide and an application to urban census data. *Environment and Planning*, **A, 21**, 1511–23.

Griffith, D.A., Haining, R.P. and Arbia, G. (1994). Heterogeneity of attribute sampling error in spatial data sets. *Geographical Analysis*, **26**, 300–20.

Griffith, D.A. and Layne, L.J. (1999). *A Casebook for Spatial Statistical Data Analysis – A Compilation of Different Thematic Data Sets*. Oxford: Oxford University Press.

Grigg, D.B. (1967). Regions, models and classes. *Models in Geography*, eds. Chorley, R.J. and Haggett, P., pp. 461–509. London: Methuen.

Grimson, R.C. (1991). A versatile test for clustering and a proximity analysis of neurons. *Methods of Information in Medicine*, **30**, 299–303.

Gumpertz, M.L., Graham, J.M. and Ristaino, J.B. (1997). Autologistic model of spatial pattern of phytophthora epidemic in bell pepper: effects of soil variables on disease presence. *Journal of Agricultural, Biological and Environmental Statistics*, **2**, 131–56.

Guptill, S. C. (1994). Synchronization of discrete geospatial databases. *Advances in GIS Research. Proceedings of the 6th International Symposium on Spatial Data Handling*, eds Waugh, T.C. and Healey, R.G., IGU Commission on GIS and the Association for Geographic Information, Vol. **2**, pp. 945–56.

Guptill, S.C. and Morrison, J.L. (1995). *Elements of Spatial Data Quality*. Oxford: Elsevier Science.

Guyon, X. (1995). *Random Fields on a Network*. New York: Springer-Verlag.

Hagerstand, T. (1967). *Innovation Diffusion as a Spatial Process*. Chicago, IL: Chicago University Press.

Haining, R.P. (1978). The moving average model for spatial interaction. *Transactions of the Institute for British Geographers*, **NS3**, 202–25.

Haining, R. (1980). Spatial autocorrelation problems. *Geography and the Urban Environment*, eds. Herbert, D.T. and Johnston, R.J., pp. 1–44. New York: Wiley.

Haining, R.P. (1983a). Anatomy of a price war. *Nature*, **304**, 679–80.

Haining, R.P. (1983b). Modelling intra-urban price competition: an example of gasoline pricing. *Journal of Regional Science*, **23**, 517–28.

Haining, R.P. (1985). The spatial structure of competition and equilibrium price dispersion. *Geographical Analysis*, **17**, 231–42.

Haining, R.P. (1987). Small area aggregate income models: theory and methods with an application to urban and rural income data for Pennsylvania. *Regional Studies*, **21**, 519–30.

Haining, R.P. (1988). Estimating spatial means with an application to remotely sensed data. *Communications in Statistics, Theory and Methods*, **17**, 573–97.

Haining, R.P. (1990a). The use of added variable plots in regression modelling with spatial data. *Professional Geographer*, **42**, 336–44.

Haining, R.P. (1990b). *Spatial Data Analysis in the Social and Environmental Sciences*. Cambridge: Cambridge University Press.

Haining, R.P. (1991). Bivariate correlation with spatial data. *Geographical Analysis*, **23**, 210–27.

Haining, R.P. (1991a). Estimation with heteroscedastic and correlated errors: a spatial analysis of intra-urban mortality data. *Papers in Regional Science*, **70**, 223–41.

Haining, R.P. (1994). Designing spatial data analysis modules for GIS. *Spatial Analysis and GIS*, eds. Fotheringham, S. and Rogerson, P. London: Taylor & Francis.

Haining, R.P. (1994a). Diagnostics of regression modelling in spatial econometrics. *Journal of Regional Science*, **34**, 325–41.

Haining, R.P. (1995). Data problems in spatial econometric modelling. *New Directions in Spatial Econometrics*, eds. Anselin, L. and Florax, R.J.G., pp. 156–71. Berlin: Springer.

Haining, R. P. and Arbia, G. (1993). Error propagation through map operations. *Technometrics*, **35**, 293–305.

Haining, R.P. and Cliff, A. (2002). Using a scan statistic to map the incidence of an infectious disease: measles in the USA 1962–1995. GEOMED 2000.

Haining, R.P., Griffith, D.A. and Bennett, R.J. (1983). Simulating two dimensional autocorrelated surfaces. *Geographical Analysis*, **15**, 247–55.

Haining, R.P., Griffith, D.A. and Bennett, R.J. (1984). A statistical approach to the problem of missing data using a first order Markov model. *Professional Geographer*, **36**, 338–48.

Haining, R.P., Griffith, D.A. and Bennett, R.J. (1989). Maximum likelihood estimation with missing spatial data and with an application to remotely sensed data. *Communications in Statistics: Theory and Methods*, **18**, 1875–94.

Haining, R.P., Plummer, P. and Sheppard, E. (1996). Spatial price equilibrium in interdependent markets: price and sales configuration. *Papers of the Regional Science Association*, **75**, 41–64.

Haining, R., Wise, S. and Blake, M. (1994). Constructing regions for small area analysis: material deprivation and colorectal cancer. *Journal of Public Health Medicine*, **16**, 429–38.

Haining, R.P., Wise, S.M. and Signoretta, P.E. (2000). Providing scientific visualization for spatial data analysis: criteria and an assessment of SAGE. *Journal of Geographical Systems*, **2**, 121–40.

Haining, R.P., Wise, S.M. and Ma, J. (1998). Exploratory spatial data analysis in a geographic information system environment. *The Statistician*, **47**, 457–69.

Hampel, F.R., Ronchetti, E.M., Rousseeuw, P.J. and Stahel, W.A. (1986). *Robust Statistics*. New York: Wiley.

Hansen, K.M. (1991). Head-banging: robust smoothing in the plane. *IEEE Transactions on Geoscience and Remote Sensing*, **29**, 369–78.

Hansen, M.H., Madow, W.G. and Tepping, B.J. (1983). An evaluation of model dependent and probability sampling inferences in sample surveys (with comments). *Journal of the American Statistical Association*, **78**, 776–807.

Harris, R.J. and Longley, P.A. (2000). New data and approaches for urban analysis: modelling residential densities. *Transactions in GIS*, **4**, 217–34.

Haslett, J., Bradley, R., Craig, P., Unwin, A. and Wills, G. (1991). Dynamic graphics for exploring spatial data with applications to locating global and local anomalies. *The American Statistician*, **45**, 234–42.

Haslett, J., Wills, G. and Unwin, A. (1990). SPIDER, an interactive tool for the analysis of spatially distributed data. *International Journal of Geographical Information Systems*, **4**, 285–96.

Hawkins, D.M., Bradu, D. and Kass, G. (1984). Location of several outliers in multiple regression using elemental sets. *Technometrics*, **26**, 197–208.

Heagerty, P.J. and Lele, S.R. (1998). A composite likelihood approach to binary spatial data. *Journal of the American Statistical Association*, **93**, 1099–111.

Hearnshaw, H.M. and Unwin, D.J. (1994). *Visualization in Geographic Information Systems*. New York: Wiley.

Hepple, L. (1979). Bayesian analysis of the linear model with spatial dependence. *Exploratory and Explanatory Statistical Analysis of Spatial Data*, eds. Bartels, C.P.A. and Ketellapper, R.H., pp. 179–99. Boston, MA: Martinus Nijhoff.

Hepple, L.W. (1996). Directions and opportunities in spatial econometrics. *Spatial Analysis Modelling in a GIS Environment*, eds. Longley, P. and Batty, M., pp. 231–46. Cambridge: GeoInformation International.

Hepple, L.W. (1998). Exact testing for spatial autocorrelation among regression residuals. *Environment and Planning*, **A, 30**, 85–108.

Heuvelink, G.B.M. (1993). *Error Propagation in Environmental Modelling with GIS*. London: Taylor & Francis.

Heuvelink, G.B.M. (1999). Aggregation and error propagation in GIS. *Spatial Accuracy Assessment: Land Information Uncertainties in Natural Resources*, eds. Lowell, K. and Jaton, A., pp. 219–25. Chelsea, Michigan: Ann Arbor Press.

Heuvelink, G.B.M., Burrough, P.A. and Stein, A. (1989). Propagation of errors in spatial modelling with GIS. *International Journal of Geographical Information Systems*, **3**, 303–22.

Hill, P.W. and Goldstein, H. (1998). Multi-level modelling of educational data with cross classification and missing identification of units. *Journal of Educational and Behavioural Statistics*, **23**, 117–28.

Hirschfield, A. and Bowers, K.J. (1997). The effect of social cohesion on levels of recorded crime in disadvantaged areas. *Urban Studies*, **34**, 1275–95.

Hirschfield, A. and Bowers, K. (2001). *Mapping and Analysing Crime Data: Lessons from Research and Practice*. London: Taylor & Francis.

Hjalmars, U., Kulldorff, M., Gustafsson, G. and Nagarwalla, N. (1996). Childhood leukaemia in Sweden: using GIS and a spatial scan statistic for cluster detection. *Statistics in Medicine*, **15**, 707–15.

Hoaglin, D.C., Mosteller, F. and Tukey, J.W. (1983). *Understanding Robust and Exploratory Data Analysis*. New York: Wiley.

Hoaglin, D.C., Mosteller, F. and Tukey, J.W. (1985). *Exploring Data Tables, Trends and Shapes*. New York: Wiley.

Hodder, I. (1977). Some new directions in the spatial analysis of archaeological data at the regional scale (macro). *Spatial Archaeology*, ed. Clarke, D.L., pp. 223–351. London: Academic Press.

Hole, D.J. and Lamont, D.W. (1992). Problems in the interpretation of small area analysis of epidemiological data: the case of cancer incidence in the West of Scotland. *Journal of Epidemiology and Community Health*, **46**, 305–10.

Holmes, J.H. and Haggett, P. (1977). Graph theory interpretation of flow matrices: a note on maximization procedures for identifying significant linkages. *Geographical Analysis*, **9**, 388–99.

Holt, D., Steel, D.G. and Tranmer, M. (1996). Area homogeneity and the modifiable areal unit problem. *Geographical Systems*, **3**, 181–200.

Holt, D., Steel, D.G., Tranmer, M. and Wrigley, N. (1996). Aggregation and ecological effects in geographically based data. *Geographical Analysis*, **28**, 244–61.

Horn, M.E.T. (1995). Solution techniques for large regional partitioning problems. *Geographical Analysis*, **27**, 230–48.

Horton, C.W., Hempkins, W.B. and Hoffman, A.A.J. (1964). A statistical analysis of some aeromagnetic maps from the Northwestern Canadian Shield. *Geophysics*, **4**, 582–601.

Hubert, L.J., Golledge, R.G. and Costanzo, C.M. (1981). Generalized procedures for evaluating spatial autocorrelation. *Geographical Analysis*, **13**, 224–33.

Hubert, L.J., Golledge, R.G., Costanzo, C.M. and Gale, N. (1985). Measuring association between spatially defined variables: an alternative procedure. *Geographical Analysis*, **17**, 36–46.

Huffer, F.W. and Wu, H. (1998). Markov Chain Monte Carlo for autologistic regression models with application to the distribution of plant species. *Biometrics*, **54**, 509–24.

Hughes, J.P. and Lettenmaier, D.P. (1981). Data requirements for kriging: estimation and network design. *Water Resources Research*, **17**, 1641–50.

Isaaks, E.H. and Srivastava, R.M. (1989). *An Introduction to Applied Geostatistics*. Oxford: Oxford University Press.

Isard, W. (1960). *Methods of Regional Analysis*. New York: Technology Press of MIT and Wiley.

Jaakkola, O. (1998). Multi-scale categorical databases with automatic generalization transformations based on map algebra. *Cartography and Geographic Information Systems*, **25**, 195–207.

Jacquez, G.M. (1994). Cuzick and Edward's test when exact locations are unknown. *American Journal of Epidemiology*, **140**, 58–64.

Jacquez, G. (1995). The map comparison problem: tests for the overlap of geographic boundaries. *Statistics in Medicine*, **14**, 2343–61.

Jacquez, G.M., Maruca, S. and Fortin, M.-J. (2000). From fields to objects: a review of geographic boundary analysis. *Journal of Geographical Systems*, **2**, 221–41.

James, W. and Stein, C. (1960). Estimation with quadratic loss. *Proceedings of the Fourth Berkeley Symposium*, vol. **1**, ed. Neyman, J., pp. 361–80. Berkeley, CA: University of California Press.

Jarman, B. (1993). Identification of underprivileged areas. *British Medical Journal*, **286**, 1705–8.

Johnston, R.J. (1986). The neighbourhood effect revisited: spatial science or political regionalism. *Environment and Planning*, **D, 4**, 41–55.

Johnstone, J. (1978). Social class, social areas and delinquency. *Sociology and Social Research*, **63**, 49–72.

Jolley, D., Jarman, B. and Elliott, P. (1992). Socio-economic confounding. *Geographical and Environmental Epidemiology: Methods for Small Area Studies*, eds. Elliott, P., Cuzich, J., English, D. and Stern, R. Oxford: Oxford University Press.

Jones III, J.P. and Casetti, E. (1992). *Applications of the Expansion Method*. London: Routledge.

Jones, K. and Duncan, C. (1996). People and Places: the multi level model as a general framework for the quantitative analysis of geographical data. *Spatial Analysis: Modelling in a GIS Environment*, eds Longley, P. and Batty, M., pp. 79–104. Cambridge: GeoInformation International.

Jones, K., Gould, M.I. and Watt, R. (1998). Multiple contexts as cross classified models: the Labour vote in the British General Election of 1992. *Geographical Analysis*, **30**, 65–93.

Jones, M. and Sibson, R. (1987). What is projection pursuit? (with discussion). *Journal of the Royal Statistical Society*, **B, 150**, 1–36.

Journel, A.E. (1983). Non parametric estimation of spatial distributions. *Journal of the International Association of Mathematical Geology*, **15**, 445–68.

Kafadar, K. (1994). Choosing among two dimensional smoothers in practice. *Computational Statistics and Data Analysis*, **18**, 419–39.

Kafadar, K. (1996). Smoothing geographical data, particularly rates of disease. *Statistics in Medicine*, **15**, 2539–60.

Kafadar, K. (1999). Simultaneous smoothing and adjusting mortality rates in US counties: melanoma in white females and white males. *Statistics in Medicine*, **18**, 3167–88.

Kahn, H. and Sempos, C.T. (1989). *Statistical Methods in Epidemiology*. Oxford: Oxford University Press.

Kainz, W. (1995). Logical consistency. *Elements of Spatial Data Quality*, eds. Guptill, S.C. and Morrison, J.L., pp. 109–37. Oxford: Elsevier Science.

Kaiser, M.S. and Cressie, N. (1997). Modeling Poisson variables with positive spatial dependence. *Statistics and Probability Letters*, **35**, 423–32.

Kauth, R.J. and Thomas, G.S. (1976). The tassled cap: a graphic description of the spectral-temporal development of agricultural crops as seen by LANDSAT. Proceedings of LARS 1976, Symposium on Machine Processing of Remote Sensed Data. Purdue University.

Kelejian, H.H. and Robinson, D.P. (1995). Spatial correlation: a suggested alternative to the autoregressive model. *New Directions in Spatial Econometrics*, eds. Anselin, L. and Florax, R.J.G.M., pp. 75–93. Berlin: Springer.

Kelsall, J.E. and Diggle, P.J. (1995). Non-parametric estimation of spatial variation in relative risk. *Statistics in Medicine*, **14**, 2335–42.

Kennedy, S. (1989). The small number problem and the accuracy of spatial databases. *Accuracy of Spatial Databases*, eds. Goodchild, M. and Gopal, S., pp. 187–96. London: Taylor & Francis.

Kennedy, S. and Tobler, W.R. (1983). Geographic interpolation. *Geographical Analysis*, **15**, 151–6.

Kershaw, K.A. (1973). *Quantitative and Dynamic Plant Ecology*, second edition, London: Arnold.

Kiefer, J. and Wynn, H.P. (1981). Optimum balanced block and Latin square designs for correlated observations. *Annals of Statistics*, **9**. 737–57.

King, G. (1997). *A Solution to the Ecological Inference Problem*. Princeton, NJ: Princeton University Press.

King, G. (1999). The future of ecological inference research: a comment on Freedman et al. *Journal of the American Statistical Association*, **94**, 352–5.

King, G. (2000). Geography, statistics and ecological inference. *Annals of the Association of American Geographers*, **90**, 601–6.

Kingman, J.F.C. (1975). Markov models for spatial variation. *The Statistician*, **24**, 167–74.

Knack, S. and Keefer, P. (1997). Does social capital have an economic payoff? A cross country investigation. *The Quarterly Journal of Economics*, **112**, 1251–88.

Knorr-Held, L. and Besag, J.E. (1998). Modelling risk from a disease in space and time. *Statistics in Medicine*, **17**, 2045–60.

Knorr-Held, L. and Rasser, G. (2000). Bayesian detection of clusters and discontinuities in disease maps. *Biometrics*, 56, 13–21.

Koenker, R. and Bassett, G. (1982). Robust tests for heteroskedasticity based on regression quantilers. *Econometrica*, **50**, 43–61.

Koop, J.C. (1990). Systematic sampling of two dimensional surfaces and related problems. *Communications in Statistics: Theory and Methods*, **19**, 1701–50.

Krishna Iyer, P.V.A. (1949). The first and second moments of some probability distributions arising from points on a lattice and their applications. *Biometrika*, **36**, 135–41.

Krug, T. and Martin, R.J. (1990). Information loss on the mean for spatial processes when some values are missing. *Communications in Statistics: Theory and Methods*, **20**, 2168–95.

Krugman, P. (1991). *Geography and Trade*. London: MIT Press.

Krugman, P. (1995). *Development, Geography and Economic Theory*. London: MIT Press.

Krugman, P. (1996). Urban concentration: the role of increasing returns and transport costs. *International Regional Science Review*, **19**, 5–30.

Krugman, P. (1998). What's new about the New Economic Geography? *Oxford Review of Economic Policy*, **14**, 7–17.

Kulldorff, M. (1997). A spatial scan statistic. *Communications in Statistics: Theory and Methods*, **26**, 1481–96

Kulldorff, M. (1998). Statistical methods for spatial epidemiology: tests for randomness. *GIS and Health*, eds. Gatrell, A. and Löytönen, M. pp. 49–62. London: Taylor & Francis.

Kulldorff, M. and Nagarwalla, N. (1995). Spatial disease clusters: detection and inference. *Statistics in Medicine*, **14**, 799–810.

Labovitz, M.L. and Masuoka, E.J. (1984). The influence of auotcorrelation in signature extraction: an example from a geobotanical investigation of Cotter Basin, Montana. *International Journal of Remote Sensing*, **5**, 315–32.

Langford, M., Maguire, D.J. and Unwin, D.J. (1991). The areal interpolation problem: estimating population using remote sensing in a GIS framework. *Handling Geographical Information: Methodology and Potential Applications*, eds. Masser, I. and Blakemore, M., pp. 55–77. Harlow: Longman.

Lawson, A.B., Biggeri, A.B., Boehning, D., Lesaffre, E., Viel, J-F., Clark, A., Schlattmann, P. and Divino, F. (2000). Disease mapping models: an empirical evaluation. *Statistics in Medicine*, **19**, 2217–42.

Lawson, A.B. and Clark, A. (2002). Spatial mixture relative risk models applied to disease mapping. *Statistics in Medicine*, **21**, 359–70.

Lawson, A.B. and Williams, F.L. (2001). *An Introductory Guide to Disease Mapping*. Chichester: Wiley.

Leamer, E.E. (1978). *Specification Searches: Ad Hoc Inference with Non-experimental Data*. New York: Wiley.

Lee, J., Snyder, P.K. and Fisher, P.F. (1992). Modelling the effect of data errors on feature extraction from digital elevation models. *Photogrammetric Engineering and Remote Sensing*, **58**, 1461–7.

Lee, P. (1999a). Where are the socially excluded: continuing debates in the identification of poor neighbourhoods. *Regional Studies*, **33**, 483–9.

Lee, P. (1999b). Where are the deprived? Measuring deprivation in cities and regions. *Statistics in Society: The Arithmetic of Politics*, eds. Dorling, D. and Simpson, S., London: Arnold.

Lee, S-I. (2001). Developing a bivariate spatial association measure: an integration of Pearson's r and Moran's I. *Journal of Geographical Systems*, **3**, 369–86.

Lee, Gallo, J. and Ertur, C. (2002). Exploratory spatial data analysis in the distribution of regional per capita GDP in Europe. *Papers in Regional Science* (forthcoming).

Legendre, P., Oden, N.L., Sikal, R.R., Vaudor, A. and Kim, J. (1990). Approximate analysis of variance of spatially autocorrelated regional data. *Journal of Classification*, **7**, 53–75.

Leonard, T. (1983). Some philosophies of inference and modelling. *Scientific Inference, Data Analysis and Robustness*, eds. Box. G.E.P., Leonard, T. and Wu, C.-F., pp. 9–23. New York: Academic Press.

Little, R.J.A. and Rubin, D.B. (1987). *Statistical Analysis with Missing Data*. New York: Wiley.

Lloyd, O.L. (1995). The exploration of the possible relationship between deaths, births and air pollution in Scottish towns. *The Added Value of Geographical Information Systems in Public and Environmental Health*. eds. deLepper, M.J.C., Scholten, H.J. and Stern, R.M., pp. 167–80. Dordrecht: Kluwer Academic.

Longford, N.T. (1999). Multivariate shrinkage estimation of small area means and proportions. *Journal of the Royal Statistical Society*, **A**, **162**, 227–45.

Longley, P.A. and Batty, M. (1996). *Spatial Analysis: Modelling in a GIS Environment*. Cambridge: GeoInformation International.

Longley, P.A., Goodchild, M.F., Maguire, D.J. and Rhind, D.W. (1999). *Geographical Information Systems*, second edition, Vols 1 and 2. New York: Wiley.

Longley, P.A., Goodchild, M.F., Maguire, D.J. and Rhind, D.W. (2001). *Geographical Information Systems and Science*. Chichester: Wiley.

Lopez, A.D. (1992). Mortality Data. *Geographical and Environmental Epidemiology: Methods for Small Area Studies*, eds. Elliott, P.J., Cuzick, J., English, D. et al., pp. 37–50. Oxford: Oxford University Press.

Losch, A. (1939). *The Economics of Location* (English translation, 1957). New York: J.Wiley & Sons.

Lovett, A., Haynes, R., Bentham, G., Gale, S., Brainard, J. and Suennenberg, G. (1998). Improving health needs assessment using patient register information in a GIS. *GIS and Health*, eds. Gatrell, A. and Löytönen, M., pp. 191–203. London: Taylor & Francis.

Lowell, K. and Jaton, A. (1999). *Spatial Accuracy Assessment: Land Information Uncertainties in Natural Resources*. Chelsea, MI: Ann Arbor Press.

McBratney, A.B. and Webster, R. (1986). Choosing functions for semi variograms of soil properties and fitting them to sampling estimates. *Journal of Soil Science*, **37**, 617–39.

McLain, D.H. (1974). Drawing contours from arbitrary data points. *Computer Journal*, **17**, 318–24.

MacDougall, E.B. (1992). Exploratory analysis, dynamic statistical visualization and geographic information systems. *Cartography and Geographic Information Systems*, **19**, 237–46.

MacEachren, A.M. (1995). *How Maps Work: Representation, Visualization and Design*. New York: The Guilford Press.

MacEachren, A.M. and Kraak, M-J. (1997). Exploratory cartographic visualization: advancing the agenda. *Computers and Geosciences*, **23**, 335–43.

MacEachren, A.M. and Monmonier, M. (1992). Introduction (to special edition on Geographic Visualization). *Cartography and Geographic Information Systems*, **19**, 197–200.

MacLeod, M., Graham, E., Johnston, M., Dibben, C. and Morgan, I. (1999). How does relative deprivation affect health? *ESRC Health Variations Newsletter*, **3**, January, Lancaster University.

Macmillan, W. (2001). Redistricting in a GIS environment: an optimisation algorithm using switching points. *Journal of Geographical Systems*, **3**, 167–80.

MacQueen, J. (1967). Some methods for classification and analysis of multivariate observations. *Proceedings of the 5th Berkeley Symposium on Mathematical Statistics and Probability*, **1**, 281–97.

Maffini, G., Arno, M. and Bitterlich, W. (1989). Observations and comments on the generation and treatment of error in digital GIS data. *Accuracy in Spatial Databases*, eds. Goodchild, M. and Gopal. S., pp. 55–67. London: Taylor & Francis.

Majure, J., Cook, D., Cressie, N. Kaiser, M., Lahiri, S. and Symanzik, J. (1996). Spatial CDF estimation and visualization with appliction to forest health monitoring. *Computing Science and Statistics*, **27**, 93–101.

Majure, J.J. and Cressie, N. (1997). Dynamic graphics for exploring spatial dependence in multivariate spatial data. *Geographical Systems*, **4**, 131–58.

Manton, K.G. and Stallard, E. (1981). Methods for the analysis of mortality risks across heterogeneous small populations: estimation of space-time gradients in cancer mortality in North Carolina counties 1970–75. *Demography*, **18**, 217–30.

Mardia, K.V. and Marshall, R.J. (1984). Maximum likelihood estimation of models for residual covariance in spatial regression. *Biometrika*, **70**, 135–46.

Mark, D.M. (1999). Spatial representation: a cognitive view. *Geographical Information Systems: Volume 1 Principles and Technical Issues*, eds. Longley, P.A., Goodchild, M.F., Maguire, D.J. and Rhind, D.W., pp. 81–9. New York: Wiley.

Markoff, J. and Shapiro, G. (1973). The linkage of data describing overlapping geographical units. *Historical Methods Newsletter*, **7**, 34–46.

Martens, P.L. (1993). An ecological model of socialisation in explaining offending. *Integrating Individual and Ecological Aspects of Crime*, eds. Farrington, D.P., Sampson, R.J. and Wikstrom, P.-O.H. Stockholm: National Council for Crime Prevention.

Martin, D.J., Senior, M.L. and Williams, H.C.W.L. (1994). On measures of deprivation and the spatial allocation of resources for primary health care. *Environment and Planning*, **A**, **26**, 1911–29.

Martin, D.J. (1998). Optimizing census geography: the separation of collection and output geographies. *International Journal of Geographical Information Science*, **12**, 673–85.

Martin, D.J. (1999). Spatial representation: the social scientists' perspective. *Geographical Information Systems: Volume 1. Principles and Technical Issues*, second edition, eds. Longley, P.A., Goodchild, M.F., Maguire, D.J. and Rhind, D.W., pp. 71–89. New York: Wiley.

Martin, R.L. (1999). The new 'geographical turn' in economics: some critical reflections. *Cambridge Journal of Economics*, **23**, 68–91.

Martin, R.L. and Sunley, P. (1996). Paul Krugman's geographical economics and its implications for regional development theory: a critical assessment. *Economic Geography*, **72**, 259–92.

Martin, R.J. (1984). Exact maximum likelihood for incomplete data from a correlated Gaussian process. *Communications in Statistics: Theory and Methods*, **13**, 1275–88.

Martin, R.J. (1987). Some comments on correction techniques for boundary effects and missing value techniques. *Geographical Analysis*, **19**, 273–82.

Martin, R.J. (1989). Information loss due to incomplete data from a spatial Gaussian one-parameter first-order conditional process. *Communications in Statistics: Theory and Methods*, **18**, 4631–45.

Maruca, S.L. and Jacquez, G.M. (2002). Area-based tests for association between spatial patterns. *Journal of Geographical Systems*, **4**, 69–83.

Matern, B. (1986). Spatial Variation. *Lecture Notes in Statistics*, **36**. Berlin: Springer-Verlag.

Matheron, G. (1976). A simple substitute for conditional expectation: the disjunctive kriging. *Advanced Geostatistics in the Mining Industry*, eds. Guarascio, M., David, M. and Huijbrechts, C., pp. 221–36. Dordrecht: Reidel.

Matula, D.W. and Sokal, R.R. (1980). Properties of Gabriel graphs relevant to geographic variation and the clustering of points in the plane. *Geographical Analysis*, **12**, 205–22.

Mawby, R.I. (1989). Policing and the criminal area. *The Geography of Crime*, eds. Evans, D.J. and Herbert, D.T., London: Routledge.

McGuire, P.G. (2000). The New York Police Department – COMPSTAT process. *Analyzing Crime Patterns: Frontiers of Practice*, eds. Goldsmith, V., McGuire, P.G., Mollenkopf, J.H. and Ross, T.A., pp. 11–22. Thousand Oaks: Sage.

Mencken, F.C. and Barnett, C. (1999). Murder, non-negligent manslaughter and spatial autocorrelation in mid-South counties. *Journal of Quantitative Criminology*, **15**, 407–22.

Messner, S.F., Anselin, L., Baller, R.D., Hawkins, D.F., Deane, G. and Tolnay, S.E. (1999). The spatial patterning of county homicide rates: an application of exploratory spatial data analysis. *Journal of Quantitative Criminology*, **15**, 423–50.

Miesch, A.T. (1975). Variograms and variance components in geochemistry and ore evaluation. *Quantitiative Studies in the Geological Sciences*, ed. Whitten, E.H.T., pp. 333–40. Colorado: Geological Society of America.

Mikhail, E.M. (1976). *Observations and Least Squares*. New York: IEP-Dun-Donnely.

Milner, A. (1959). A centric systematic area sample treated as a random sample. *Biometrics*, **15**, 270–97.

Minister for the Cabinet Office (1999). White Paper: Modernising Government. London: HMSO. Also: www.cabinet-office.gov.uk/seu/exec.summary/summary.htm

Mollie, A. (1996). Bayesian mapping of disease. *Markov Chain Monte Carlo in Practice: Interdisciplinary Statistics*, pp. 359–79. London: Chapman & Hall.

Monmonier, M.S. (1989). Geographic brushing: exhancing exploratory analysis of the scatterplot matrix. *Geographical Analysis*, **21**, 81–4.

Monmonier, M.S. (1991). *How to Lie with Maps*. Chicago, IL.: University of Chicago Press.

Monmonier, M.S. (1992). Authoring graphic scripts: experiences and principles. *Cartography and Geographic Information Systems*, **19**, 247–60.

Monmonier, M.S. and MacEachren, A.M. (1992). Graphic visualization. *Cartography and Geographic Information Systems*, Special Edition, **19**, 197–260.

Moran, C.J. and Bui, E.N. (2000). Stochastic and partly-determined statistical perturbation of points, polygons and surface to assess certainty of environmental modelling. In *Accuracy 2000, Proceedings of the 4th International Symposium on Spatial Accuracy Assessment in Natural Resource and Environmental Science*, Amsterdam, July, pp. 493–500. Delft University Press.

Moran, P.A.P. (1948). The interpretation of statistical maps. *Journal of the Royal Statistical Society*, **B, 10**, 243–51.

Moran, P.A.P. (1950). Notes on continuous stochastic phenomena. *Biometrika*, **37**, 17–23.

Moreno, R. and Trehan, B. (1997). Location and the growth of nations. *Journal of Economic Growth*, **2**, 399–418.

Morganstern, H. (1982). Uses of ecologic analysis in epidemiologic research. *American Journal of Public Health*, **72**, 1336–44.

Morphet, C.S. (1993). The mapping of small-area census data – a consideration of the role of enumeration district boundaries. *Environment and Planning*, **A, 25**, 1267–77.

Morrison, D.F. (1967). *Multivariate Statistical Methods*. New York: McGraw Hill.

Mrozinski, R.D. and Cromley, R.G. (1999). Singly and doubly constrained methods of areal interpolation for vector-based GIS. *Transactions in GIS*, **3**, 285–301.

Mungiole, M., Pickle, L.W. and Simonson, K.H. (1999). Application of a weighted head banging algorithm to mortality data maps. *Statistics in Medicine*, **18**, 3201–9.

Neprash, J.A. (1934). Some problems in the correlation of spatially distributed variables. *Journal of the American Statistical Association*, **29**, 167–8.

Nester, M.R. (1996). An applied statistician's creed. *Applied Statistician*, **45**, 401–10.

Ning, X. and Haining, R.P. (2002). Spatial pricing in interdependent markets: a case study of petrol retailing in Sheffield. Submitted to *Environment and Planning*, A.

Norcliffe, G. (1977). *Inferential Statistics for Geographers*. London: Hutchinson.

O'Connell, P.E., Gurney, R.J., Jones, D.A., Miller, J.B., Nicholas, C.A. and Senior, M.R. (1979). A case study of rationalization of a rain guage network in SW England. *Water Resources Research*, **15**, 1813–22.

Oden, N. (1995). Adjusting Moran's I for population density. *Statistics in Medicine*, **14**, 17–26.

Oden, N.L., Sokal, R.R., Fortin, M-J. and Goebl, H. (1993). Categorical Wombling: detecting regions of significant change in spatially located categorical variables. *Geographical Analysis*, **25**, 315–36.

Okabe, A. and Tagashira, N. (1996). Spatial aggregation bias in a regression model containing a distance variable. *Geographical Systems*, **3**, 77–100.

Oliver, M.A. and Webster, R. (1986). Semi-variograms for modelling the spatial pattern of land form and soil properties. *Earth Surface Processes and Land forms*, **11**, 491–504.

Oliver, M.A. and Webster, R. (1989). A geostatistical basis for spatial weighting in multivariate classification. *Mathematical Geology*, **21**, 15–35.

Olson, J.M. (1976). Non contiguous area cartograms. *Professional Geographer*, **28**, 371–80.

Olson, J.M. (1987). Review of 'Elements of Graphing Data'. *The American Cartographer*, **14**, 88–89.

Openshaw, S. (1978). An optimal zoning approach to the study of spatially aggregated data. In *Spatial Representation and Spatial Interaction*, eds. Masser, I. and Brown, P.J.B., pp. 93–113. Leiden: Martinus Nijhoff.

Openshaw, S. (1995). The future of the Census. *Census User's Handbook*, ed. Openshaw, S., pp. 389–411. Cambridge: GeoInformation International.

Openshaw, S. (1996). Developing GIS-relevant zone based spatial analysis methods. *Spatial Analysis: Modelling in a GIS Environment*, eds. Longley, P. and Batty, M., pp. 55–78. Cambridge: GeoInformation International.

Openshaw, S. and Abrahart, R. (1996). Geocomputation. *Geocomputation '96: Proceedings of First International Conference on Geocomputation, University of Leeds*, pp. 665–6. Leeds: Department of Geography.

Openshaw, S., Charlton, M., Wymer, C. and Craft, A. (1987). A mark 1 Geographical Analysis Machine for the automated analysis of point data sets. *International Journal of Geographical Information Systems*, **1**, 335–58.

Openshaw, S. and Rao, L. (1995). Algorithms for re-engineering 1991 Census geography. *Environment and Planning*, **A**, **27**, 425–46.

Openshaw, S. and Wymer, C. (1995). Classifying and regionalizing Census data. *Census User's Handbook*, ed. Openshaw, S., pp. 239–70. Cambridge: GeoInformation International.

Orchard, R. and Woodbury, M. (1972). The missing information principle: theory and application. *Proceedings of the 6th Berkeley Symposium on Mathematical Statistics and Probability*, eds. LeCam, L., Neyman, J. and Scott, E., Vol. **1**, pp. 697–715. Berkeley, CA: University of California Press.

Ord, J.K. (1975). Estimation methods for models of spatial interaction. *Journal of the American Statistical Association*, **70**, 120–6.

Orford, S., Dorling, D. and Harris, R. (1998). Review of visualization in the social sciences: a state of the art survey and report. Advisory Group on Computer Graphics. *Technical Report Series*, No. **41**, July ISSN 1356–9066.

Pace, R.K., Barry, R., Clapp, J.M. and Rodriguez, M. (1998). Spatiotemporal autoregressive models of neighbourhood effects. *The Journal of Real Estate Finance and Economics*, **17**, 15–33.

Pelto, C.R., Elkins, T.A. and Boyd, H.A. (1968). Automatic contouring of irregularly spaced data. *Geophysics*, **33**, 424–30.

Pereira, J.M.C., Carreiras, J.M.B. and Perestrello de Vasconcelos, M.J. (1998). Exploratory data analysis of the spatial distribution of wildfire in Portugal, 1980–89. *Geographical Systems*, **5**, 355–90.

Perruchet, C. (1983). Constrained agglomerative hierarchical classification. *Pattern Recognition*, **16**, 213–17.

Phipps, M. (1989). Dynamical behaviour of cellular automata under the constraint of neighbourhood coherence. *Geographical Analysis*, **21**, 197–215.

Pickett, K.E. and Pearl, M. (2001). Multi-level analyses of neighbourhood socio-economic context and health outcomes: a critical review. *Journal of Epidemiology and Community Health*, **55**, 111–22.

Plummer, P., Haining, R. and Sheppard, E. (1998). Spatial pricing in interdependent markets: testing assumptions and modeling price variation. A case study of gasoline retailing in St Cloud, Minnesota. *Environment and Planning*, **A**, **30**, 67–84.

Pocock, S.J., Cook, D.G. and Beresford, S.A.A. (1981). Regression of area mortality rates on explanatory variables: what weighting is appropriate. *Applied Statistician*, **30**, 286–96.

Pocock, S.J., Cook, D.G. and Shaper, A.G. (1982). Analyzing geographic variation in cardiovascular mortality: methods and results. *Journal of the Royal Statistical Society*, **A**, **145**, 313–41.

Porter, M.E. (1998). *The Competitive Advantage of Nations*. London: MacMillan.

Portugali, J., Benenson, I. and Omer, I. (1994). Sociospatial residential dynamics: stability and instability within a self-organizing city. *Geographical Analysis*, **26**, 321–40.

Putnam, R.D. (1993). *Making Democracy Work: Civic Traditions in Modern Italy*. Princeton, NJ: Princeton University Press.

Raper, J.F. (1999). Spatial representation: the scientist's perspective. *Geographic Information Systems*, second edition, eds. Longley, P.A., Goodchild, M.F., Maguire, D.J. and Rhind, D.W., pp. 61–70. New York: Wiley.

Raper, J.F. (2001). Defining spatial socio-economic units: retrospective and prospective. *Life and Motion of Socio-economic Units*, eds. Frank, A., Raper, J. and Cheylan, J-P., pp. 13–20. London: Taylor & Francis.

Raybould, S. and Walsh, S. (1995). Road traffic accidents involving children in North-East England. *The Added Value of Geographical Information Systems in Public and Environmental Health*. eds. deLepper, M.J.C., Scholten, H.J. et al., pp. 181–8. Dordrecht: Kluwer Academic.

Rey, S.J. (2001). Spatial empirics for economic growth and convergence. *Geographical Analysis*, **33**, 195–214.

Rey, S.J. and Montouri, B.D. (1999). US regional income convergence: a spatial econometric perspective. *Regional Studies*, **33**, 143–56.

Richards, J.A. (1986). *Remote Sensing Digital Image Analysis*. Berlin: Springer-Verlag.

Richardson, H.W. (1970). *Regional Economics*. London: MacMillan.

Richardson, S. (1992). Statistical methods for geographical correlation studies. *Geographical and Environmental Epidemiology: Methods for Small Area Studies*, eds. Elliot, P., Cuzich, J., English, D. and Stern, R., pp. 181–204. Oxford: Oxford University Press.

Richardson, S. and Hémon, D. (1981). On the variance of the sample correlation between two independent lattice processes. *Journal of Applied Probability*, **18**, 943–8.

Richardson, S., Monfort, C., Green, M., Draper, G. and Muirhead, C. (1995). Spatial variation of natural radiation and childhood leukaemia incidence in Great Britain. *Statistics in Medicine*, **14**, 2487–501.

Riedwyl, H. and Schuepbach, M. (1994). Parquet diagram to plot contingency tables. *Advances in Statistical Software*, ed. Faulbaum, F., pp. 293–9. Stuttgart: Gustav Fischer.

Ripley, B.D. (1981). *Spatial Statistics*. New York: Wiley.

Ripley, B.D. (1988). *Statistical Inference for Spatial Processes*. Cambridge: Cambridge University Press.

Robinson, W.S. (1950). Ecological correlations and the behaviour of individuals. *American Sociological Review*, **15**, 351–7.

Rodriguez-Iturbe, I. and Mejia, J.M. (1974). The design of rainfall networks in time and space. *Water Resources Research*, **10**, 713–28.

Rogerson, P.A. (1999). The detection of clusters using a spatial version of the chi-square goodness of fit statistic. *Geographical Analysis*, **31**, 130–47.

Rosenthal, R. (1978). How often are our numbers wrong? *American Psychologist*, **33**, 1005–8.

Rossiter, D.J. and Johnston, R.J. (1981). Program GROUP: the identification of all possible solutions to a constituency-delimitation problem. *Environment and Planning*, **A**, **13**, 231–8.

Rossmo, D.K. (2000). *Geographic Profiling*. Boca Raton: CRC Press.

Rushton, G. (1998). Improving the geographic basis of health surveillance using GIS. *GIS and Health*, eds. Gatrell, A. and Löytönen, M., pp. 63–79. London: Taylor & Francis.

Rushton, G. and Lolonis, P. (1996). Exploratory spatial analysis of birth defect rates in an urban population. *Statistics in Medicine*, **15**, 717–26.

Sadahiro, Y. (1999). Accuracy of areal interpolation: a comparison of alternative methods. *Journal of Geographical Systems*, **1**, 323–46.

Sadahiro, Y. (2000). Accuracy of count data estimated by the point-in-polygon method. *Geographical Analysis*, **32**, 64–89.

Salgé, F. (1995). *Elements of Spatial Data Quality*, eds. Guptill, S.C. and Morrison, J.L., pp. 139–51. Oxford: Elsevier Science.

Sampson, R.J., Raudenbush, S.W. and Earls, F. (1997). Neighborhoods and violent crime: a multi-level study of collective efficacy. *Science*, **277**, 918–24.

Savelieva, E. Kanevski, M., Demyanov, V., Chernov, S. and Maignan, M. (1998). Conditional stochastic co-simulations of the Chernobyl fallout. *geoENV II – Geostatistics for Environmental Applications*, eds. Gómez-Hernández, J., Soares, A. and Froidevaux, R., pp. 453–65. Dordrecht: Kluwer Academic.

Schaeffer, F. (1953). Exceptionalism in Geography. *Annals of the Association of American Geographers*, **43**, 226–49.

Schulman, J., Selvin, S. and Merrill, D.W. (1988). Density equalised map projections: a method for analysing clustering around a fixed point. *Statistics in Medicine*, **7**, 491–505.

Shaw, C.R. and McKay, H.D. (1942). *Juvenile Delinquency and Urban Areas*. Chicago, IL: Chicago University Press.

Shepard, D.S. (1983). Computer mapping: the symap interpolation algorithm. *Spatial Statistics and Models*, eds. Gaile, G.L. and Willmott, C.J., pp. 55–79. Dordrecht: Reidel.

Sheppard, E., Haining, R. and Plummer, P. (1992). Spatial pricing in interdependent markets. *Journal of Regional Science*, **32**, 55–75.

Short, N.M. (1999). Remote Sensing Tutorial. NASA publ. http://rst.gsfc.nasa.gov/Front/tofc.html.

Sibson, R. (1981). A brief description of natural neighbour interpolation. *Interpreting Multivariate Data*, ed. Barnett, V., pp. 21–36. Chichester: Wiley.

Silverman, B.W. (1986). *Density Estimation of Statistics and Data Analysis*. Andover, Hants: Routledge, Chapman & Hall.

Simpson, C.H. (1951). The interpretation of interaction in contingency tables. *Journal of the Royal Statistical Society*, **B**, **13**, 242–9.

Smans, M. and Estive, J. (1992). Practical approaches to disease mapping. *Geographical and Environmental Epidemiology: Methods for Small Area Studies*, eds. Elliott, P., Cuzick, J., English, E. and Stern, R., pp. 141–9. Oxford: Oxford University Press.

Sokal, R.R., Oden, N.L., Thompson, B.A. and Kim, J. (1993). Testing for regional differences in means: distinguishing inherent from spurious spatial autocorrelation by restricted randomization. *Geographical Analysis*, **25**, 199–210.

Sooman, A. and Macintyre, S. (1995). Health and perceptions of the local environment in socially contrasting neighbourhoods in Glasgow. *Health and Place*, **1**, 15–26.

Spence, N.A. (1968). A multivariate uniform regionalization of British counties on the basis of employment data for 1961. *Regional Studies*, **2**, 87–104.

Spiegelhalter, D.J., Best, N.G. Carlin, B.P. and van der Linde, A. (2001). Bayesian measures of model complexity and fit. Technical report, Medical Research Council Biostatistics Unit, Cambridge, UK (www.mrc-bsu.cam.ac.uk/Publications/preslid.shtml).

Stehman, S. (1996). Estimating the Kappa coefficient and its variance under stratified random sampling. *Photogrammetric Engineering and Remote Sensing*, **62**, 401–7.

Stephan, F. (1934). Sampling errors and interpretations of social data ordered in time and space. *Journal of the American Statistical Association*, **29**, Suppl. 165–6.

Stone, R.A. (1988). Investigations of excess environmental risks around putative sources: statistical problems and a proposed test. *Statistics in Medicine*, **7**, 649–60.

Student (1914). The elimination of spurious correlation due to position in time and space. *Biometrika*, **10**, 179–80.

Swartz, C. (2000). The spatial analysis of crime. *Analyzing Crime Patterns: Frontiers of Practice*, eds. Goldsmith, V., McGuire, P.G., Mollenkopf, J.H. and Ross, R.A., pp. 33–46. Thousand Oaks: Sage.

Swerdlow, A.J. (1992). Cancer incidence data for adults. *Geographical and Environmental Epidemiology: Methods for Small Area Studies*, eds. Elliott, P.J., Cuzick, J., English D. and Stern, R., pp. 51–62. Oxford: Oxford University Press.

Switzer, P. (2000). Multiple simulation of spatial fields. *Proceedings of the 4th International Symposium on Spatial Accuracy Assessment in Natural Resources and Environmental Sciences*, pp. 629–35. Delft: Delft University Press.

Switzer, P. (2000). Probabilistic exploration strategies. Spatial archaeometry: using spatial statistical methods in archaeology and cultural heritage safeguard research. Report of the Interdisciplinary Workshop on spatial statistical methods in archaeology and cultural heritage research, Pescara, Italy, July.

Symarzik, J., Majure, J. and Cook, D. (1996). Dynamic graphics in a GIS: a bidirectional link between Arc View 2.0 and Xgobi. *Computing Science and Statistics*, **27**, 299–303.

Tango, T. (1995). A class of tests for detecting 'general' and 'focused' clustering of rare diseases. *Statistics in Medicine*, **7**, 649–60.

Taylor, J.R. (1982). *An Introduction to Error Analysis*. Mill Valley, CA: University Science Books.

Taylor, P. (1969). The location variable in taxonomy. *Geographical Analysis*, **1**, 181–95.

Theobald, D.M. (1989). Accuracy and bias issues in surface representation. *Accuracy in Spatial Databases*, eds. Goodchild, M. and Gopal, S., pp. 99–105. London: Taylor & Francis.

Thompson, D. (1998). The National Health Service breast cancer screening programme in Sheffield: service delivery and uptake. Ph.D. Thesis, University of Sheffield.

Tiefelsdorf, M. and Boots, B. (1995). The exact distribution of Moran's I. *Environment and Planning*, **A**, **27**, 985–99.

Tinkler, K. (1972). The physical interpretation of eigenfunctions of dichotomous matrices. *Transactions of the Institute of British Geographers*, **55**, 17–46.

Tjostheim, D. (1978). A measure of association for spatial variables. *Biometrika*, **65**, 109–14.

Tobler, W. (1979). Smooth pycnophylactic interpolation for geographical regions. *Journal of the American Statistical Association*, **74**, 519–36 (with discussion).

Tobler, W. (1989). Frame independent spatial analysis. *Accuracy of Spatial Databases*, eds. Goodchild, M. and Gopal, S., pp. 115–22. London: Taylor & Francis.

Tobler, W.R. and Kennedy, S. (1985). Smooth multi-dimensional interpolation. *Geographical Analysis*, **17**, 251–7.

Tobler, W.R. and Lau, J. (1978). Isopleth mapping using histosplines. *Geographical Analysis*, **10**, 273–9.

Townsend, P., Phillimore, P. and Beattie, A. (1988). *Health and Deprivation: Inequality and the North*. London: Croom Helm.

Tranmer, M. and Steel, D.G. (1998). Using census data to investigate the causes of the ecological fallacy. *Environment and Planning*, **A**, **30**, 817–31.

Tufte, E.R. (1983). *The Visual Display of Quantitative Information*. Cheshire, CT: Graphics Press.

Tufte, E.R. (1990). *Envisioning Information*. Chesire, CT: Graphics Press.

Tukey, J.W. (1977). *Exploratory Data Analysis*. Reading: Addison-Wesley.

Tukey, J.W. (1979). Statistical mapping: what should not be plotted. *The Collected Works of John W. Tukey, Volume V: Graphics, 1965–1985*, pp. 109–21. Belmont, CA: Wadsworth (1988).

Tukey, P.A. and Tukey, J.W. (1981). Graphical display of data sets in three or more dimensions. *Interpreting Multivariate Data*, ed. Barnett, V., New York: Wiley.

Turnbull, B.W., Iwano, E.J., Burnett, W.S., Howe, H.L. and Clark, L.C. (1990). Monitoring for clusters of disease: application to leukaemia incidence in upstate New York. *American Journal of Epidemiology*, **132**, S136–S143.

Ulm, K. (1990). A simple method to calculate the confidence interval of a standardized mortality ratio. *American Journal of Epidemiology*, **131**, 373–5.

Unwin, A., Hawkins, G., Hofmann, H. and Siegl, B. (1996). Interactive graphics for data sets with missing values – MANET. *Journal of Computational and Graphical Statistics*, **5**, 113–22.

Unwin, D.J. (1995). Geographical Information Systems and the problem of 'error and uncertainty'. *Progress in Human Geography*, **19**, 549–58.

Unwin, D.J. (1997). Graphics, visualiztion and the social sciences. *Advisory Group on Computer Graphics, Technical Report Series*, **No. 33**, ISSN 1356–9066, pp. 103–8. Burleigh Court, Loughborough University.

Unwin, D.J. and Wrigley, N. (1987). Control point distribution in trend surface modelling revisited: an application of the concept of leverage. *Transactions, Institute of British Geographers*, **12**, 147–60.

Unwin, D.J. and Wrigley, N. (1987a). Towards a general theory of control point distribution effects in trend surface models. *Computers and Geosciences*, **13**, 351–5.

Upton, G.J.G. (1985). Distance weighted geographic interpolations. *Environment and Planning*, **A**, **17**, 667–71.

Upton, G.J.G. (1991). Rectangular cartograms, spatial autocorrelation and interpolation. *Papers in Regional Science*, **70**, 287–302.

Upton, G.J.G. and Fingleton, B. (1985). *Spatial data analysis by example. Vol. 1, Point pattern and quantitative data*. Chichester: Wiley.

Upton, G.J.G. and Fingleton, B. (1989). *Spatial Data Analysis by Example: Volume 2. Categorical and Directional Data*. Chichester: Wiley.

Venables, A.J. (1999). But why does geography matter, and which geography matters? *International Regional Science Review*, **22**(2), 238–41.

Veregin, H. (1994). Integration of simulation modelling and error propagation for the buffer operation in GIS. *Photogrammetric Engineering and Remote Sensing*, **60**, 427–35.

Veregin, H. (1995). Developing and testing of an error propagation model for GIS overlay operations. *International Journal of Geographical Information Systems*, **9**, 595–619.

Veregin, H. and Hargitai, P. (1995). An evaluation matrix for geographical data quality. *Elements of Spatial Data Quality*, eds. Guptill, S.C. and Morrison, J.L., pp. 167–188. Oxford: Elsevier Science.

Verly, G., David, M., Journel, A.G. and Marechal, A. (1984). *Geostatistics for Natural Resources Characterization*. Dordrecht: Reidel.

Viel, J-F., Arveux, P., Baverel, J. and Cahn, J-Y. (1999). Soft-tissue sarcoma and Non-Hodgkin's Lymphoma clusters around a municipal solid waste incinerator with high dioxin emission levels. *American Journal of Epidemiology*, **152**, 13–19.

Visvalingam, M. (1983). Operational definitions of area based social indicators. *Environment and Planning*, **A, 15**, 831–9.

Wakefield, J. and Elliott, P. (1999). Issues in the statistical analysis of small area health data. *Statistics in Medicine*, **18**, 2377–99.

Waldhor, T. (1996). The spatial autocorrelation coefficient Moran's I under heteroscedasticity. *Statistics in Medicine*, **15**, 887–92.

Waldrop, M.W. (1992). *Complexity*. New York: Simon & Schuster.

Wang, J., Wise, S. and Haining, R. (1997). An integrated regionalization of earthquake, flood and drought hazards in China. *Transactions in Geographic Information Systems*, **2**, 25–44.

Webster, M.R. (1977). *Quantitative and Numerical Methods in Soil Classification and Survey*. Oxford: Clarendon Press.

Webster, M.R. (1985). Quantitative analysis of soil in the field. *Advances in Soil Sciences*, **3**, 1–70.

Webster, M.R. and Burgess, T.M. (1981). Optimal interpolation and isarithmic mapping of soil properties. III Changing drift and universal kriging. *Journal of Soil Sciences*, **32**, 505–24.

Webster, R. and Burrough, P.A. (1972). Computer based soil mapping of small areas from sample data. II Classification smoothing. *Journal of Soil Science*, **23**, 222–34.

Webster, R. and Oliver, M.A. (2001). *Geostatistics for Environmental Scientists*. Chichester: John Wiley.

Webster, R., Oliver, M.A., Muir, K.R. and Mann, J.R. (1994). Kriging the local risk of a rare disease from a register of diagnoses. *Geographical Analysis*, **26**, 168–85.

Wegener, M. (2000). Spatial models and GIS. *Spatial Models and GIS*, eds. Fotheringham, A.S. and Wegener, M., pp. 3–20. London: Taylor & Francis.

Wegman, E. (1995). Huge data sets and the frontiers of computational feasibility. *Journal of Computational and Graphical Statistics*, **4**, 281–95.

Wegman, E.J., Posta, W.L. and Solka, J.L. Image grand tour. (www.galaxy.gmu.edn/stats/center.html)

Weisberg, S. (1985). *Applied Linear Regression*. New York: John Wiley.

Welch, R., Jordan, T.R. and Ehlers, M. (1985). Comparative evaluations of the geodetic accuracy and cartographic potential of LANDSAT-4 and LANDSAT-5 Thematic Mapper image data. *Photogrammetric Engineering and Remote Sensing*, **51**, 1799–812.

White, H. (1980). A heteroskedasticity – consistent covariance matrix estimator and a direct test for heteroskedasticity. *Econometrica*, **48**, 817–38.

White, R. and Engelen, G. (1994). Cellular dynamics and GIS: modelling spatial complexity. *Geographical Systems*, **1**, 237–53.

Whittle, P. (1954). On stationary processes in the plane. *Biometrika*, **41**, 434–49.

Wikström, P.-O.H. (1990). Delinquency and the urban structure. *Crime and Measures against Crime in the City*, ed. Wikström, P.-O.H., Stockholm: National Council for Crime Prevention.

Wikström, P.-O.H. (1991). *Urban Crime, Criminals and Victims: The Swedish Experience in an Anglo-American Comparative Perspective*. New York: Springer-Verlag.

Wikström, P-O. and Loeber, R. (2000). Do disadvantaged neighbourhoods cause well-adjusted children to become adolescent delinquents? A study of male juvenile serious offending, individual risk and protective factors and neighbourhood context. *Criminology*, **38**, 1109–42.

Wilhelm, A. and Sander, M. (1998). Interactive statistical analysis of dialect features. *The Statistician*, **47**, 445–55.

Wilkinson, C., Grundy, C., Landon, M. and Stevenson, S. (1998). GIS in public health. *GIS and Health*, eds. Gatrell, A.C. and Löytönen, M., pp. 179–89. London: Taylor & Francis.

Wilkinson, R.G. (1996). *Unhealthy Societies*. London: Routledge.

Wilson, W.J. (1997). *When Work Disappears: The World of the New Urban Poor*. New York: Alfred Knopf.

Wise, S.M., Haining, R.P. and Ma, J. (1997). Regionalization tools for the exploratory spatial analysis of health data. *Recent Developments in Spatial Analysis: Spatial Statistics, Behavioural Modelling and Neuro-Computing*, eds. Fischer, M. and Getis, A., pp. 83–100. Berlin: Springer-Verlag.

Wise, S.M., Haining, R.P. and Signoretta, P.E. (1999). Scientific visualization and the exploratory analysis of area data. *Environment and Planning*, **A**, **31**, 1823–38.

Womble, W.H. (1951). Differential systematics. *Science*, **114**, 315–22.

Wong, D. and Amrhein, C. (1996). Research on the MAUP: old wine in a new bottle or real breakthrough? *Geographical Systems*, **3**, 73–76.

Wray, N.R., Alexander, F.E., Muirhead, C.R., Pukkala, E., Schmidtmann, I. and Stiller, C. (1999). A comparison of some simple methods to identify geographical areas with excess incidence of a rare disease such as childhood leukaemia. *Statistics in Medicine*, **18**, 1501–16.

Wright, D.L. Stern, H.S. and Cressie, N. (2002). Loss functions for estimation of extrema with an application to disease mapping. Paper presented at the Spatial Econometrics workshop, Toulouse, France, June 15, 2002 and submitted to the *Canadian Journal of Statistics*.

Wrigley, N., Holt, T., Steel, D. and Tranmer, M. (1996). Analysing, modelling and resolving the ecological fallacy. *Spatial Analysis: Modelling in a GIS Environment*, eds. Longley, P. and Batty, M., pp. 23–40. Cambridge: GeoInformation International.

Yichun Xie. (1996). A generalized model for cellular urban dynamics. *Geographical Analysis*, **28**, 350–73.

Youden, W.J. and Mehlich, A. (1937). Selection of efficient methods for soil sampling. *Contributions to the Boyce Thompson Institute of Plant Research*, **9**, 59–70.

Yule, G.U. (1926). Why do we sometimes get nonsense correlations between two time series? *Journal of the Royal Statistical Society*, **89**, 1–69.

Zhang, J. and Kirby, R.P. (2000). A geostatistical approach to modelling positional errors in vector data. *Transactions in GIS*, **4**, 145–59.

Index